Springer Texts in Business and Economics

For further volumes:
http://www.springer.com/series/10099

Gebhard Kirchgässner • Jürgen Wolters
Uwe Hassler

Introduction to Modern Time Series Analysis

Second Edition

 Springer

Gebhard Kirchgässner
SIAW-HSG
University of St. Gallen
St. Gallen
Switzerland

Jürgen Wolters
Institute for Statistics and Econometrics
FU Berlin
Berlin
Germany

Uwe Hassler
Applied Econometrics and
  International Economic Policy
Goethe University Frankfurt
Frankfurt
Germany

ISSN 2192-4333   ISSN 2192-4341 (electronic)
ISBN 978-3-642-44029-8 ISBN 978-3-642-33436-8 (eBook)
DOI 10.1007/978-3-642-33436-8
Springer Heidelberg New York Dordrecht London

© Springer-Verlag Berlin Heidelberg 2013
Softcover reprint of the hardcover 2nd edition 2013
This work is subject to copyright. All rights are reserved by the Publisher, whether the whole or part of the material is concerned, specifically the rights of translation, reprinting, reuse of illustrations, recitation, broadcasting, reproduction on microfilms or in any other physical way, and transmission or information storage and retrieval, electronic adaptation, computer software, or by similar or dissimilar methodology now known or hereafter developed. Exempted from this legal reservation are brief excerpts in connection with reviews or scholarly analysis or material supplied specifically for the purpose of being entered and executed on a computer system, for exclusive use by the purchaser of the work. Duplication of this publication or parts thereof is permitted only under the provisions of the Copyright Law of the Publisher's location, in its current version, and permission for use must always be obtained from Springer. Permissions for use may be obtained through RightsLink at the Copyright Clearance Center. Violations are liable to prosecution under the respective Copyright Law.
The use of general descriptive names, registered names, trademarks, service marks, etc. in this publication does not imply, even in the absence of a specific statement, that such names are exempt from the relevant protective laws and regulations and therefore free for general use.
While the advice and information in this book are believed to be true and accurate at the date of publication, neither the authors nor the editors nor the publisher can accept any legal responsibility for any errors or omissions that may be made. The publisher makes no warranty, express or implied, with respect to the material contained herein.

Printed on acid-free paper

Springer is part of Springer Science+Business Media (www.springer.com)

Preface to the Second Edition

In preparing this second and enlarged edition, a third author has joined the team. Still, the scope of the book has not changed. We try to provide a rigorous understanding of the theory and methods of univariate and multivariate time series analysis. At the same time, the main objective is the development of empirical skills with a special emphasis on the link to economic applications. Therefore, we strengthened the specific feature of our book that now contains 63 examples, most of them using real data sets. The computations for the empirical examples were performed by means of EViews, Version 7.2. Note that previous versions partly result in (slightly) different numbers for parameters, standard errors and test statistics. The same is likely to hold true with other computer programmes or future versions of EViews. Since the empirical examples are central to the book, we now provide all data sets contained in EViews files on the homepage of UWE HASSLER.

For this second edition we have updated some of the time series analysed in the examples, while other data sets containing historical series taken from the literature remain unchanged. The major change of this enlarged edition, however, consists of additional material. First, the new *Chapter 7* covers nonstationary panel data analysis. This accommodates that during the last decade many of the time series techniques treated in our book have been carried to the panel situation where series from several, possibly correlated units are investigated. Second, the final chapter on conditional heteroscedasticity has been supplemented by a section on multivariate ARCH models accounting for time-varying conditional correlation. Third, some subsections have been added (see *Section 2.2.2* on temporal aggregation), while others have been enlarged (see *Section 5.5.1* on fractional integration). Finally, we removed typos from the first edition and improved the exposition where this seemed necessary.

We wish to thank all those who have helped us with this second edition. It is our pleasure to mention, in particular, FLORIAN HABERMACHER. TERESA KÖRNER, and GABRIELA SCHMID. They have made valuable contributions towards improving the presentation but, of course, are not responsible for any remaining deficiencies. Moreover, we are indebted to Dr. MARTINA BIHN and RUTH MILEWSKI from Springer for their kind collaboration.

St Gallen/Berlin/Frankfurt, August 2012

GEBHARD KIRCHGÄSSNER   JÜRGEN WOLTERS   UWE HASSLER

Preface to the First Edition

Econometrics has been developing rapidly over the past four decades. This is not only true for microeconometrics which more or less originated during this period, but also for time series econometrics where the cointegration revolution influenced applied work in a substantial manner. Economists have been using time series for a very long time. Since the 1930s when econometrics became an own subject, researchers have mainly worked with time series. However, economists as well as econometricians did not really care about the statistical properties of time series. This attitude started to change in 1970 with the publication of the textbook *Time Series Analysis, Forecasting and Control* by GEORGE E.P. BOX and GWILYM M. JENKINS. The main impact, however, stems from the work of CLIVE W.J. GRANGER starting in the 1960s. In 2003 together with ROBERT F. ENGLE, he received the Nobel Prize in Economics for his work.

This textbook provides an introduction to these recently developed methods in time series econometrics. Thus, it is assumed that the reader is familiar with a basic knowledge of calculus and matrix algebra as well as of econometrics and statistics at the level of introductory textbooks. The book aims at advanced Bachelor and especially Master students in economics and applied econometrics but also at the general audience of economists using empirical methods to analyse time series. For these readers, the book is intended to bridge the gap between methods and applications by also presenting a lot of empirical examples.

A book discussing an area in rapid development is inevitably incomplete and reflects the interests and experiences of the authors. We do not include, for example, the modelling of time-dependent parameters with the Kalman filter as well as Markov Switching Models, panel unit roots and panel cointegration. Moreover, frequency domain methods are not treated either.

Earlier versions of the different chapters were used in various lectures on time series analysis and econometrics at the Freie Universität Berlin, Germany, and the University of St. Gallen, Switzerland. Thus, the book has developed over a number of years. During this time span, we also learned a lot from our students and we do hope that this has improved the presentation in the book.

We would like to thank all those who have helped us in producing this book and who have critically read parts of it or even the whole manuscript. It is our pleasure to mention, in particular, MICHAEL-DOMINIK BAUER, ANNA CISLAK, LARS P. FELD, SONJA LANGE, THOMAS MAAG, ULRICH K. MÜLLER, GABRIELA SCHMID, THORSTEN UEHLEIN, MARCEL R. SAVIOZ, and ENZO WEBER. They have all made valuable contributions towards improving the presentation but, of course, are not responsible for any remaining deficiencies. Our special thanks go to MANUELA KLOSS-MÜLLER who edited the text in English. Moreover, we are indebted to Dr. WERNER A. MÜLLER and MANUELA EBERT from Springer for their kind collaboration.

St Gallen/Berlin, April 2007

GEBHARD KIRCHGÄSSNER   JÜRGEN WOLTERS

Contents

Preface V

1 **Introduction and Basics** 1
   1.1 The Historical Development of Time Series Analysis 2
   1.2 Graphical Representations of Economic Time Series 5
   1.3 The Lag Operator 10
   1.4 Ergodicity and Stationarity 12
   1.5 The Wold Decomposition 21
   References 22

2 **Univariate Stationary Processes** 27
   2.1 Autoregressive Processes 27
      2.1.1 First Order Autoregressive Processes 27
      2.1.2 Second Order Autoregressive Processes 40
      2.1.3 Higher Order Autoregressive Processes 49
      2.1.4 The Partial Autocorrelation Function 52
      2.1.5 Estimating Autoregressive Processes 56
   2.2 Moving Average Processes 58
      2.2.1 First Order Moving Average Processes 58
      2.2.2 MA(1) and Temporal Aggregation 62
      2.2.3 Higher Order Moving Average Processes 65
   2.3 Mixed Processes 68
      2.3.1 ARMA(1,1) Processes 69
      2.3.2 ARMA(p,q) Processes 75
   2.4 Forecasting 78
      2.4.1 Forecasts with Minimal Mean Squared Errors 78
      2.4.2 Forecasts of ARMA(p,q) Processes 81
      2.4.3 Evaluation of Forecasts 85
   2.5 The Relation between Econometric Models and
         ARMA Processes 89
   References 90

3 Granger Causality ... 95
- 3.1 The Definition of Granger Causality ... 97
- 3.2 Characterisation of Causal Relations in Bivariate Models ... 99
  - 3.2.1 Characterisation of Causal Relations Using the Autoregressive and Moving Average Representations ... 99
  - 3.2.2 Characterisation of Causal Relations Using the Residuals of the Univariate Processes ... 101
- 3.3 Causality Tests ... 104
  - 3.3.1 The Direct Granger Procedure ... 104
  - 3.3.2 The Haugh-Pierce Test ... 108
  - 3.3.3 The Hsiao Procedure ... 112
- 3.4 Applying Causality Tests in a Multivariate Setting ... 116
  - 3.4.1 The Direct Granger Procedure with More Than Two Variables ... 116
  - 3.4.2 Interpreting the Results of Bivariate Tests in Systems With More Than Two Variables ... 119
- 3.5 Concluding Remarks ... 120
- References ... 122

4 Vector Autoregressive Processes ... 127
- 4.1 Representation of the System ... 129
- 4.2 Granger Causality ... 138
- 4.3 Impulse Response Analysis ... 140
- 4.4 Variance Decomposition ... 146
- 4.5 Concluding Remarks ... 151
- References ... 152

5 Nonstationary Processes ... 155
- 5.1 Forms of Nonstationarity ... 155
- 5.2 Trend Elimination ... 161
- 5.3 Unit Root Tests ... 165
  - 5.3.1 The Dickey-Fuller Test ... 167
  - 5.3.2 The Augmented Dickey-Fuller Test ... 170
  - 5.3.3 The Phillips-Perron Test ... 173
  - 5.3.4 Unit Root Tests and Structural Breaks ... 178
  - 5.3.5 A Test with the Null Hypothesis of Stationarity ... 180
- 5.4 Decomposition of Time Series ... 183
- 5.5 Further Developments ... 190
  - 5.5.1 Fractional Integration ... 191
  - 5.5.2 Seasonal Integration ... 193

	5.6	Deterministic versus Stochastic Trends in Economic Time Series 196
	References 198

6 Cointegration 205
6.1 Definition and Properties of Cointegrated Processes 209
6.2 Cointegration in Single Equation Models: Representation, Estimation and Testing 211
6.2.1 Bivariate Cointegration 211
6.2.2 Cointegration with More Than Two Variables 214
6.2.3 Testing Cointegration in Static Models 215
6.2.4 Testing Cointegration in Dynamic Models 221
6.3 Cointegration in Vector Autoregressive Models 225
6.3.1 The Vector Error Correction Representation 225
6.3.2 The Johansen Approach 228
6.3.3 Analysis of Vector Error Correction Models 237
6.4 Cointegration and Economic Theory 242
References 244

7 Nonstationary Panel Data 251
7.1 Issues with Panel Data 252
7.1.1 Omitted Variable Bias 252
7.1.2 Estimation and Testing 253
7.1.3 Mixed Panel Evidence 255
7.2 Panel Unit Root Tests 258
7.2.1 First Generation Tests 258
7.2.2 Second Generation Tests 259
7.2.3 The Null Hypothesis of Stationarity 262
7.3 The Combination of Significance 263
7.3.1 The Inverse Normal Method 263
7.3.2 Bonferroni-Type Tests 265
7.4 Panel Cointegration 267
7.4.1 Single Equation Approaches 267
7.4.2 System Approaches 273
7.5 Concluding Remarks 274
References 275

8 Autoregressive Conditional Heteroscedasticity 281
8.1 ARCH Models 285
8.1.1 Definition and Representation 285
8.1.2 Unconditional Moments 288
8.1.3 Temporal Aggregation 289

8.2 Generalised ARCH Models 292
    8.2.1 GARCH Models 292
    8.2.2 The GARCH(1,1) Process 294
    8.2.3 Nonlinear Extensions 297
8.3 Estimation and Testing 299
8.4 Multivariate Models 301
    8.4.1 VAR-Type Models 302
    8.4.2 Correlation Models 304
8.5 ARCH/GARCH Models as Instruments of Financial
    Market Analysis 305
References 307

Index of Names and Authors 311

Subject Index 315

1 Introduction and Basics

A time series is defined as a set of quantitative observations arranged in chronological order. We generally assume that time is a discrete variable. Time series have always been used in the field of econometrics. Already at the outset, JAN TINBERGEN (1939) constructed the first econometric model for the United States and thus started the scientific research programme of empirical econometrics. At that time, however, it was hardly taken into account that chronologically ordered observations might depend on each other. The prevailing assumption was that, according to the classical linear regression model, the residuals of the estimated equations are stochastically independent from each other. For this reason, procedures were applied which are also suited for cross section or experimental data without any time dependence.

DONALD COCHRANE and GUY H. ORCUTT (1949) were the first to notice that this practice might cause problems. They showed that if residuals of an estimated regression equation are positively autocorrelated, the variances of the regression parameters are underestimated and, therefore, the values of the F and t statistics are overestimated. This problem could be solved, at least for the frequent case of first order autocorrelation, by transforming the data adequately. Almost at the same time, JAMES DURBIN and GEOFFREY S. WATSON (1950/51) developed a test procedure which made it possible to identify first order autocorrelation. The problem seemed to be solved (more or less), and, until the 1970's, the issue was hardly ever raised in the field of empirical econometrics.

This did not change until GEORGE E.P. BOX and GWILYM M. JENKINS (1970) published a textbook on time series analysis that received considerable attention. First of all, they introduced univariate models for time series which simply made systematic use of the information included in the observed values of time series. This offered an easy way to predict the future development of this variable. Today, the procedure is known as *Box-Jenkins Analysis* and is widely applied. It became even more popular when CLIVE W.J. GRANGER and PAUL NEWBOLD (1975) showed that simple forecasts which only considered information given by one single time series often outperformed the forecasts based on large econometric models consisting sometimes of many hundreds of equations.

In fact, at that time, many procedures applied in order to analyse relations between economic variables were not really new. Partly, they had already been used in other sciences, in particular, for quite a while, in the experimental natural sciences. Some parts of their theoretical foundations had also been known for a considerable time. From then on, they have been used in economics, too, mainly because of two reasons. Up to then, contrary to the natural sciences there had not been enough economic observations available to even consider the application of these methods. Moreover, at the beginning of the 1970's, electronic computers became available which were quite powerful compared to earlier times and which could manage numerical problems comparatively easy. Since then, the development of new statistical procedures and larger, more powerful computers as well as the availability of larger data sets has advanced the application of time series methods which help to deal with economic issues.

Before we discuss modern (parametric) time series procedures in this chapter, we give a brief historical overview (*Section 1.1*). In *Section 1.2*, we demonstrate how different transformations can show the properties of time series. In *Section 1.3*, we show how the lag operator can be used as a simple but powerful instrument for modelling economic time series.

Certain conditions have to be fulfilled in order to make statistical inference based on time series data. It is essential that some properties of the underlying data generating process, in particular expectation, variance and covariances, between elements of these series, are not time dependent, i.e. that the observed time series are stationary. Therefore, the exact definition of stationarity is given in *Section 1.4*. which also introduces the autocorrelation function as an important statistical instrument for describing (time) dependencies between the elements of a time series. Finally, in *Section 1.5*, we introduce *Wold's Decomposition*, a general representation of a stationary time series. Thus, this chapter mainly covers some notions and tools necessary to understand the later chapters of this textbook.

1.1 The Historical Development of Time Series Analysis

Time series already played an important role in the early natural sciences. Babylonian astronomy used time series of the relative positions of stars and planets to predict astronomical events. Observations of the planets' movements provided the basis of the laws JOHANNES KEPLER discovered.

The analysis of time series helps to detect regularities in the observations of a variable and derive 'laws' from them, and/or exploit all information included in this variable to better predict future developments. The basic methodological idea behind these procedures, which were also valid for the Babylonians, is that it is possible to decompose time series into a finite number of independent but not directly observable components that develop regularly and can thus be calculated in advance. For this procedure, it is necessary that there are different independent factors which have an impact on the variable.

In the middle of the 19th century, this methodological approach to astronomy was taken up by the economists CHARLES BABBAGE and WILLIAM STANLEY JEVONS. The decomposition into unobserved components that depend on different causal factors, as it is usually employed in the classical time series analysis, was developed by WARREN M. PERSONS (1919). He distinguished four different components:

- a long-run development, the trend,

- a cyclical component with periods of more than one year, the business cycle,

- a component that contains the ups and downs within a year, the seasonal cycle, and

- a component that contains all movements which neither belong to the trend nor to the business cycle nor to the seasonal component, the *residual*.

Under the assumption that the different non-observable factors are independent, their additive overlaying generates the time series which we can, however, only observe as a whole. In order to get information about the data generating process, we have to make assumptions about its unobserved components. The classical time series analysis assumes that the systematic components, i.e. trend, business cycle and seasonal cycle, are not influenced by stochastic disturbances and can thus be represented by deterministic functions of time. Stochastic impact is restricted to the residuals, which, on the other hand, do not contain any systematic movements. It is therefore modelled as a series of independent or uncorrelated random variables with expectation zero and constant variance, i.e. as a pure random process.

However, since the 1970's, a totally different approach has increasingly been applied to the statistical analysis of time series. The purely descriptive procedures of classical time series analysis were abandoned and, instead, results and methods of probability theory and mathematical statistics have been employed. This has led to a different assessment of the role of stochastic movements with respect to time series. Whereas the classical approach regards these movements as residuals without any significance for the structure of time series, the modern approach assumes that there are stochastic impacts on all components of a time series. Thus, the 'law of movement' of the whole time series is regarded as a stochastic process, and the time series to be analysed is just one realisation of the data generating process. Now the focus is on stochastic terms with partly rather complex dependence structures.

The first steps in this direction were taken by the Russian statistician EVGENIJ EVGENIEVICH SLUTZKY and the British statistician GEORGE UDNY YULE at the beginning of the last century. Both of them showed that time series with cyclical properties similar to economic (and other) time series can be generated by constructing weighted or unweighted sums or differences of pure random processes. EVGENIJ EVGENIEVICH SLUTZKY and GEORGE UDNY YULE developed moving average and autoregressive processes as models to represent time series. HERMAN WOLD (1938) systematised and generalised these approaches in his doctoral thesis. Their widespread practical usage is due to GEORGE E.P BOX and GWILYM M. JENKINS (1970), who developed methods to implement these models empirically. They had abandoned the idea of different components and assumed that there was a common stochastic model for the whole generation process of time series. Firstly, this method identifies a specific model on the basis of certain statistical figures. Secondly, the parameters of this model are estimated. Thirdly, the specification of the model is checked by statistical tests. If specification errors become obvious, the specification has to be changed and the parameters have to be re-estimated. This procedure is re-iterated until it generates a model that satisfies the given criteria. This model can finally be used for forecasts.

Recently, the idea of decomposing a time series has been taken up again, particularly for the modelling of seasonal variations. However, contrary to the classical approach, it is now assumed that all components of a time series can be represented by simple stochastic models. The procedure for the seasonal adjustment of time series used by EUROSTAT is, for example, based on such an approach.

Moreover, since the 1980's the possible nonstationarity of time series has increasingly been taken into consideration. Nonstationarity might not only be caused by deterministic but also by stochastic trends and, furthermore, the nonstationarity of time series is no longer simply eliminated through the application of filters in order to continue within the framework of stationary models. Nonstationarity is rather explicitly taken into account when constructing models, as long as this is possible and seems to make sense. Accordingly, after this introduction of the basic principles, we will first deal with models of stationary time series and then turn to the modelling of nonstationary time series.

1.2 Graphical Representations of Economic Time Series

When investigating (economic) time series, it is generally useful to start with graphical representations to detect those properties of the series which can be seen by simply looking at the plot of a time series. In this context, it is important to consider different transformations of the time series to be analysed, as, for example, its levels, its changes and its relative changes.

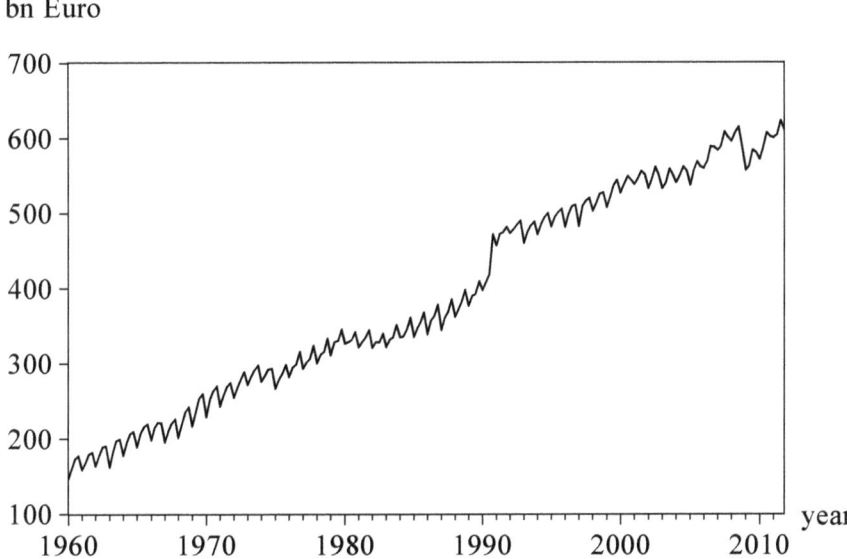

Figure 1.1: Real Gross Domestic Product of the Federal Republic of Germany in billions of Euro, 1960 – 2011

Figure 1.1 shows the real Gross Domestic Product (GDP) of the Federal Republic of Germany in billions of Euros from the first quarter of 1960 to the fourth quarter of 2011, in prices of 1995. Up to 1999, the data stem from the National Accounts of the Federal Republic of Germany issued by the German Institute of Economic Research (DIW) in Berlin. From 2000 onwards, the series is based on data published by the German Bundesbank based on data of the Federal Statistical Office. This time series increases in the long run, i.e. it has a positive trend. On the other hand, it shows well-pronounced short-run movements which take place within one year. These are seasonal variations. There are two remarkable shifts in the series. The first one is due to the German Unification: from the third quarter of 1990 on, the series is based on data for the unified Germany while the earlier data are based on the former West Germany only. The second one is due to the big financial and economic crisis which caused a drop of the German

6 Introduction and Basics

GDP of about 10 per cent from the third quarter in 2008 to the first quarter in 2009.

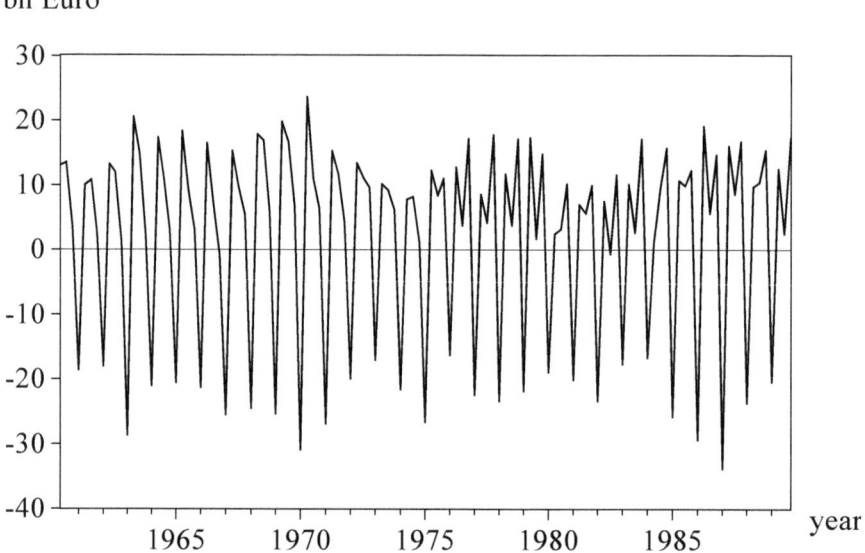

Figure 1.2: Quarterly Changes of the Real Gross Domestic Product (ΔGDP) of the Federal Republic of Germany, 1960 – 1989

When changes from quarter to quarter are analysed, i.e. $\Delta GDP_t = GDP_t - GDP_{t-1}$, where t is the time index, *Figure 1.2* shows that the trend is eliminated by this transformation while the seasonal variations remain. (Because of the structural break due to the German Unification, we only consider the West German data from 1960 to 1989.) The resulting values fluctuate around zero with almost constant amplitude. Moreover, the seasonal component shows a break: up to 1974, the annual minimum is almost always located in the first quarter, from 1975 onwards in the fourth quarter.

If the relative changes from quarter to quarter are to be observed, we take the quarterly growth rates. In percentage points, these are usually calculated as

$$(1.1) \qquad \overline{qgr_t} = \frac{GDP_t - GDP_{t-1}}{GDP_{t-1}} \cdot 100 .$$

However, the problem with this representation is that there is an asymmetry with respect to positive and negative changes: A rise from 100 to 125 is seen as an increase of 25 percent, whereas a decline from 125 to 100 is seen as a decrease of 'only' 20 percent. This can lead to considerable problems if average growth rates are calculated for time series with strongly pronounced fluctuations. In an extreme case this might lead to the calculation of positive average growth rates in spite of a negative trend. In order to avoid this, 'continuous' growth rates are usually employed today, which are calculated (again in percentage points) as

(1.1') $\qquad \text{qgr}_t = (\ln(\text{GDP}_t) - \ln(\text{GDP}_{t-1})) \cdot 100.$

Here, $\ln(\cdot)$ denotes the natural logarithm. In the following, we will always use this definition. As the approximation $\ln(1 + x) \approx x$ is valid for small values of x, the differences between (1.1) and (1.1') can generally be neglected for small growth rates.

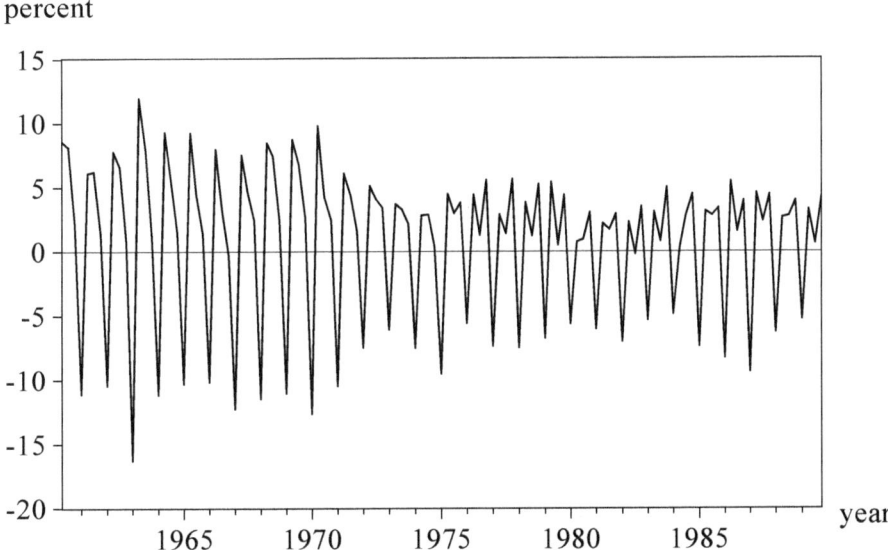

Figure 1.3: Quarterly Growth Rates of the Real Gross Domestic Product (qgr) of the Federal Republic of Germany, 1960 – 1989

Figure 1.3 shows that the growth rates, too, reflect a seasonal pattern. In 1975, this pattern is clearly disrupted. However, contrary to *Figure 1.2*, the amplitude and thus the relative importance of the seasonal variation has obviously been declining over time.

8 Introduction and Basics

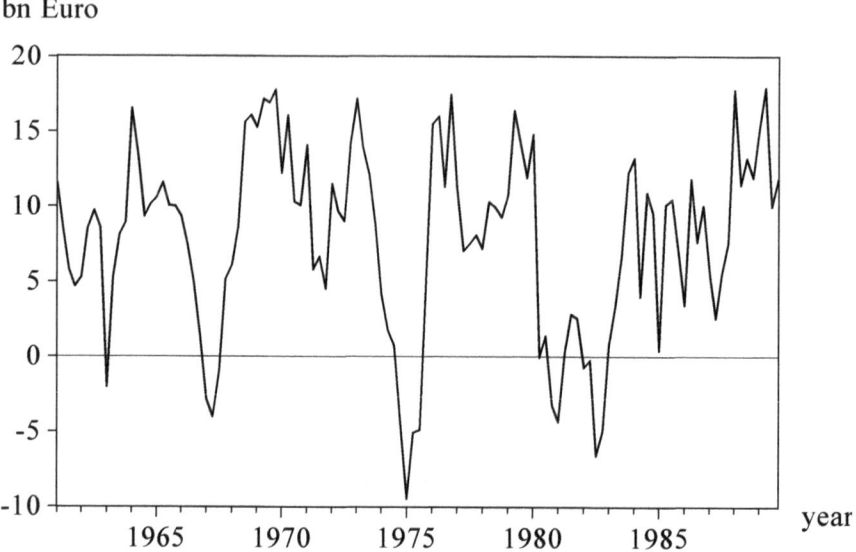

Figure 1.4: Annual Changes of the Real Gross Domestic Product ($\Delta_4 GDP$) of the Federal Republic of Germany, 1961 – 1989

If seasonal variations are to be eliminated, changes should be related to the same quarter of the preceding year and not to the preceding quarter. With $\Delta_4 GDP_t = GDP_t - GDP_{t-4}$, *Figure 1.4* shows the annual changes in the German Gross Domestic Product compared to the same quarter of the previous year. This series does no longer show any seasonal variations. These changes are mostly positive; they are only negative during recessions. This is particularly true for 1967, when Germany faced its first 'real' recession after the Second World War, as well as for the recessions in 1975 and 1981/82 which followed the two oil price shocks.

The annual growth rates, i.e. the corresponding relative annual changes (in percent), are, however, more revealing. They are presented in *Figure 1.5* and can be calculated as

$$\text{agr}_t = (\ln(GDP_t) - \ln(GDP_{t-4})) \cdot 100.$$

The sixties and seventies are characterised by highly fluctuating growth rates between -3.5 and just below 10 percent. In the seventies, the big recession of 1975 can clearly be recognised as well as the recession in the early eighties. Subsequently, real growth rates were positive, but at a lower level than before, between zero and just under five percent.

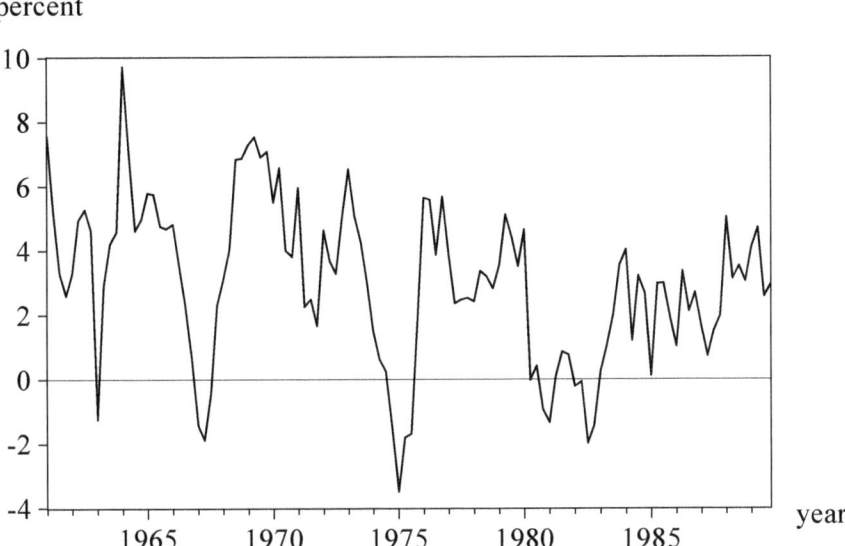

Figure 1.5: Annual Growth Rates of Real Gross Domestic Product (agr) of the Federal Republic of Germany, 1960 – 1989

A further possibility to eliminate seasonal variations without eliminating the trend is given by the following transformation:

$$\text{GDPS}_t = \frac{1}{4}(\text{GDP}_t + \text{GDP}_{t-1} + \text{GDP}_{t-2} + \text{GDP}_{t-3}).$$

Four consecutive values of the time series are added and, in order to avoid a shift in the level, divided by 4. Thus, we get an (unweighted) moving average of order four, i.e. with four elements. *Figure 1.6* shows the series GDP and GDPS for the period from 1961 to 2011. The latter indicates the long-term development, the so-called smooth component of the Gross Domestic Product around which the actual values fluctuate. The smooth component clearly indicates four (normal) recessions: in the late 1960's, the mid 1970's, the early 1980's and the last one after 1992. It also shows the structural break caused by the German Unification and the shift caused by the financial and economic crisis. But while the German Unification clearly caused a shift in the level of the series, the one caused by the financial and economic crisis might just be the result of the big recession from which the German economy had not yet fully recovered in 2011. Thus, it might have no long-run impact on the level of the series. It is also obvious that these shifts are partly smoothed and thus 'averaged away'. This example clearly shows that different ways of transforming one and the same time series can reveal the different kinds of information contained in it.

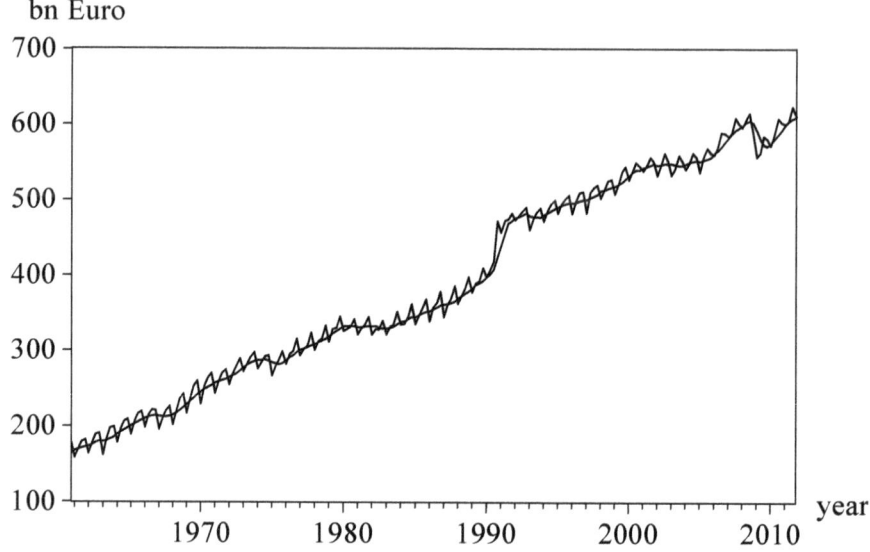

Figure 1.6: 'Smooth Component' and actual values of the Real Gross Domestic Product of the Federal Republic of Germany, 1961 – 2011

1.3 The Lag Operator

We introduce the *lag operator* L to show the relation between the differences and the moving average. Let x be a time series. If we apply the lag operator on this series, all values are delayed by one period, i.e.

(1.2) $\qquad Lx_t = x_{t-1}.$

If we apply the lag operator to x_{t-1}, we get x_{t-2} because of relation (1.2), and we can indicate

$$Lx_{t-1} = L(Lx_t) = L^2 x_t = x_{t-2}.$$

By generalising we get

(1.3) $\qquad L^k x_t = x_{t-k}, \quad k = ..., -1, 0, 1, 2, ... \, .$

For $k = 0$ we get the identity $L^0 x_t = x_t$. Usually, instead of L^0 we just write '1'. For $k > 0$ the series is shifted k periods backwards, and for $k < 0$ $|k|$ periods forward. For example: $L^{-3} x_t = x_{t+3}$. Furthermore, the usual rules for powers apply. Thus, we can write the following:

$$L^m x_{t-n} = L^m(L^n x_t) = L^{m+n} x_t = x_{t-(m+n)}.$$

1.3 The Lag Operator

The following notation results from using the lag operator for the first differences:

(1.4) $$\Delta x_t = x_t - x_{t-1} = (1-L)x_t.$$

For fourth differences it holds that

(1.5) $$\Delta_4 x_t = x_t - x_{t-4} = (1-L^4)x_t,$$

while growth rates as compared to the same quarter of the preceding year can be written as

(1.6) $$\Delta_4 \ln(x_t) = \ln(x_t) - \ln(x_{t-4}) = (1-L^4)\ln(x_t).$$

Finally, the unweighted moving average of order four can be written as

(1.7) $$xs_t = \frac{1}{4}(x_t + x_{t-1} + x_{t-2} + x_{t-3}) = \frac{1}{4}(1 + L + L^2 + L^3)x_t.$$

Quite generally, a polynomial of order p in the lag operator can be represented as

$$\alpha(L)x_t = (1 - \alpha_1 L - \alpha_2 L^2 - \ldots - \alpha_p L^p)x_t$$
$$= x_t - \alpha_1 x_{t-1} - \alpha_2 x_{t-2} - \ldots - \alpha_p x_{t-p}.$$

Trivially, there can be no delay if we apply the lag operator on a constant δ, i.e. it holds that

$$\alpha(L)\delta = (1 - \alpha_1 - \alpha_2 - \ldots - \alpha_p)\delta.$$

Thus, the value of the lag polynomial is the sum of all its coefficients in this case. We get the same result if we substitute L by $L^0 = 1$:

(1.8) $$\alpha(1) = 1 - \sum_{i=1}^{p} \alpha_i.$$

Relations (1.4) to (1.7) show the great advantage of the lag operator: transformations can be represented independently from the special time series, simply by a polynomial in the lag operator. Moreover, the same operations as with common polynomials (in real or complex variables) can be performed with lag polynomials, especially multiplication and division. For the multiplication the commutative law holds, i.e.

$$\alpha(L)\beta(L) = \beta(L)\alpha(L).$$

Such polynomials of the lag operator are also called 'linear filters'. If we multiply the first difference filter (1.4) with the moving average of third order (1.7) multiplied by four, we get the filter of fourth difference (1.5) because of

$$(1-L)(1+L+L^2+L^3) = (1-L^4).$$

This reveals that, as the long-term component is eliminated by the first difference filter and the seasonal component by the moving average, both components are eliminated from a time series by the product of those two filters, the filter of fourth differences.

1.4 Ergodicity and Stationarity

Formal models for time series are developed on the basis of probability theory. Let the T-dimensional vector of random variables $x_1, x_2, \ldots, x_T$ be given with the corresponding multivariate distribution. This can also be interpreted as a series of random variables $\{x_t\}_{t=1}^{T}$, as *stochastic process* or as *data generating process* (DGP). Let us now consider a sample of this process of length T. Consequently, the real numbers $\{x_1^{(1)}, x_2^{(1)}, \ldots, x_T^{(1)}\}$ are just one possible result of the underlying data generating process. Even if we were able to observe this process infinitely long, $\{x_t^{(1)}\}_{t=1}^{\infty}$ would be just one realisation of this stochastic process. It is obvious, however, that there is not just one realisation of such a process, but, in principle, an arbitrary number of realisations which all have the same statistical properties as they all result from the same data generating process.

In the following, a *time series* is considered as one realisation of the underlying stochastic process. We can also regard the stochastic process as the entirety of all of its possible realisations. To make the notation as simple as possible, we will not distinguish between the process itself and its realisation. This can be taken out of the context.

Stochastic processes of the dimension T can be completely described by a T-dimensional distribution function. This is, however, not a practicable procedure. We rather concentrate on the first and second order moments, i.e. on the mean (or expected value)

$$E[x_t], \, t = 1, 2, \ldots, T,$$

the T variances

$$V[x_t] = E[(x_t - E[x_t])^2], \quad t = 1, 2, \ldots, T,$$

as well as the T(T-1)/2 covariances

1.4 Ergodicity and Stationarity

$$Cov[x_t, x_s] = E[(x_t - E[x_t])(x_s - E[x_s])], \quad t < s.$$

Quite often, these are denoted as autocovariances because they are covariances between random variables of the same stochastic process. If the stochastic process has a multivariate normal distribution, its distribution function is fully described by its moments of first and second order. This holds, however, only in this special case.

As we usually have only one time series, i.e. just one realisation of the stochastic process in practical applications, we have to make additional assumptions in order to be able to perform statistical inference. For example, to be able to estimate the expected value, the variance and the covariances of the stochastic process $\{x_t\}$, there should be more than one realisation of this random variable available for a given point in time t.

The assumption of *ergodicity* means that the sample moments which are calculated on the basis of a time series with a finite number of observations converge (in some sense) for $T \to \infty$ against the corresponding moments of the population. This concept is only meaningful, however, if we can assume that, for example, the expectations $E[x_t] = \mu$ and the variances $V[x_t] = \sigma_x^2$ are constant for all t.

More precisely, a DGP is said to be *mean ergodic* if

$$\lim_{T \to \infty} E\left[\left(\frac{1}{T}\sum_{t=1}^{T} x_t - \mu\right)^2\right] = 0,$$

and variance ergodic if

$$\lim_{T \to \infty} E\left[\left(\frac{1}{T}\sum_{t=1}^{T} (x_t - \mu)^2 - \sigma_x^2\right)^2\right] = 0.$$

These conditions are 'consistency properties' for dependent random variables and cannot be tested. Therefore, they have to be assumed.

A stochastic process has to be in statistical equilibrium in order to be ergodic, i.e. it has to be stationary. Two different kinds of stationarity can be distinguished. If we assume that the common distribution function of the stochastic process does not change by a shift in time, the process is said to be *strictly stationary*. As this concept is difficult to apply in practice, we only consider *weak stationarity* or *stationarity in the second moments*. We first define stationarity for the corresponding moments of the stochastic process $\{x_t\}$:

(i) *Mean Stationarity*: A process is mean stationary if $E[x_t] = \mu_t = \mu$ is constant for all t.

14 Introduction and Basics

(ii) *Variance Stationarity*: A process is variance stationary if $V[x_t] = E[(x_t - \mu_t)^2] = \sigma_x^2 = \gamma(0)$ is constant and finite for all t.

(iii) *Covariance Stationarity*: A process is covariance stationary if $Cov[x_t, x_s] = E[(x_t - \mu_t)(x_s - \mu_s)] = \gamma(|s-t|)$ is only a function of the time distance between the two random variables and does not depend on the actual point in time t.

(iv) *Weak Stationarity*: As variance stationarity immediately results from covariance stationarity for s = t, a stochastic process is weakly stationary when it is mean and covariance stationary.

Because we only assume this kind of stationarity in the following, we will mostly drop the adjective weak.

Example 1.1

We call the stochastic process $\{u_t\}$ a *pure random or a white noise process*, if it has the following properties: $E[u_t] = 0$ and $V[u_t] = \sigma^2$ for all t, as well as $Cov[u_t, u_s] = E[u_t u_s] = 0$ for all $t \neq s$. Apparently, this process is weakly stationary. The random variables all have mean zero and variance σ^2 and are uncorrelated with each other.

Example 1.2

Let the stochastic process $\{x_t\}$ be defined as

(E1.1) $$x_t = \begin{cases} u_1 & \text{for} \quad t = 1, \\ x_{t-1} + u_t & \text{for} \quad t = 2, 3, \ldots, \end{cases}$$

where $\{u_t\}$ is a pure random process. This stochastic process, a *random walk without drift*, can also be written as

(E1.2) $$x_t = \sum_{j=1}^{t} u_j \ .$$

Let us assume that we generate $\{u_t\}$ by flipping a fair coin. We get heads with probability 0.5 (in this case, our random variable has the value +1) and tails with probability 0.5 (in this case, our random variable has the value -1). Let us start, for example, with $x_0 = 0$ for t = 0. Then it is easy to see that all possible realisations (time series) of this random walk can only take values within the area in *Figure 1.7* which is limited by the two angle bisectors. If each flip results in heads (tails), the corresponding time series would take the value +1 (-1) for t = 1, the value +2 (-2) for t = 2, and so on.

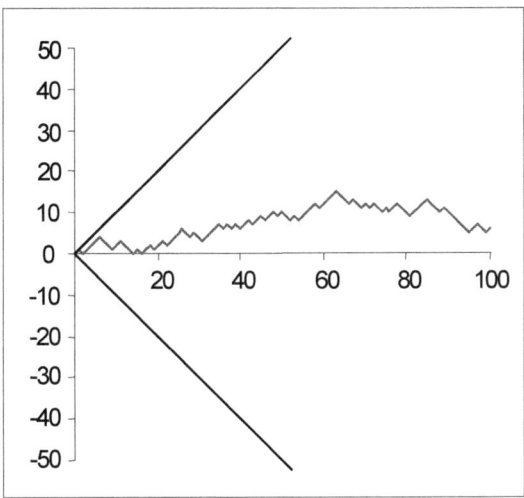

Figure 1.7: Example of a Random Walk where only the steps +1 and –1 are possible

Which moments of first and second order does the stochastic process as defined in (E1.1) have? Due to (E1.2) and the properties of a pure random process it holds that

$$E[x_t] = E\left[\sum_{j=1}^{t} u_j\right] = \sum_{j=1}^{t} E[u_j] = 0,$$

$$V[x_t] = V\left[\sum_{j=1}^{t} u_j\right] = \sum_{j=1}^{t} V[u_j] = t\sigma^2, \text{ and}$$

$$Cov[x_t, x_s] = E\left[\left(\sum_{j=1}^{t} u_j\right)\left(\sum_{i=1}^{s} u_i\right)\right] = \sum_{j=1}^{t}\sum_{i=1}^{s} E[u_j u_i] = \min(t,s)\sigma^2.$$

Thus, the random walk without drift is mean stationary, but neither variance nor covariance stationary and, consequently, also not weakly stationary. The random walk without drift is an important element of a category of nonstationary stochastic processes which, as will be shown later, are well suited to describe the development of economic time series.

It is impossible to evaluate the strength of dependence of random variables of a stochastic process by using autocovariances as these are not normalised and, therefore, dependent on the applied measurement units. If the covariances are normalised with the respective variances, the result is a term which is independent of the applied measurement unit, the autocorrelation function. For weakly stationary processes this is given by

$$\text{(1.9)} \quad \rho(\tau) = \frac{E[(x_t - \mu)(x_{t+\tau} - \mu)]}{E[(x_t - \mu)^2]} = \frac{\gamma(\tau)}{\gamma(0)}, \quad \tau = ..., -1, 0, 1, ...,$$

and has the following properties:

(i) $\quad\quad\quad\quad\quad\quad\quad \rho(0) = 1$,

(ii) $\quad\quad\quad\quad\quad\quad\quad \rho(\tau) = \rho(-\tau)$, and

(iii) $\quad\quad\quad\quad\quad\quad\quad |\rho(\tau)| \leq 1$, for all τ.

Because of (i) and the symmetry (ii) it is sufficient to know the autocorrelation function or the autocorrelogram for $\tau = 1, 2, ...$.

Due to the ergodicity assumption, mean, variance and autocovariances of stationary processes can be estimated in the following way:

$$\hat{\mu} = \frac{1}{T}\sum_{t=1}^{T} x_t,$$

$$\hat{\gamma}(0) = \frac{1}{T}\sum_{t=1}^{T}(x_t - \hat{\mu})^2,$$

$$\hat{\gamma}(\tau) = \frac{1}{T}\sum_{t=1}^{T-\tau}(x_t - \hat{\mu})(x_{t+\tau} - \hat{\mu}), \quad \tau = 1, 2, ..., T-1.$$

These are consistent estimators of μ, $\gamma(0)$ und $\gamma(\tau)$. The consistent estimator of the autocorrelation function is given by

$$\text{(1.10)} \quad \hat{\rho}(\tau) = \frac{\sum_{t=1}^{T-\tau}(x_t - \hat{\mu})(x_{t+\tau} - \hat{\mu})}{\sum_{t=1}^{T}(x_t - \hat{\mu})^2} = \frac{\hat{\gamma}(\tau)}{\hat{\gamma}(0)}, \quad \tau = 1, 2, ..., T-1.$$

This estimator is asymptotically unbiased. For white noise processes, its variance can be approximated by $1/T$ and is asymptotically normally distributed. Due to this, approximate pointwise 95 percent confidence intervals of $\pm 2/\sqrt{T}$ are often indicated for the estimated autocorrelation coefficients.

According to MAURICE STEVENSON BARTLETT (1946), the variance of autocorrelation coefficients of stochastic processes in which all autocorrelation coefficients disappear from the index value $k + 1$ on, $\rho(\tau) = 0$ for $\tau > k$, is approximately given by

$$V[\hat\rho(\tau)] \approx \frac{1}{T}\left(1+2\sum_{j=1}^{k}\rho(j)^2\right), \quad \tau > k.$$

In order to evaluate estimated time series models, it is important to know whether the residuals of the model really have the properties of a pure random process, in particular, whether they are uncorrelated. Thus, the null hypothesis to be tested is

$$H_0: \rho(\tau) = 0 \text{ for } \tau = 1, 2, ..., m, \, m < T.$$

The first possibility to check this is to apply the 95 percent confidence limits $\pm 2/\sqrt{T}$ valid under the null hypothesis to every estimated correlation coefficient. Under H_0 at most 5 percent of $\hat\rho(\tau)$ may lie outside these limits.

To make a global statement, i.e. to test the common hypothesis whether a given number of m autocorrelation coefficients are zero altogether, GEORGE E. P. BOX and DAVID A. PIERCE (1970) have developed the following test statistic:

(1.11) $$Q^* = T \sum_{j=1}^{m} \hat\rho(j)^2.$$

Under the null hypothesis it is asymptotically χ^2 distributed with m-k degrees of freedom, k being the number of estimated parameters.

As – strictly applied – the distribution of this test statistic holds only asymptotically, GRETA M. LJUNG and GEORGE E. P. BOX (1978) proposed the following modification for small samples,

(1.12) $$Q = T(T+2) \sum_{j=1}^{m} \frac{\hat\rho(j)^2}{T-j},$$

which is also asymptotically χ^2 distributed with m-k degrees of freedom.

It should be intuitively clear that the null hypothesis of non-autocorrelation of the residuals should be rejected if some of the $\hat\rho(j)$ are too large, i.e. if Q^* or Q is too large, or – to be more precise – if they are larger than the corresponding critical values of the χ^2 distribution with m-k degrees of freedom for a specified significance level.

An alternative to these testing procedures is the Lagrange-Multiplier Test (LM Test) developed by TREVOR S. BREUSCH (1978) and LESLIE G. GODFREY (1978). Like for the Q (Q^*) test the null hypothesis is

H_0: The residuals are not autocorrelated,

which is tested against the alternative that the residuals follow an autoregressive or a moving average process of order m. The test can be performed with an auxiliary regression. The estimated residuals are regressed on the explanatory variables of the main model and on the lagged residuals, up to order m. The test statistic which is χ^2 distributed with m degrees of freedom is given by T times the multiple correlation coefficient R^2 of the auxiliary regression, with T being the number of observations. Alternatively, an F test can be used for testing the combined significance of the lagged residuals in the auxiliary regression.

Compared to the Durbin-Watson test which is used in traditional econometrics for testing autocorrelation of the residuals of an estimated model, the Q (Q*) as well as the LM test have two major advantages: firstly, they can check for autocorrelation of any order, and not only of first order. Secondly, the results are also correct if there are lagged endogenous variables in the regression equation, whereas in such cases the results of the Durbin-Watson test are biased in favour of the null hypothesis.

The fact that the residuals are not autocorrelated does not imply that they are independently and/or normally distributed; absence of autocorrelation does only imply stochastic independence if the variables are normally distributed. It is, however, often assumed that they are normally distributed, as the usual testing procedures are based on this assumption. Whether this is actually true depends on the higher moments of the distribution. Especially the third and fourth moments are important,

$$E[(x_t - E[x_t])^i], \quad i = 3, 4.$$

The third moment is necessary to determine the skewness of the distribution which can be estimated by

$$\hat{S} = \frac{1}{T} \frac{\sum_{t=1}^{T}(x_t - \hat{\mu})^3}{\sqrt{\hat{\gamma}(0)^3}}.$$

For symmetric distributions (as the normal distribution) the theoretical value of the skewness is zero. The kurtosis which is based on the forth moment can be estimated by

$$\hat{K} = \frac{1}{T} \frac{\sum_{t=1}^{T}(x_t - \hat{\mu})^4}{\hat{\gamma}(0)^2}.$$

For the normal distribution it holds that K = 3. Values larger than three indicate that the distribution has 'fat tails': the density of a distribution in the centre and at the tails, i.e. outside the usual ± 2σ limits, is higher and in the areas in between smaller than the density of a normal distribution. This holds, for example, for the t distribution. Such fat tails are typical for high frequency financial market data.

Using the skewness S and the kurtosis K, CARLOS M. JARQUE and ANIL K. BERA (1980) proposed a test for normality. It can be applied directly on the time series itself (or on its differences). Usually, however, it is applied to check estimated regression residuals. The test statistic

$$JB \;=\; \frac{T}{6}\left(\hat{S}^2 + \frac{1}{4}(\hat{K}-3)^2\right)$$

is χ^2 distributed with 2 degrees of freedom. T is again the sample size. The hypothesis that the variable is normally distributed is rejected whenever the values of the test statistic are larger than the corresponding critical values.

Example 1.3

The price development in efficient markets as, for example, stock prices or exchange rates, can often be represented by a random walk. An example is the exchange rate between the Swiss Franc and the U.S. Dollar. Monthly data of this series are shown in *Figure 1.8a* for the period from January 1974 to December 2011. Below this, continuous monthly returns corresponding to (1.1') are presented. They behave like a pure random process. This can be seen from the correlogram: none of the estimated correlation coefficients which are presented in Figure 1.8c is significantly different from zero. (The dashed lines in *Figure 1.8c* represent the approximate 95 percent confidence limits.) Moreover, neither the Ljung-Box Q test nor the Breusch-Godfrey LM test indicate autocorrelation: For m = 2 and m = 12 the test statistics are Q(2) = 0.767, Q(12) = 11.813, LM(2) = 0.749, LM(12) = 13.608. (The critical values of the χ^2 distribution with 2 degrees of freedom are 4.605 and with 12 degrees of freedom 18.549, both at the 10 percent significance level.) On the other hand, the hypothesis of normality has to be rejected at the 0.1 percent level since JB = 18.178. (The critical value of the χ^2 distribution with 2 degrees of freedom at the 0.1 percent level is 13.816). The reason for this is the kurtosis with a value of 3.964.

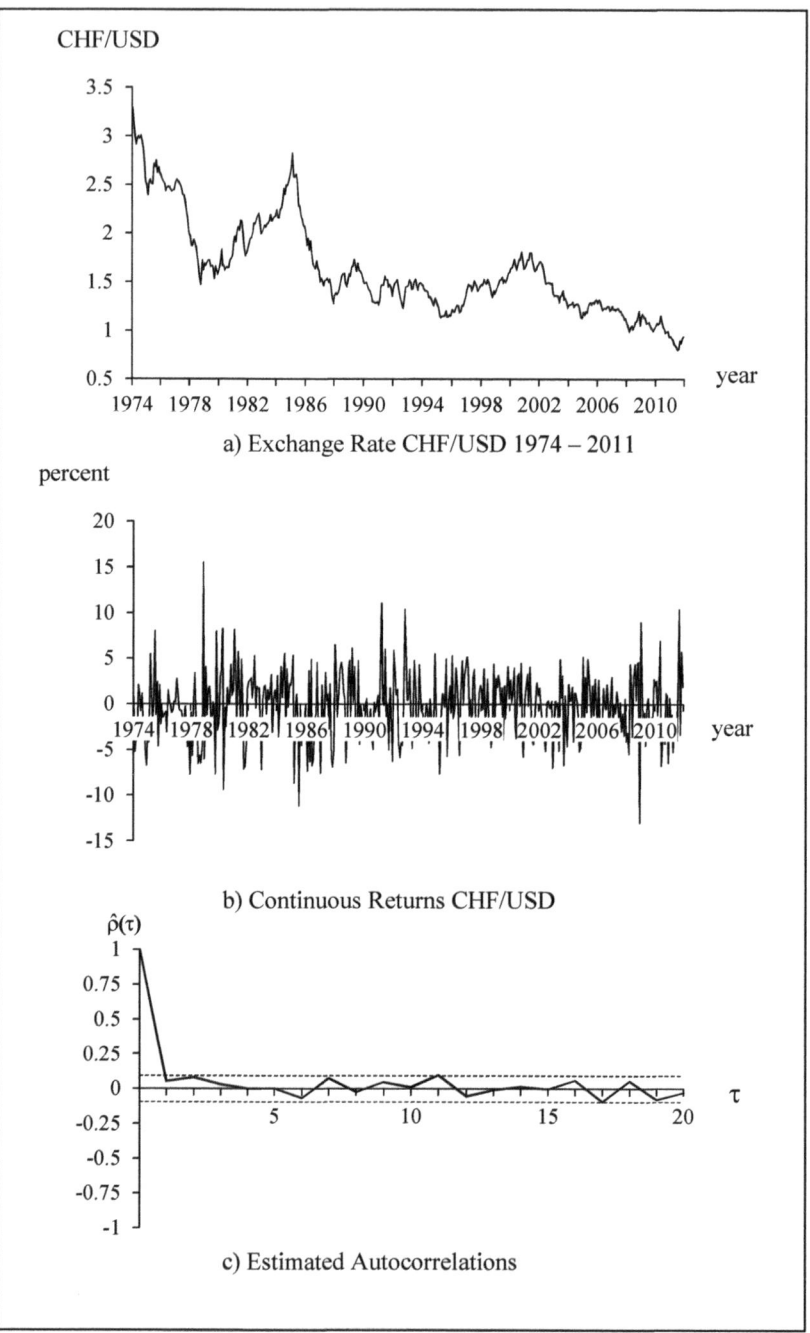

Figure 1.8: Exchange Rate Swiss Franc U.S. Dollar, Monthly data, January 1974 to December 2011

1.5 The Wold Decomposition

Before we deal with special models of stationary processes, a general property of such processes is discussed: the *Wold Decomposition*. This decomposition traces back to HERMAN WOLD (1938). It exists for every covariance stationary, purely non-deterministic stochastic process: After subtracting the mean function, each of such processes can be represented by a linear combination of a series of uncorrelated random variables with zero mean and constant variance, which are the errors made in forecasting x_t on the basis of a linear function of lagged x.

Purely non-deterministic means that all additive deterministic components of a time series have to be subtracted in advance. By using its own lagged values, any deterministic component can be perfectly predicted in advance. This holds, for example, for a constant mean, as well as for periodic, polynomial, or exponential series in t. Thus, one can write:

$$(1.13) \quad x_t - \mu_t = \sum_{j=0}^{\infty} \psi_j u_{t-j} \quad \text{with} \quad \psi_0 = 1 \quad \text{and} \quad \sum_{j=0}^{\infty} \psi_j^2 < \infty.$$

There, u_t is a pure random process, i.e. it holds that

$$E[u_t] = 0 \quad \text{and} \quad E[u_t\, u_s] = \begin{cases} \sigma^2 & \text{for} \quad t = s \\ 0 & \text{otherwise} \end{cases}.$$

The quadratic convergence of the series of the ψ_j guarantees the existence of second moments of the process. There is no need of any distributional assumption for this decomposition to hold. Especially, there is no need of the u_t to be independent, it is sufficient that they are uncorrelated.

For the mean we get

$$E[x_t - \mu_t] = E\left[\sum_{j=0}^{\infty} \psi_j u_{t-j}\right] = \sum_{j=0}^{\infty} \psi_j E[u_{t-j}] = 0,$$

i.e., it holds that

$$E[x_t] = \mu_t.$$

The variance can be calculated as follows:

$$V[x_t] = E[(x_t - \mu_t)^2] = E\left[(u_t + \psi_1 u_{t-1} + \psi_2 u_{t-2} + \ldots)^2\right].$$

Because of $E[u_t u_{t-j}] = 0$ for $j \neq 0$, this can be simplified to

$$V[x_t] = E[u_t^2] + \psi_1^2 E[u_{t-1}^2] + \psi_2^2 E[u_{t-2}^2] + \ldots$$

$$= \sigma^2 \sum_{j=0}^{\infty} \psi_j^2 = \gamma(0).$$

Thus, the variance is finite and not time dependent. Correspondingly, with $\tau > 0$ we get the time independent autocovariances

$$\begin{aligned}
\text{Cov}[x_t, x_{t+\tau}] &= E[(x_t - \mu_t)(x_{t+\tau} - \mu_{t+\tau})] \\
&= E[(u_t + \psi_1 u_{t-1} + \ldots + \psi_\tau u_{t-\tau} + \psi_{\tau+1} u_{t-\tau-1} + \ldots) \\
&\quad \cdot (u_{t+\tau} + \psi_1 u_{t+\tau-1} + \ldots + \psi_\tau u_t + \psi_{\tau+1} u_{t-1} + \ldots)] \\
&= \sigma^2 (1 \cdot \psi_\tau + \psi_1 \psi_{\tau+1} + \psi_2 \psi_{\tau+2} + \ldots) \\
&= \sigma^2 \sum_{j=0}^{\infty} \psi_j \psi_{\tau+j} = \gamma(\tau) < \infty,
\end{aligned}$$

with $\psi_0 = 1$. It becomes clear that the autocovariances are only functions of the time difference, i.e. the distance between two random variables. Thus, all conditions of covariance stationarity are fulfilled. Because of (1.9) the autocorrelation function is given by:

$$\rho(\tau) = \frac{\sum_{j=0}^{\infty} \psi_j \psi_{\tau+j}}{\sum_{j=0}^{\infty} \psi_j^2}, \quad \tau = 1, 2, \ldots.$$

All stationary models discussed in the following chapters can be represented on the basis of the Wold Decomposition (1.13). However, this representation is, above all, interesting for theoretical reasons: in practice, applications of models with an infinite number of parameters are hardly useful.

References

An introduction to the **history of time series analysis** is given by

MARC NERLOVE, DAVID M. GRETHER and JOSÉ L. CARVALHO, *Analysis of Economic Time Series: A Synthesis*, Academic Press, New York et al. 1979, pp. 1 – 21.

The first estimated **econometric model** was presented in

JAN TINBERGEN, *Statistical Analysis of Business Cycle Theories, Vol. 1: A Method and Its Application to Business Cycle Theory, Vol. 2: Business Cycles in the*

United States of America, 1919 – 1932, League of Nations, Economic Intelligence Service, Geneva 1939.

That **autocorrelation of the residuals** can cause problems for the statistical estimation and testing of econometric models was first noticed by

DONALD COCHRANE and GUY H. ORCUTT, Application of Least Squares Regression to Relationships Containing Autocorrelated Error Terms, *Journal of the American Statistical Association* 44 (1949), pp. 32 – 61.

In this article, one can also find the transformation to eliminate first order autocorrelation which was named after these two authors. With this transformation and the testing procedure proposed by

JAMES DURBIN and GEOFFREY S. WATSON, Testing for Serial Correlation in Least Squares Regression, I, *Biometrika* 37 (1950), pp. 409 – 428; II, *Biometrika* 38 (1951), pp. 159 – 178,

econometricians believed to cope with these problems.

However, **methods of time series analysis** had already been **applied** earlier to investigate economic time series.

WARREN M. PERSONS, Indices of Business Conditions, *Review of Economic Statistics* 1 (1919), pp. 5 – 107,

was the first to distinguish different components of economic time series. Such procedures are still applied today. For example, the seasonal adjustment procedure SEATS, which is used by EUROSTAT and which is described in

AUGUSTIN MARAVALL and VICTOR GOMEZ, The Program SEATS: ‚Signal Extraction in ARIMA Time Series', Instruction for the User, European University Institute, Working Paper ECO 94/28, Florence 1994,

is based on such an approach.

The **more recent development** of time series analysis has been initiated by the textbook of

GEORGE E.P. BOX and GWILYM M. JENKINS, *Time Series Analysis: Forecasting and Control*, Holden Day, San Francisco et al. 1970; 2nd enlarged edition 1976.

This book mainly proposes the **time domain for the analysis of time series** and focuses on univariate models. The **theoretical basis** of this approach is the **decomposition theorem** for stationary time series shown by

HERMAN WOLD, *A Study in the Analysis of Stationary Time Series*, Almquist and Wicksell, Stockholm 1938.

An argument in favour of the application of this time series approach is that short-term predictions thus generated are often considerably better than predictions generated by the use of large econometric models. This was shown, for example, by

CLIVE W.J. GRANGER and PAUL NEWBOLD, Economic Forecasting: The Atheist's Viewpoint, in: G.A. RENTON (ed.), *Modelling the Economy*, Heinemann, London 1975, pp. 131 – 148.

Besides analyses in the time domain there is also the possibility to analyse time series in the **frequency domain**. See, for example,

CLIVE W.J. GRANGER and MICHIO HATANAKA, *Spectral Analysis of Economic Time Series*, Princeton University Press, Princeton N.J. 1964.

Extensive **surveys on modern methods of time series analysis** are given by

JAMES D. HAMILTON, *Time Series Analysis*, Princeton University Press, Princeton N.J. 1994, and

HELMUT LÜTKEPOHL, *New Introduction to Multiple Time Series Analysis*, Springer, Berlin et al., 2005.

In JAMES D. HAMILTON's book one can also find remarks on the relation between ergodicity and stationarity (pp. 45ff.).

Textbooks focusing on the application of these methods are

WALTER ENDERS, *Applied Econometric Time Series*, Wiley, New York, 3rd edition 2010, as well as

HELMUT LÜTKEPOHL and MARKUS KRÄTZIG (eds.), *Applied Time Series Econometrics*, Cambridge University Press, Cambridge et al. 2004.

For a **deeper discussion of stochastic processes** see, for example,

ARIS SPANOS, *Statistical Foundations of Econometric Modelling*, Cambridge University Press, Cambridge (England) et al. 1986, pp. 130ff, or

EMANUEL PARZEN, *Stochastic Processes*, Holden-Day, San Francisco 1962.

The **test statistic for the variance of single estimated autocorrelation coefficients** is given by

MAURICE STEVENSON BARTLETT, On the Theoretical Specification and Sampling Properties of Auto-Correlated Time Series, *Journal of the Royal Statistical Society (Supplement)* 8 (1946), pp. 24 – 41.

The **statistic for testing a given number of autocorrelation coefficients** was developed by

GEORGE E.P. BOX and DAVID A. PIERCE, Distribution of Residual Autocorrelations in Autoregressive Moving Average Time Series Models, *Journal of the American Statistical Association* 65 (1970), pp. 1509 – 1526,

while the modification for small samples is due to

GRETA M. LJUNG and GEORGE E.P. BOX, On a Measure of Lack of Fit in Time Series Models, *Biometrika* 65 (1978), pp. 297 – 303.

The **Lagrange-Multiplier test** for residual autocorrelation has been developed by

TREVOR S. BREUSCH, Testing for Autocorrelation in Dynamic Linear Models, *Australian Economic Papers* 17 (1978), pp. 334 – 355, and by

LESLIE G. GODFREY, Testing Against General Autoregressive and Moving Average Error Models When Regressors Include Lagged Dependent Variables, *Econometrica* 46 (1978), S. 1293 – 1302.

The **test on normal distribution** presented above has been developed by

CARLOS M. JARQUE and ANIL K. BERA, Efficient Tests for Normality, Homoscedasticity and Serial Independence of Regression Residuals, *Economics Letters* 6 (1980), pp. 255 – 259.

2 Univariate Stationary Processes

As mentioned in the introduction, the publication of the textbook by GEORGE E.P. BOX and GWILYM M. JENKINS in 1970 opened a new road to the analysis of economic time series. This chapter presents the Box-Jenkins Approach, its different models and their basic properties in a rather elementary and heuristic way. These models have become an indispensable tool for short-run forecasts. We first present the most important approaches for statistical modelling of time series. These are autoregressive (AR) processes (*Section 2.1*) and moving average (MA) processes (*Section 2.2*), as well as a combination of both types, the so-called ARMA processes (*Section 2.3*). In *Section 2.4* we show how this class of models can be used for predicting the future development of a time series in an optimal way. Finally, we conclude this chapter with some remarks on the relation between the univariate time series models described in this chapter and the simultaneous equations systems of traditional econometrics (*Section 2.5*).

2.1 Autoregressive Processes

We know autoregressive processes from traditional econometrics: Already in 1949, DONALD COCHRANE and GUY H. ORCUTT used the first order autoregressive process for modelling the residuals of a regression equation. We will start with this process, then treat the second order autoregressive process and finally show some properties of autoregressive processes of an arbitrary but finite order.

2.1.1 First Order Autoregressive Processes

Derivation of Wold's Representation

A *first order autoregressive process*, an AR(1) process, can be written as an inhomogeneous stochastic first order difference equation,

(2.1) $$x_t = \delta + \alpha\, x_{t-1} + u_t,$$

where the inhomogeneous part $\delta + u_t$ consists of a constant term δ and a pure random process u_t. Let us assume that for $t = t_0$ the initial value x_{t_0} is given. By successive substitution in (2.1) we get

$$x_{t_0+1} = \delta + \alpha x_{t_0} + u_{t_0+1}$$

$$x_{t_0+2} = \delta + \alpha x_{t_0+1} + u_{t_0+2}$$
$$= \delta + \alpha(\delta + \alpha x_{t_0} + u_{t_0+1}) + u_{t_0+2}$$
$$= \delta + \alpha\delta + \alpha^2 x_{t_0} + \alpha u_{t_0+1} + u_{t_0+2}$$

$$x_{t_0+3} = \delta + \alpha x_{t_0+2} + u_{t_0+3}$$

$$x_{t_0+3} = \delta + \alpha\delta + \alpha^2\delta + \alpha^3 x_{t_0} + \alpha^2 u_{t_0+1} + \alpha u_{t_0+2} + u_{t_0+3}$$

$$\vdots$$

$$x_{t_0+\tau} = (1 + \alpha + \alpha^2 + \ldots + \alpha^{\tau-1})\delta + \alpha^\tau x_{t_0}$$
$$+ \alpha^{\tau-1} u_{t_0+1} + \alpha^{\tau-2} u_{t_0+2} + \ldots + \alpha u_{t_0+\tau-1} + u_{t_0+\tau},$$

or

$$x_{t_0+\tau} = \alpha^\tau x_{t_0} + \frac{1-\alpha^\tau}{1-\alpha}\delta + \sum_{j=0}^{\tau-1} \alpha^j u_{t_0+\tau-j}.$$

For $t = t_0 + \tau$, we get

(2.2) $$x_t = \alpha^{t-t_0} x_{t_0} + \frac{1-\alpha^{t-t_0}}{1-\alpha}\delta + \sum_{j=0}^{t-t_0-1} \alpha^j u_{t-j}.$$

The development and thus the properties of this process are mainly determined by the assumptions on the initial condition x_{t_0}.

The case of a *fixed (deterministic) initial condition* is given if x_0 is assumed to be a fixed (real) number, for example for $t_0 = 0$, i.e. no random variable. Then we can write:

$$x_t = \alpha^t x_0 + \frac{1-\alpha^t}{1-\alpha}\delta + \sum_{j=0}^{t-1} \alpha^j u_{t-j}.$$

This process consists of time dependent deterministic and stochastic parts. Thus, it can never be weakly stationary, since first and second order moments are time dependent. It is, however, *asymptotically stationary* because the time dependence vanishes for $t_0 \to -\infty$.

We can imagine the case of *stochastic initial conditions* as (2.1) being generated along the whole time axis, i.e. $-\infty < t < \infty$. If we observe the process only for positive values of t, the initial value x_0 is a random variable which is generated by this process. Formally, the process with stochastic initial conditions results from (2.2) if the solution of the homogeneous difference equation has disappeared. This is only possible if $|\alpha| < 1$. Therefore, in the following, we restrict α to the interval $-1 < \alpha < 1$. If $\lim_{t_0 \to -\infty} x_{t_0}$ is bounded, (2.2) for $t_0 \to -\infty$ converges to

$$(2.3) \qquad x_t = \frac{\delta}{1-\alpha} + \sum_{j=0}^{\infty} \alpha^j u_{t-j} .$$

The time dependence has disappeared. According to *Section 1.5*, the AR(1) process (2.1) has the Wold representation (2.3) with $\psi_j = \alpha^j$ and $|\alpha| < 1$. This results in the convergence of

$$\sum_{j=0}^{\infty} \psi_j^2 = \sum_{j=0}^{\infty} \alpha^{2j} = \frac{1}{1-\alpha^2} .$$

Thus, assuming stochastic initial conditions, the process (2.1) is weakly stationary.

The Lag Operator

Equation (2.3) can also be derived from relation (2.1) by using the lag operator defined in *Section 1.3*:

$$(2.1') \qquad (1 - \alpha L)x_t = \delta + u_t .$$

If we solve for x_t we get

$$(2.4) \qquad x_t = \frac{\delta}{1-\alpha L} + \frac{1}{1-\alpha L} u_t .$$

The expression $1/(1 - \alpha L)$ can formally be expanded to a geometric series,

$$\frac{1}{1-\alpha L} = 1 + \alpha L + \alpha^2 L^2 + \alpha^3 L^3 + \dots .$$

Thus, we get

$$\begin{aligned} x_t &= (1 + \alpha L + \alpha^2 L + \dots)\delta + (1 + \alpha L + \alpha^2 L + \dots)u_t \\ &= (1 + \alpha + \alpha^2 + \dots)\delta + u_t + \alpha u_{t-1} + \alpha^2 u_{t-2} + \dots , \end{aligned}$$

and because of $|\alpha| < 1$

$$x_t = \frac{\delta}{1-\alpha} + \sum_{j=0}^{\infty} \alpha^j u_{t-j}.$$

The first term could have been derived immediately if we substituted the value '1' for L in the first term of (2.4). (See also relation (1.8) on p. 11).

Calculation of Moments

Due to representation (2.3), the first and second order moments can be calculated. As $E[u_t] = 0$ holds for all t, we get for the mean

$$E[x_t] = E\left[\frac{\delta}{1-\alpha} + \sum_{j=0}^{\infty} \alpha^j u_{t-j}\right]$$

$$E[x_t] = \frac{\delta}{1-\alpha} + \sum_{j=0}^{\infty} \alpha^j E[u_{t-j}] = \frac{\delta}{1-\alpha} = \mu$$

i.e. the mean is constant. It is different from zero if and only if $\delta \neq 0$. Because of $1 - \alpha > 0$, the sign of the mean is determined by the sign of δ. For the variance we get

$$V[x_t] = E\left[\left(x_t - \frac{\delta}{1-\alpha}\right)^2\right] = E\left[\left(\sum_{j=0}^{\infty} \alpha^j u_{t-j}\right)^2\right]$$

$$= E[(u_t + \alpha u_{t-1} + \alpha^2 u_{t-2} + \ldots)^2]$$

$$= E[u_t^2 + \alpha^2 u_{t-1}^2 + \alpha^4 u_{t-2}^2 + \ldots + 2\alpha u_t u_{t-1} + 2\alpha^2 u_t u_{t-2} + \ldots]$$

$$= \sigma^2(1 + \alpha^2 + \alpha^4 + \ldots),$$

because $E[u_t u_s] = 0$ for $t \neq s$ and $E[u_t u_s] = \sigma^2$ for $t = s$. Applying the summation formula for the geometric series, and because of $|\alpha| < 1$, we get the constant variance

$$V[x_t] = \frac{\sigma^2}{1-\alpha^2}.$$

The covariances can be calculated as follows:

$$\text{Cov}[x_t, x_{t-\tau}] = E\left[\left(x_t - \frac{\delta}{1-\alpha}\right)\left(x_{t-\tau} - \frac{\delta}{1-\alpha}\right)\right]$$

$$
\begin{aligned}
&= E[(u_t + \alpha\, u_{t-1} + \ldots + \alpha^\tau u_{t-\tau} + \ldots) \\
&\qquad \cdot (u_{t-\tau} + \alpha\, u_{t-\tau-1} + \alpha^2 u_{t-\tau-2} + \ldots)] \\
&= E[(u_t + \alpha\, u_{t-1} + \ldots + \alpha^{\tau-1} u_{t-\tau+1} \\
&\qquad + \alpha^\tau (u_{t-\tau} + \alpha\, u_{t-\tau-1} + \alpha^2 u_{t-\tau-2} + \ldots)) \\
&\qquad \cdot (u_{t-\tau} + \alpha\, u_{t-\tau-1} + \alpha^2 u_{t-\tau-2} + \ldots)] \\
&= \alpha^\tau E[(u_{t-\tau} + \alpha u_{t-\tau-1} + \alpha^2 u_{t-\tau-2} + \ldots)^2] \, .
\end{aligned}
$$

Thus, we get

$$
\text{Cov}[x_t, x_{t-\tau}] = \alpha^\tau V[x_{t-\tau}] = \alpha^\tau \frac{\sigma^2}{1-\alpha^2} \, .
$$

The autocovariances are only a function of the time difference τ and not of time t, and we can write:

(2.5) $$\gamma(\tau) = \alpha^\tau \frac{\sigma^2}{1-\alpha^2}, \quad \tau = 0, 1, 2, \ldots \, .$$

Therefore, the AR(1) process with $|\alpha| < 1$ and stochastic initial conditions is weakly stationary.

An Alternative Method for the Calculation of Moments

Under the condition of weak stationarity, i.e. for $|\alpha| < 1$ and stochastic initial conditions, the mean of x_t is constant. If we apply the expectation operator on equation (2.1), we get:

$$E[x_t] = E[\delta + \alpha\, x_{t-1} + u_t] = \delta + \alpha\, E[x_{t-1}] + E[u_t] \, .$$

Because of $E[u_t] = 0$ and $E[x_t] = E[x_{t-1}] = \mu$ for all t we can write

$$E[x_t] = \mu = \frac{\delta}{1-\alpha} \, .$$

If we consider the deviations from the mean,

$$\tilde{x}_t = x_t - \mu$$

and substitute this in relation (2.1), we get:

$$\tilde{x}_t + \mu = \delta + \alpha\, \tilde{x}_{t-1} + \alpha\mu + u_t \, .$$

From this it follows that

$$\tilde{x}_t = \delta + \mu(\alpha - 1) + \alpha \, \tilde{x}_{t-1} + u_t$$

$$= \delta + \frac{\delta}{1-\alpha}(\alpha - 1) + \alpha \, \tilde{x}_{t-1} + u_t$$

(2.6) $$\tilde{x}_t = \alpha \, \tilde{x}_{t-1} + u_t \, .$$

This is the AR(1) process belonging to (2.1) with $E[\tilde{x}_t] = 0$.

If we multiply equation (2.6) with $\tilde{x}_{t-\tau}$ for $\tau \geq 0$ and take expectations we can write:

(2.7) $$E[\tilde{x}_{t-\tau} \, \tilde{x}_t] = \alpha \, E[\tilde{x}_{t-\tau} \, \tilde{x}_{t-1}] + E[\tilde{x}_{t-\tau} \, u_t] \, .$$

Because of (2.3) we get

$$\tilde{x}_{t-\tau} = u_{t-\tau} + \alpha \, u_{t-\tau-1} + \alpha^2 \, u_{t-\tau-2} + \dots \, .$$

This leads to

(2.8) $$E[\tilde{x}_{t-\tau} \, u_t] = \begin{cases} \sigma^2 & \text{for } \tau = 0 \\ 0 & \text{for } \tau > 0 \end{cases} .$$

Because of the stationarity assumption and because of the (even) symmetry of the autocovariances, $\gamma(\tau) = \gamma(-\tau)$, equation (2.7) results in

$$\tau = 0: \quad E[\tilde{x}_t^2] = \alpha \, E[\tilde{x}_t \, \tilde{x}_{t-1}] + \sigma^2,$$

or

$$\gamma(0) = \alpha \, \gamma(1) + \sigma^2,$$

$$\tau = 1: \quad E[\tilde{x}_t \, \tilde{x}_{t-1}] = \alpha \, E[\tilde{x}_{t-1}^2],$$

or

$$\gamma(1) = \alpha \, \gamma(0) \, .$$

This leads to the variance of the AR(1) process

$$\gamma(0) = \frac{\sigma^2}{1-\alpha^2} \, .$$

For $\tau \geq 1$ (2.7) implies

$$\gamma(1) = \alpha \, \gamma(0)$$

$$\gamma(2) = \alpha \, \gamma(1) = \alpha^2 \, \gamma(0)$$

$$\gamma(3) = \alpha\,\gamma(2) = \alpha^3\,\gamma(0)$$
$$\vdots$$
$$\gamma(\tau) = \alpha\,\gamma(\tau-1) = \alpha^\tau\,\gamma(0).$$

Thus, the covariances can be calculated from the linear homogeneous first order difference equation

$$\gamma(\tau) - \alpha\,\gamma(\tau-1) = 0$$

with the initial value $\gamma(0) = \sigma^2/(1 - \alpha^2)$.

The Autocorrelogram

Because of $\rho(\tau) = \gamma(\tau)/\gamma(0)$, the autocorrelation function (the autocorrelogram) of the AR(1) process is

(2.9) $$\rho(\tau) = \alpha^\tau, \quad \tau = 1, 2, \ldots.$$

This function converges geometrically to zero for $\tau \to \infty$, and its infinite sum equals $1/(1 - \alpha)$ since $|\alpha| < 1$. This convergence is monotone for positive and oscillating for negative values of α.

Example 2.1

For $\delta = 0$ and $\alpha \in \{0.9, 0.5, -0.9\}$, *Figures 2.1 to 2.3* each present one realisation of the corresponding AR(1) process with T = 240 observations. To generate these series, we used realisations of normally distributed pure random processes with mean zero and variance one. We always dropped the first 60 observations to eliminate the dependence of the initial values.

The realisation for $\alpha = 0.9$, presented in *Figure 2.1*, is relatively smooth. This is to be expected given the theoretical autocorrelation function because random variables with a considerable distance between each other still have high positive correlations.

The development of the realisation in *Figure 2.2* with $\alpha = 0.5$ is much less systematic. The geometric decrease of the theoretical autocorrelation function is rather fast. The fourth order autocorrelation coefficient is only 0.0625.

Contrary to this, the realisation of the AR(1) process with $\alpha = -0.9$, presented in *Figure 2.3*, follows a well pronounced zigzag course with, however, alternating positive and negative amplitudes. This is consistent with the theoretical autocorrelation function indicating that all random variables with even-numbered distance are positively correlated and those with odd-numbered distance negatively correlated.

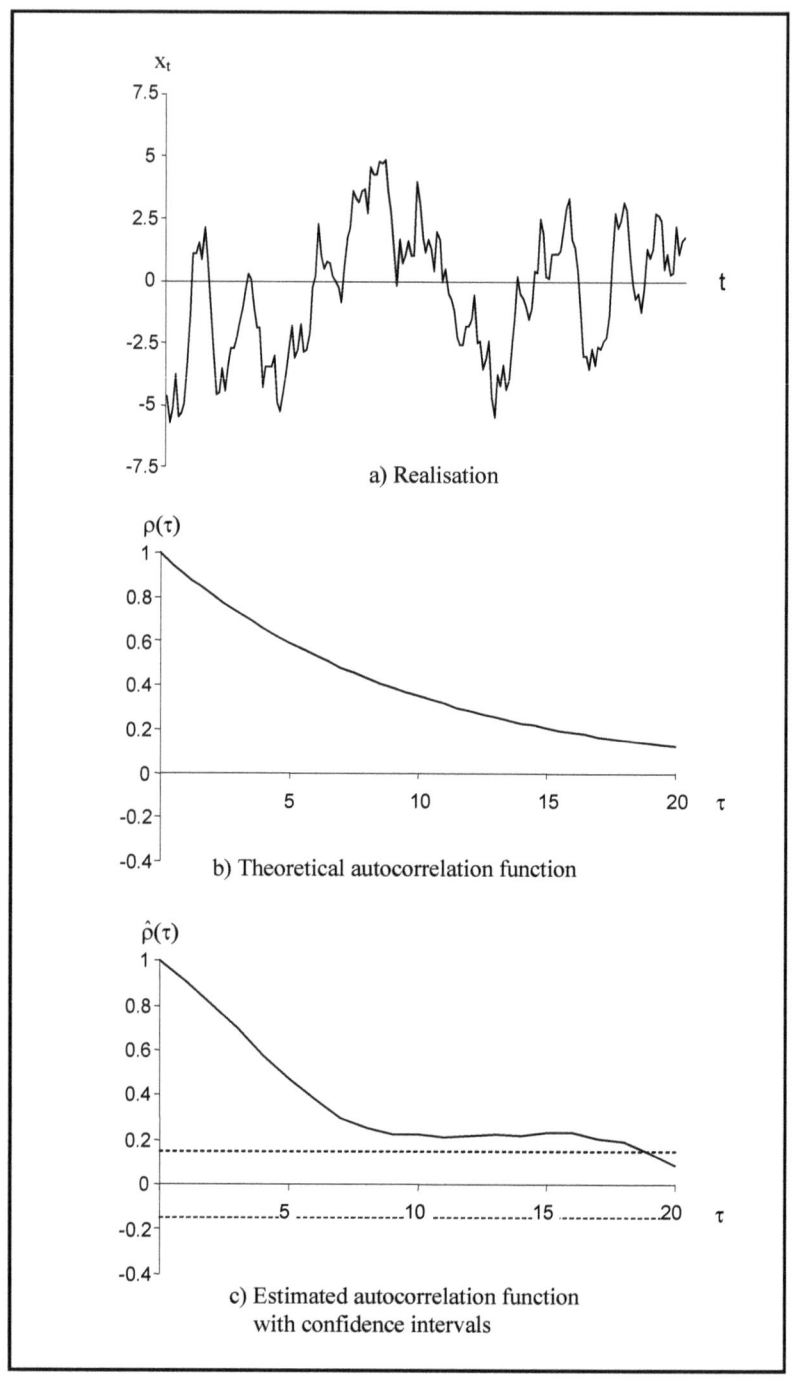

Figure 2.1: *AR(1) process with α = 0.9*

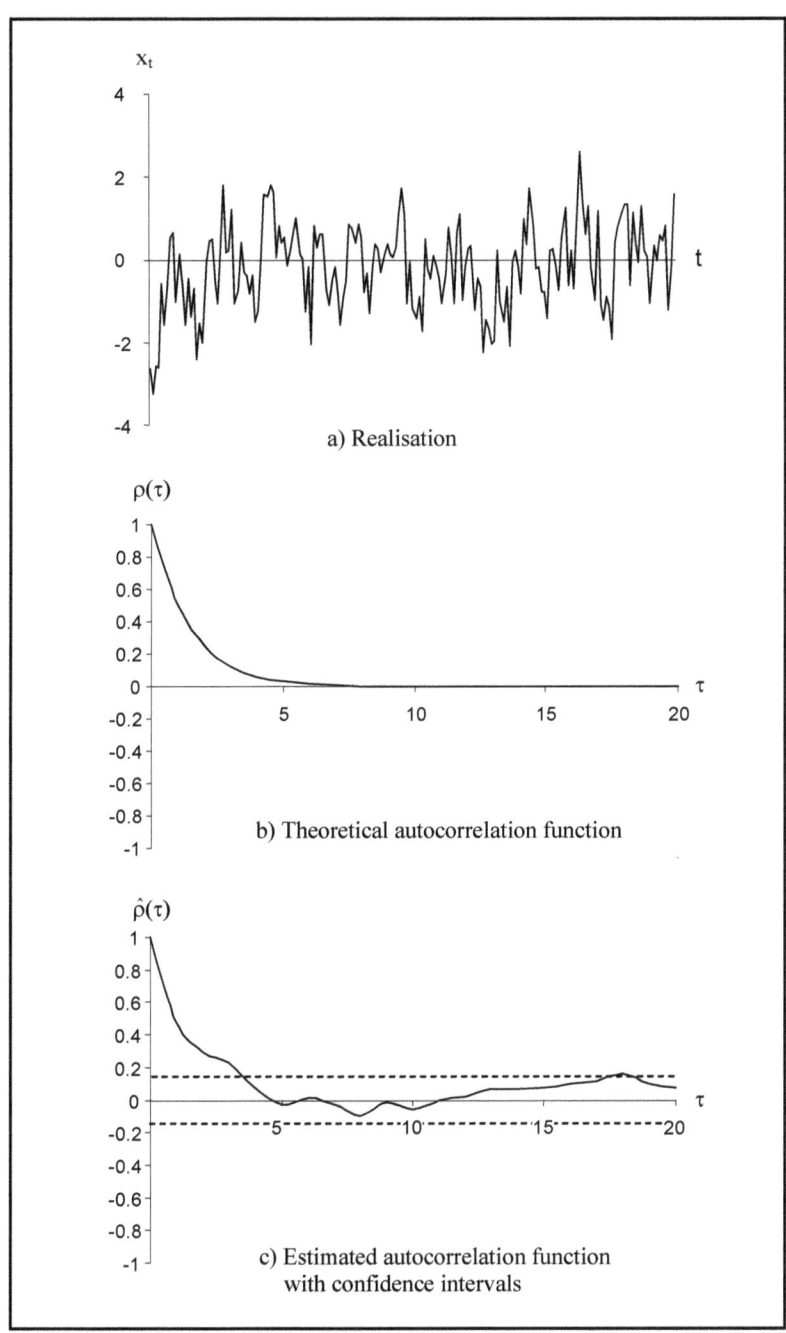

Figure 2.2: AR(1) process with α= 0.5

36   Univariate Stationary Processes

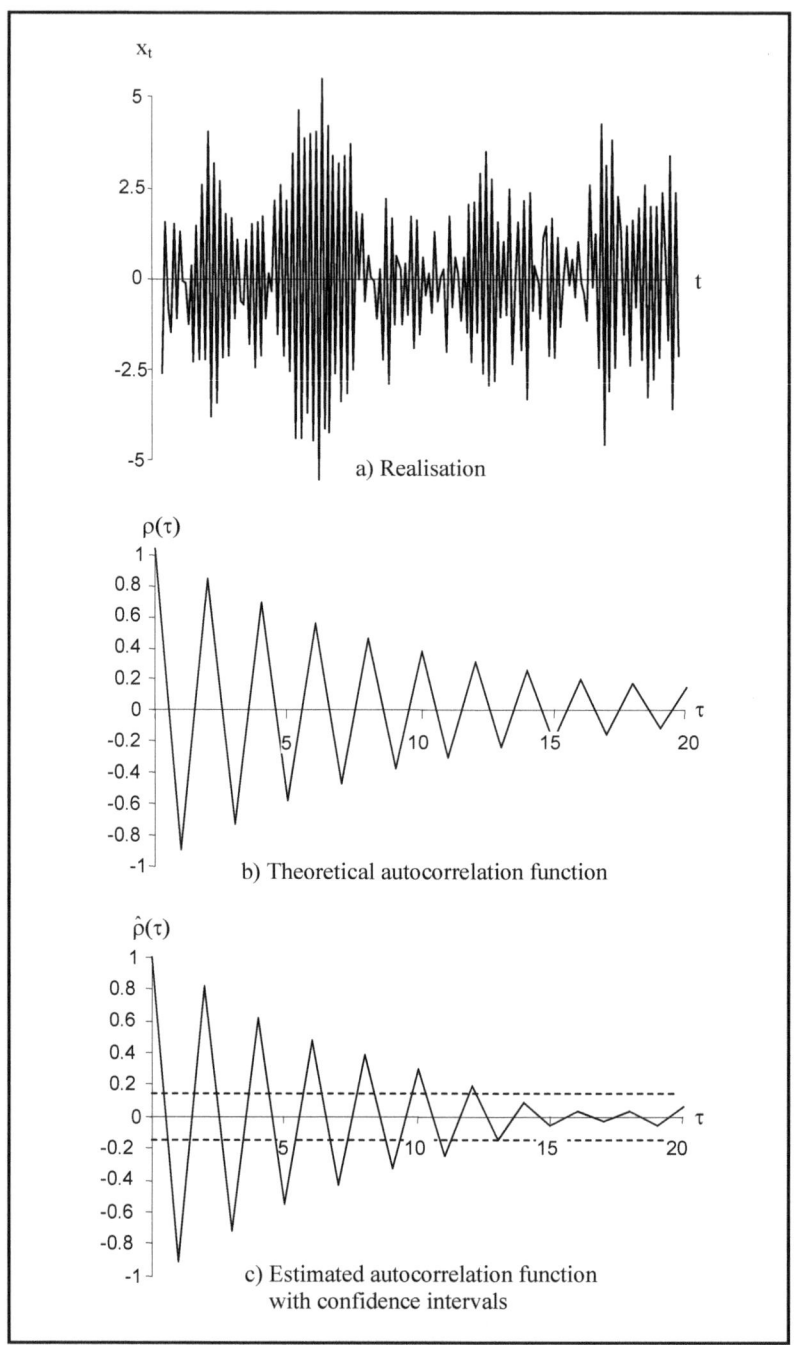

Figure 2.3: AR(1) process with α = -0.9

It generally holds that the closer the parameter α is to +1, the smoother the realisations will be. For negative values of α we get zigzag developments which are the more pronounced the closer α is to -1. For α = 0 we get a pure random process.
The autocorrelation functions estimated by means of relation (1.10) with the given realisations are also presented in *Figures 2.1 to 2.3*. The dotted parallel lines show approximate 95 percent confidence intervals for the null hypothesis assuming that the true process is a pure random process. In all three cases, the estimated functions reflect quite well the typical development of the theoretical autocorrelations.

Example 2.2

In a paper on the effect of economic development on the electoral chances of the German political parties during the period of the social-liberal coalition from 1969 to 1982, GEBHARD KIRCHGÄSSNER (1985) investigated (besides other issues) the time series properties of the popularity series of the parties constructed by monthly surveys of the Institute of Demoscopy in Allensbach (Germany). For the period from January 1971 to April 1982, the popularity series of the Christian Democratic Union (CDU), i.e. the share of voters who answered that they would vote for this party (or its Bavarian sister party, the CSU) if there were a general election by the following Sunday, is given in *Figure 2.4*. The autocorrelation and the partial autocorrelation function (which is discussed in *Section 2.1.4*) are also presented in this figure. While the autocorrelation function goes slowly towards zero, the partial autocorrelation function breaks off after $\tau = 1$. This argues for an AR(1) process.

The model has been estimated with Ordinary Least Squares (OLS), the method proposed in *Section 2.1.5* for the estimation of autoregressive models. Thus, we get:

$$CDU_t = 8.053 + 0.834\, CDU_{t-1} + \hat{u}_t,$$
$$(3.43) \quad (17.10)$$

$$\overline{R}^2 = 0.683, \quad SE = 1.586, \quad Q(11) = 12.516 \;(p = 0.326).$$

The estimated t values are given in parentheses, SE denotes the standard error of the residuals. The autocorrelogram, which is also given in *Figure 2.4*, does not indicate any higher-order process. Moreover, given the high p-value, the Ljung-Box Q statistic with 12 correlation coefficients (i.e. with 11 degrees of freedom) gives no reason to reject this model. The mean is calculated as

$$\hat{\mu} = \frac{8.053}{1 - 0.834} = 48.512 \,.$$

It shows that about 48.5 percent of the voters voted on average for the CDU during this period.

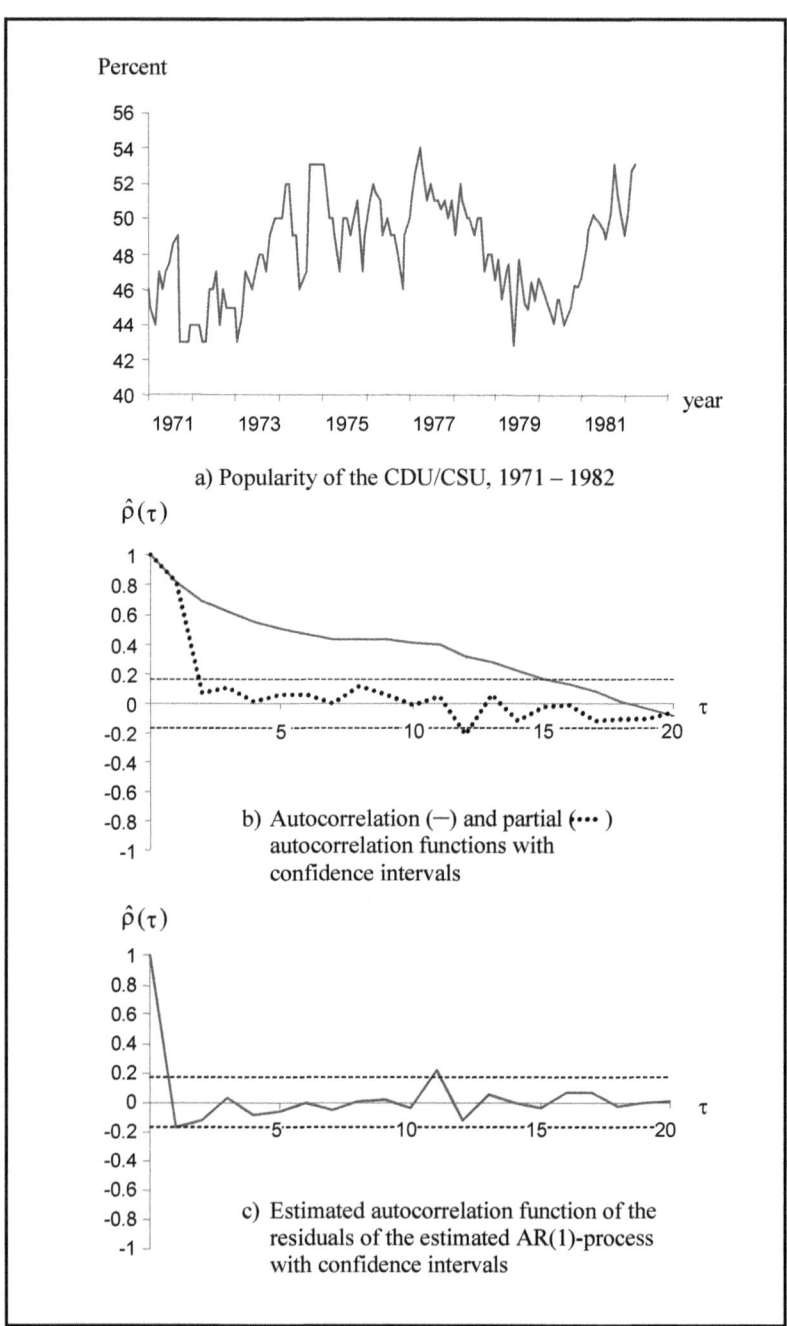

Figure 2.4: Popularity of the CDU/CSU, 1971 – 1982

Stability Conditions

Along with the stochastic initial value, the condition $|\alpha| < 1$, the so-called stability condition, is crucial for the stationarity of the AR(1) process. We can also derive the stability condition from the linear homogeneous difference equation, which is given for the process itself by

$$x_t - \alpha\, x_{t-1} = 0,$$

for its autocovariances by

$$\gamma(\tau) - \alpha\, \gamma(\tau-1) = 0$$

and for the autocorrelations by

$$\rho(\tau) - \alpha\, \rho(\tau-1) = 0.$$

These difference equations have stable solutions, i.e. $\lim_{\tau \to \infty} \rho(\tau) = 0$, if and only if their characteristic equation

(2.10) $$\lambda - \alpha = 0$$

has a solution (root) with an absolute value smaller than one, i.e. if $|\alpha| < 1$ holds. We get an equivalent condition if we do not consider the characteristic equation but the lag polynomial of the corresponding difference equations,

(2.11) $$1 - \alpha\, L = 0.$$

This implies that the solution has to be larger than one in absolute value. (Strictly speaking, L, which denotes an operator, has to be substituted by a variable, which is often denoted by 'z'. To keep the notation simple, we use L in both meanings.)

Example 2.3

Let us consider the stochastic process

(E2.1) $$y_t = x_t + v_t.$$

In this equation, x_t is a stationary AR(1) process, $x_t = \alpha\, x_{t-1} + u_t$, with $|\alpha| < 1$; v_t is a pure random process with mean zero and constant variance σ_v^2 which is uncorrelated with the other pure random process u_t with mean zero and constant variance σ_u^2.

We can interpret the stochastic process y_t as an additive decomposition of two stationary components. Then y_t itself is stationary. In the sense of MILTON FRIEDMAN (1957) we can interpret x_t as the permanent (systematic) and v_t as the transitory component.

What does the correlogram of y_t look like? As both x_t and v_t have zero mean, $E[y_t] = 0$. Multiplying (E2.1) with $y_{t-\tau}$ and taking expectations results in

$$E[y_{t-\tau}\, y_t] = E[y_{t-\tau}\, x_t] + E[y_{t-\tau}\, v_t].$$

Due to $y_{t-\tau} = x_{t-\tau} + v_{t-\tau}$, we get

$$E[y_{t-\tau}\, y_t] = E[x_{t-\tau}\, x_t] + E[v_{t-\tau}\, x_t] + E[x_{t-\tau}\, v_t] + E[v_{t-\tau}\, v_t].$$

As u_t and v_t are uncorrelated, it holds that $E[v_{t-\tau}\, x_t] = E[x_{t-\tau}\, v_t] = 0$, and because of the stationarity of the two processes, we can write

(E2.2) $$\gamma_y(\tau) = \gamma_x(\tau) + \gamma_v(\tau).$$

For $\tau = 0$ we get the variance of y_t as

$$\gamma_y(0) = \gamma_x(0) + \sigma_v^2 = \frac{\sigma_u^2}{1-\alpha^2} + \sigma_v^2.$$

For $\tau > 0$, because of $\gamma_v(\tau) = 0$ for $\tau \neq 0$, we get from (E2.2)

$$\gamma_y(\tau) = \gamma_x(\tau) = \alpha^\tau \frac{\sigma_u^2}{1-\alpha^2}.$$

Thus, we finally get

$$\rho_y(\tau) = \frac{\alpha^\tau}{1+(1-\alpha^2)\sigma_v^2/\sigma_u^2}, \quad \tau = 1, 2, ...,$$

for the correlogram of y_t. The overlay of the systematic component by the transitory component reduces the autocorrelation generated by the systematic component. The larger the variance of the transitory component, the stronger is this effect.

2.1.2 Second Order Autoregressive Processes

Generalising (2.1), the *second order autoregressive process* (AR(2)) can be written as

(2.12) $$x_t = \delta + \alpha_1 x_{t-1} + \alpha_2 x_{t-2} + u_t,$$

with u_t denoting a pure random process with variance σ^2 and $\alpha_2 \neq 0$. With the lag operator L we get

(2.13) $$(1 - \alpha_1 L - \alpha_2 L^2)\, x_t = \delta + u_t.$$

With $\alpha(L) = 1 - \alpha_1 L - \alpha_2 L^2$ we can write

(2.14) $$\alpha(L)\, x_t = \delta + u_t.$$

As for the AR(1) process, we get the Wold representation from (2.14) if we invert $\alpha(L)$; i.e. under the assumption that $\alpha^{-1}(L)$ exists and has the property

(2.15) $$\alpha(L)\,\alpha^{-1}(L) = 1$$

we can 'solve' for x_t in (2.14):

(2.16) $$x_t = \alpha^{-1}(L)\,\delta + \alpha^{-1}(L)\,u_t.$$

If we use the series expansion with undetermined coefficients for

$$\alpha^{-1}(L) = \psi_0 + \psi_1 L + \psi_2 L^2 + \dots$$

it has to hold that

$$1 = (1 - \alpha_1 L - \alpha_2 L^2)(\psi_0 + \psi_1 L + \psi_2 L^2 + \psi_3 L^3 + \dots)$$

because of (2.15). This relation is an identity only if the coefficients of L^j, $j = 0, 1, 2, \dots$, are equal on both the right and the left hand side. We get

$$\begin{aligned}
1 = \;&\psi_0 + \psi_1 L + \psi_2 L^2 + \psi_3 L^3 + \dots \\
&- \alpha_1\psi_0 L - \alpha_1\psi_1 L^2 - \alpha_1\psi_2 L^3 - \dots \\
& - \alpha_2\psi_0 L^2 - \alpha_2\psi_1 L^3 - \dots
\end{aligned}$$

Comparing the coefficients of the lag polynomials on the right- and left-hand side finally leads to

$L^0:$ $\quad\psi_0 = 1$

$L^1:$ $\quad\psi_1 - \alpha_1\psi_0 = 0 \quad\Rightarrow\quad \psi_1 = \alpha_1.$

$L^2:$ $\quad\psi_2 - \alpha_1\psi_1 - \alpha_2\psi_0 = 0 \quad\Rightarrow\quad \psi_2 = \alpha_1^2 + \alpha_2.$

$L^3:$ $\quad\psi_3 - \alpha_1\psi_2 - \alpha_2\psi_1 = 0 \quad\Rightarrow\quad \psi_3 = \alpha_1^3 + 2\alpha_1\alpha_2.$

By applying this so-called method of undetermined coefficients, we get the values ψ_j, $j = 2, 3, \dots$, from the linear homogeneous difference equation

$$\psi_j - \alpha_1\psi_{j-1} - \alpha_2\psi_{j-2} = 0$$

with the initial conditions $\psi_0 = 1$ and $\psi_1 = \alpha_1$.

The stability condition for the AR(2) process requires that, for $j \to \infty$, the ψ_j converge to zero, i.e. that the characteristic equation of (2.12),

(2.17) $$\lambda^2 - \alpha_1\lambda - \alpha_2 = 0,$$

has only roots with absolute values smaller than one, or that all solutions of the lag polynomial in (2.13),

(2.18) $$1 - \alpha_1 L - \alpha_2 L^2 = 0$$

are larger than one in modulus. Together with stochastic initial conditions, this guarantees the stationarity of the process. The stability conditions are fulfilled if the following parameter restrictions hold jointly for (2.17) and (2.18):

$$1 + (-\alpha_1) + (-\alpha_2) > 0,$$
$$1 - (-\alpha_1) + (-\alpha_2) > 0,$$
$$1 - (-\alpha_2) > 0.$$

As a constant is not changed by the application of the lag operator, the number '1' can substitute the lag operator in the corresponding terms. Thus, due to (2.16), the Wold representation of the AR(2) process is given by

(2.19) $$x_t = \frac{\delta}{1 - \alpha_1 - \alpha_2} + \sum_{j=0}^{\infty} \psi_j u_{t-j} \ , \ \psi_0 = 1.$$

Under the assumption of stationarity, the expected value of the stochastic process can be calculated directly from (2.12) since $E[x_t] = E[x_{t-1}] = E[x_{t-2}] = \mu$. We get

$$\mu = \delta + \alpha_1 \mu + \alpha_2 \mu$$

or

(2.20) $$E[x_t] = \mu = \frac{\delta}{1 - \alpha_1 - \alpha_2}.$$

As the stability conditions are fulfilled, $1 - \alpha_1 - \alpha_2 > 0$ holds, i.e. the sign of δ also determines the sign of μ.

In order to calculate the second order moments, we can assume – without loss of generality – that $\mu = 0$, which is equivalent to $\delta = 0$. Multiplying (2.12) with $x_{t-\tau}$, $\tau \geq 0$, and taking expectations leads to

(2.21) $$E[x_{t-\tau} x_t] = \alpha_1 E[x_{t-\tau} x_{t-1}] + \alpha_2 E[x_{t-\tau} x_{t-2}] + E[x_{t-\tau} u_t].$$

Because of representation (2.19), relation (2.8) holds here as well. This leads to the following equations

(2.22)
$$\begin{array}{rl} \tau = 0 \ : \ \gamma(0) & = \alpha_1 \gamma(1) + \alpha_2 \gamma(2) + \sigma^2 \\ \tau = 1 \ : \ \gamma(1) & = \alpha_1 \gamma(0) + \alpha_2 \gamma(1) \\ \tau = 2 \ : \ \gamma(2) & = \alpha_1 \gamma(1) + \alpha_2 \gamma(0) \end{array},$$

and, more generally, the following difference equation holds for the autocovariances $\gamma(\tau)$, $\tau \geq 2$,

(2.23) $\qquad \gamma(\tau) - \alpha_1 \gamma(\tau\text{-}1) - \alpha_2 \gamma(\tau\text{-}2) = 0.$

As the stability conditions hold, the autocovariances which can be recursively calculated with (2.23) are converging to zero for $\tau \rightarrow \infty$.

The relations (2.22) result in

(2.24) $\qquad V[x_t] = \gamma(0) = \dfrac{1-\alpha_2}{(1+\alpha_2)\,[(1-\alpha_2)^2 - \alpha_1^2]} \sigma^2$

for the variance of the AR(2) process, and in

$$\gamma(1) = \dfrac{\alpha_1}{(1+\alpha_2)\,[(1-\alpha_2)^2 - \alpha_1^2]} \sigma^2,$$

and

$$\gamma(2) = \dfrac{\alpha_1^2 + \alpha_2 - \alpha_2^2}{(1+\alpha_2)\,[(1-\alpha_2)^2 - \alpha_1^2]} \sigma^2,$$

for the autocovariances of order one and two.

The autocorrelations can be calculated accordingly. If we divide (2.23) by the variance $\gamma(0)$ we get the linear homogeneous second order difference equation,

(2.25) $\qquad \rho(\tau) - \alpha_1 \rho(\tau\text{-}1) - \alpha_2 \rho(\tau\text{-}2) = 0$

with the initial conditions $\rho(0) = 1$ and $\rho(1) = \alpha_1/(1-\alpha_2)$ for the autocorrelation function. Depending on the values of α_1 and α_2, AR(2) processes can generate quite different developments, and, therefore, these processes can show considerably different characteristics.

Example 2.4

Let us consider the AR(2) process

(E2.3) $\qquad x_t = 1 + 1.5\,x_{t-1} - 0.56\,x_{t-2} + u_t$

with a variance of u_t of 1. Because the characteristic equation

$$\lambda^2 - 1.5\,\lambda + 0.56 = 0$$

has the two roots $\lambda_1 = 0.8$ and $\lambda_2 = 0.7$, (E2.3) is stationary, given that we have stochastic initial conditions. The expected value of this process is

$$\mu = \frac{1}{1 - 1.5 + 0.56} = 16.\bar{6}.$$

The variance of (E2.3) can be calculated from (2.24) as $\gamma(0) = 19.31$. A realisation of this process (with 180 observations) is given in *Figure 2.5* in which the (estimated) mean was subtracted. Thus, the realisations fluctuate around zero, and the process always tends to go back to the mean. This *mean-reverting behaviour* is a typical property of stationary processes.

Due to (2.25) we get

$$\rho(\tau) - 1.5\,\rho(\tau-1) + 0.56\,\rho(\tau-2) = 0, \quad \tau = 2, 3, ...,$$
$$\text{with } \rho(0) = 1, \quad \rho(1) = 0.96$$

for the autocorrelation function. The general solution of this homogeneous difference equation is

$$\rho(\tau) = C_1 (0.8)^\tau + C_2 (0.7)^\tau,$$

where C_1 and C_2 are two arbitrary constants. Taking into account the two initial conditions we get

$$\rho(\tau) = 2.6 (0.8)^\tau - 1.6 (0.7)^\tau$$

for the autocorrelation coefficients. This development is also expressed in *Figure 2.5*. The coefficients are always positive but strictly monotonically decreasing. Initially, the estimated autocorrelogram using the given realisation is also monotonically decreasing, but, contrary to the theoretical development, the values begin to fluctuate from the tenth lag onwards. However, except for the coefficient for $\tau = 16$, the estimates are not significantly different from zero; they are all inside the approximate 95 percent confidence interval indicated by the dotted lines.

The characteristic equations of stable autoregressive processes of second or higher order can result in conjugate complex roots. In this case, the time series exhibit dampened oscillations, which are shocked again and again by the pure random process. The solution of the homogeneous part of (2.12) for conjugate complex roots can be represented by

$$x_t = d^t (C_1 \cos(ft) + C_2 \sin(ft))$$

with C_1 and C_2 again being arbitrary constants that can be determined by using the initial conditions. The dampening factor

$$d = \sqrt{-\alpha_2}$$

corresponds to the modulus of the two roots, and

$$f = \arccos\left(\frac{\alpha_1}{2\sqrt{-\alpha_2}}\right)$$

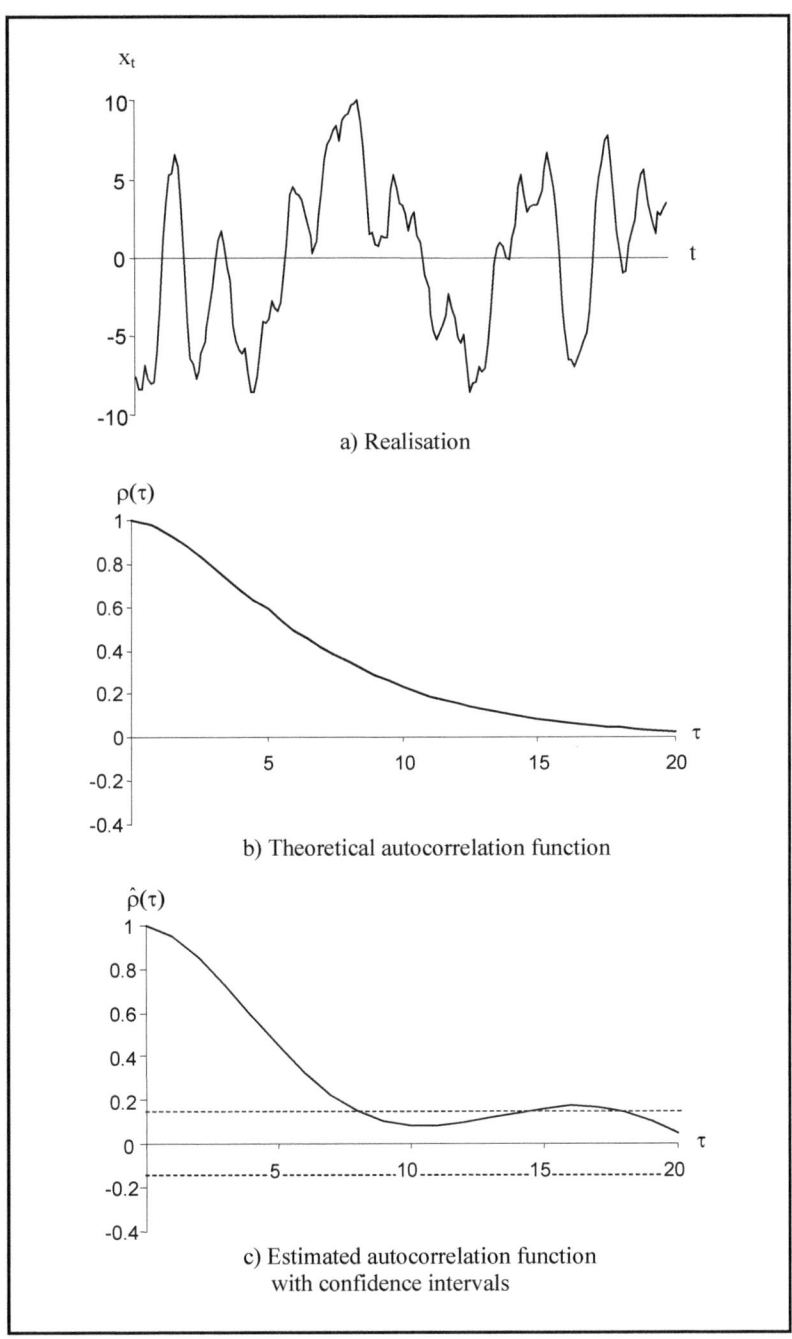

Figure 2.5: AR(2) process with $\alpha_1 = 1.5$, $\alpha_2 = -0.56$

46 Univariate Stationary Processes

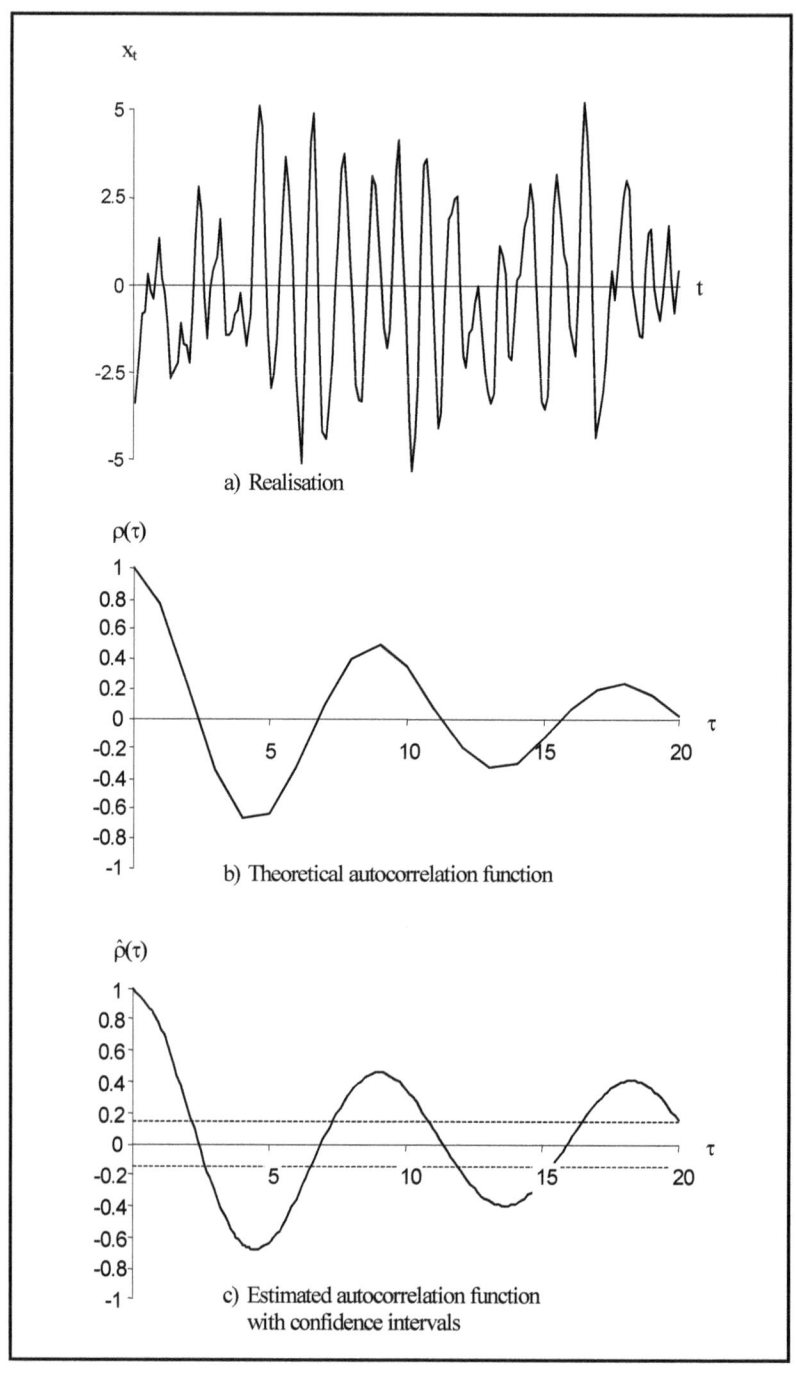

Figure 2.6: *AR(2) process with $\alpha_1 = 1.4$ and $\alpha_2 = -0.85$*

is the frequency of the oscillation. The period of the cycles is $P = 2\pi/f$. Processes with conjugate complex roots are well-suited to describe business cycle fluctuations.

Example 2.5

Consider the AR(2) process

(E2.4) $$x_t = 1.4 \, x_{t-1} - 0.85 \, x_{t-2} + u_t,$$

with a variance of u_t of 1. The characteristic equation

$$\lambda^2 - 1.4 \lambda + 0.85 = 0$$

has the two solutions $\lambda_1 = 0.7 + 0.6i$ and $\lambda_2 = 0.7 - 0.6i$. ('i' stands for the imaginary unit: $i^2 = -1$.) The modulus (dampening factor) is $d = 0.922$. Thus, (E2.4) with stochastic initial conditions and a mean of zero is stationary. According to (2.24) the variance is given by $\gamma(0) = 8.433$.

A realisation of this process with 180 observations is given in *Figure 2.6*. Its development is cyclical around its zero mean. For the autocorrelation function we get

$$\rho(\tau) - 1.4 \, \rho(\tau-1) + 0.85 \, \rho(\tau-2) = 0, \quad \tau = 2, 3, ...,$$
$$\rho(0) = 1, \quad \rho(1) = 0.76,$$

because of (2.25).

The general solution is

$$\rho(\tau) = 0.922^\tau \, (C_1 \cos(0.709 \, \tau) + C_2 \sin(0.709 \, \tau)).$$

Taking into account the two initial conditions, we get for the autocorrelation coefficients

$$\rho(\tau) = 0.922^\tau \, (\cos(0.709 \, \tau) + 0.1 \sin(0.709 \, \tau)),$$

with a frequency of $f = 0.709$.

In case of quarterly data, this corresponds to a period length of about 9 quarters. Both the theoretical and the estimated autocorrelations in *Figure 2.6* show this kind of dampened periodical behaviour.

Example 2.6

Figure 2.7 shows the development of the three month money market rate in Frankfurt (GSR) from the first quarter of 1970 to the last quarter of 1998 as well as the autocorrelation and the partial autocorrelation functions explained in *Section 2.1.4*. Whereas the autocorrelation function tends only slowly towards zero, the partial autocorrelation function breaks off after two lags. As will be shown below, this indicates an AR(2) process. For the period from 1970 to 1998, estimation with OLS results in the following:

48   Univariate Stationary Processes

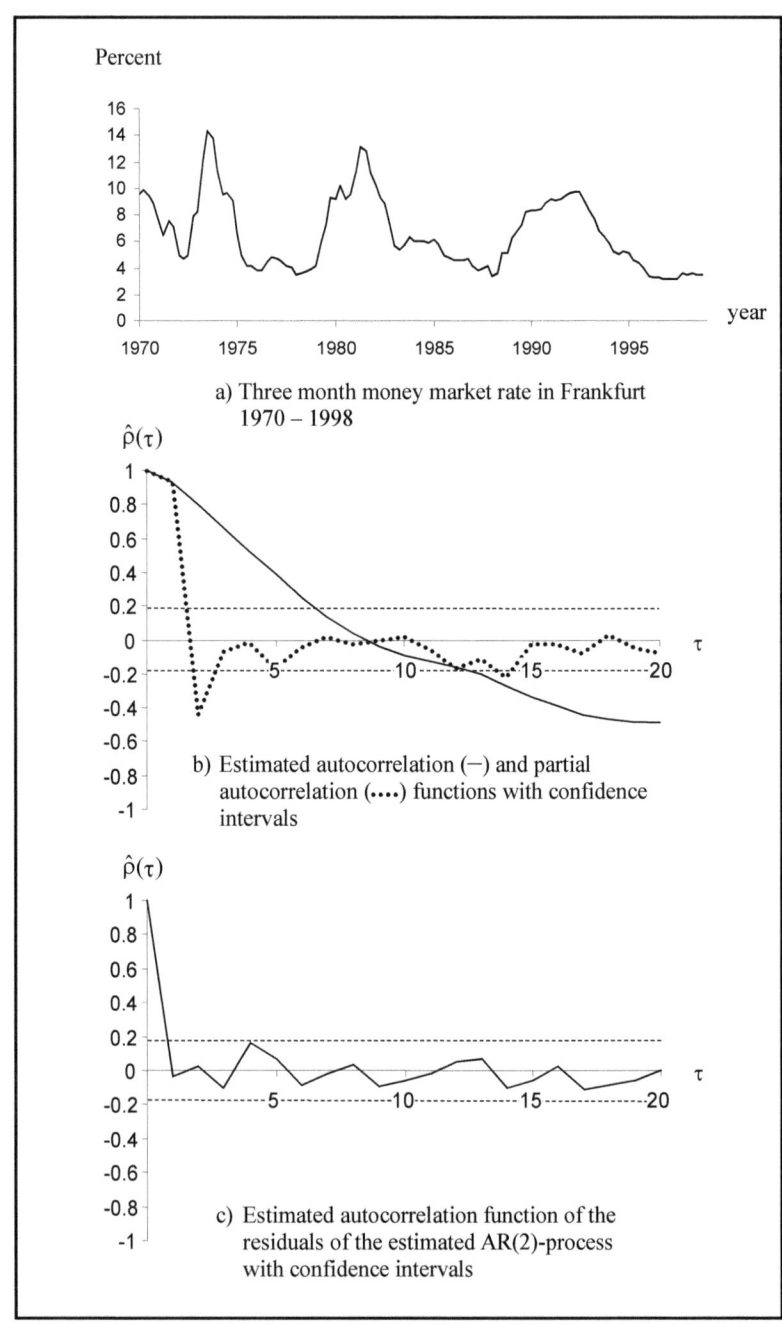

Figure 2.7:   Three month money market rate in Frankfurt, 1970 – 1998

$$GSR_t = 0.575 + 1.407\ GSR_{t-1} - 0.498\ GSR_{t-2} + \hat{u}_t.$$
$$(2.82)\quad(17.50)\qquad\quad(-6.16)$$

$$\bar{R}^2 = 0.910,\ SE = 0.812,\ Q(6) = 6.475\ (p = 0.372),$$

with t values being again given in parentheses. On the 0.1 percent level, both estimated coefficients of the lagged interest rates are significantly different from zero. The autocorrelogram of the estimated residuals (given in *Figure 2.7c*) as well as the Ljung-Box Q statistic which is calculated with 8 correlation coefficients (and 6 degrees of freedom) does not indicate any higher order process.

The two roots of the process are $0.70 \pm 0.06i$, i.e. they indicate dampened cycles. The modulus (dampening factor) is $d = 0.706$; the frequency $f = 0.079$ corresponds to a period of 79.7 quarters and therefore of nearly 20 years. Correspondingly, this oscillation cannot be detected in the estimated autocorrelogram presented in *Figure 2.7b*.

2.1.3 Higher Order Autoregressive Processes

An AR(p) process can be described by the following stochastic difference equation,

(2.26) $\qquad x_t = \delta + \alpha_1 x_{t-1} + \alpha_2 x_{t-2} + \ldots + \alpha_p x_{t-p} + u_t,$

with $\alpha_p \neq 0$, where u_t is again a pure random process with zero mean and variance σ^2. Using the lag operator we can also write:

(2.26') $\qquad (1 - \alpha_1 L - \alpha_2 L^2 - \ldots - \alpha_p L^p) x_t = \delta + u_t.$

If we assume stochastic initial conditions, the AR(p) process in (2.26) is stationary if the stability conditions are satisfied, i.e. if the characteristic equation

(2.27) $\qquad \lambda^p - \alpha_1 \lambda^{p-1} - \alpha_2 \lambda^{p-2} - \ldots - \alpha_p = 0$

only has roots with absolute values smaller than one, or if the solutions of the lag polynomial

(2.28) $\qquad 1 - \alpha_1 L - \alpha_2 L^2 - \ldots - \alpha_p L^p = 0$

only have roots with absolute values larger than one.

If the stability conditions are satisfied, we get the Wold representation of the AR(p) process by the series expansion of the inverse lag polynomial,

$$\frac{1}{1 - \alpha_1 L - \ldots - \alpha_p L^p} = 1 + \psi_1 L + \psi_2 L^2 + \ldots$$

as

(2.29) $$x_t = \frac{\delta}{1-\alpha_1 - \ldots - \alpha_p} + \sum_{j=0}^{\infty} \psi_j u_{t-j} .$$

Generalising the approach that was used to calculate the coefficients of the AR(2) process, the series expansion can again be calculated by the method of undetermined coefficients.

From (2.29) we get the constant (unconditional) expectation as

$$E[x_t] = \frac{\delta}{1-\alpha_1 - \ldots - \alpha_p} = \mu .$$

Again, similarly to the AR(1) and AR(2) cases, a necessary condition for stability is

$$1 - \alpha_1 - \alpha_2 - \ldots - \alpha_p > 0.$$

Without loss of generality we can set $\delta = 0$, i.e. $\mu = 0$, in order to calculate the autocovariances. Because of $\gamma(\tau) = E[x_{t-\tau} x_t]$, we get according to (2.26)

(2.30) $$\gamma(\tau) = E[x_{t-\tau} (\alpha_1 x_{t-1} + \alpha_2 x_{t-2} + \ldots + \alpha_p x_{t-p} + u_t)] .$$

For $\tau = 0, 1, \ldots, p$, it holds that

(2.31)
$$\begin{aligned}
\gamma(0) &= \alpha_1 \gamma(1) + \alpha_2 \gamma(2) + \ldots + \alpha_p \gamma(p) + \sigma^2 \\
\gamma(1) &= \alpha_1 \gamma(0) + \alpha_2 \gamma(1) + \ldots + \alpha_p \gamma(p-1) \\
&\vdots \\
\gamma(p) &= \alpha_1 \gamma(p-1) + \alpha_2 \gamma(p-2) + \ldots + \alpha_p \gamma(0)
\end{aligned}$$

because of the symmetry of the autocovariances and because of $E[x_{t-\tau} u_t] = \sigma^2$ for $\tau = 0$ and zero for $\tau > 0$.

This is a linear inhomogeneous equation system for given α_i and σ^2 to derive the $p + 1$ unknowns $\gamma(0), \gamma(1), \ldots, \gamma(p)$. For $\tau > p$ we get the linear homogeneous difference equation to calculate the autocovariances of order $\tau > p$:

(2.32) $$\gamma(\tau) - \alpha_1 \gamma(\tau-1) - \ldots - \alpha_p \gamma(\tau-p) = 0.$$

If we divide (2.32) by $\gamma(0)$, we get the corresponding difference equation to calculate the autocorrelations:

(2.33) $$\rho(\tau) - \alpha_1 \rho(\tau-1) - \ldots - \alpha_p \rho(\tau-p) = 0.$$

The initial conditions $\rho(1), \rho(2), \ldots, \rho(p)$ can be derived from the so-called Yule-Walker equations. We get those if we successively insert $\tau = 1, 2, \ldots, p$ in (2.33), or, if the last p equations in (2.31) are divided by $\gamma(0)$,

(2.34)
$$\begin{aligned}\rho(1) &= \alpha_1 &&+ \alpha_2\,\rho(1) &&+ \alpha_3\,\rho(2) &&+ \ldots + \alpha_p\,\rho(p-1)\\ \rho(2) &= \alpha_1\,\rho(1) &&+ \alpha_2 &&+ \alpha_3\,\rho(1) &&+ \ldots + \alpha_p\,\rho(p-2)\\ &\vdots\\ \rho(p) &= \alpha_1\,\rho(p-1) + \alpha_2\,\rho(p-2) + \alpha_3\,\rho(p-3) + \ldots + \alpha_p\end{aligned}$$

If we define $\boldsymbol{\rho}' = (\rho(1), \rho(2), \ldots, \rho(p))$, $\boldsymbol{\alpha}' = (\alpha_1, \alpha_2, \ldots, \alpha_p)$ and

$$\underset{p\times p}{R} = \begin{bmatrix} 1 & \rho(1) & \rho(2) & \cdots & \rho(p-1) \\ \rho(1) & 1 & \rho(1) & \cdots & \rho(p-2) \\ \vdots & & & & \\ \rho(p-1) & \rho(p-2) & \rho(p-3) & \cdots & 1 \end{bmatrix}$$

we can write the Yule-Walker equations (2.34) in matrix form,

(2.35) $$\boldsymbol{\rho} = R\,\boldsymbol{\alpha}.$$

If the first p autocorrelation coefficients are given, the coefficients of the AR(p) process can be calculated according to (2.35) as

(2.36) $$\boldsymbol{\alpha} = R^{-1}\boldsymbol{\rho}.$$

Equations (2.35) and (2.36) show that there is a one-to-one mapping between the p coefficients $\boldsymbol{\alpha}$ and the first p autocorrelation coefficients $\boldsymbol{\rho}$ of an AR(p) process. If there is a generating pure random process, it is sufficient to know either $\boldsymbol{\alpha}$ or $\boldsymbol{\rho}$ to identify the AR(p) process. Thus, there are two possibilities to describe the structure of an autoregressive process of order p: the parametric representation that uses the parameters $\alpha_1, \alpha_2, \ldots, \alpha_p$, and the non-parametric representation with the first p autocorrelation coefficients $\rho(1), \rho(2), \ldots, \rho(p)$. Both representations contain exactly the same information. Which representation is used depends on the specific situation. We usually use the parametric representation to describe finite order autoregressive processes (with known order).

Example 2.7

Let the fourth order autoregressive process

$$x_t = \alpha_4\, x_{t-4} + u_t, \quad 0 < \alpha_4 < 1,$$

be given, where u_t is again white noise with zero mean and variance σ^2. Applying (2.31) we get:

$$\begin{aligned}\gamma(0) &= \alpha_4\,\gamma(4) + \sigma^2,\\ \gamma(1) &= \alpha_4\,\gamma(3),\\ \gamma(2) &= \alpha_4\,\gamma(2),\end{aligned}$$

$$\gamma(3) = \alpha_4 \gamma(1),$$
$$\gamma(4) = \alpha_4 \gamma(0).$$

From these relations we get

$$\gamma(0) = \frac{\sigma^2}{1 - \alpha_4^2},$$

$$\gamma(1) = \gamma(2) = \gamma(3) = 0,$$

$$\gamma(4) = \alpha_4 \frac{\sigma^2}{1 - \alpha_4^2}.$$

As can easily be seen, only the autocovariances with lag $\tau = 4j$, $j = 1, 2, ...$ are different from zero, while all other autocovariances are zero. Thus, for $\tau > 0$ we get the autocorrelation function

$$\rho(\tau) = \begin{cases} \alpha_4^j & \text{for } \tau = 4j, \ j = 1, 2, ... \\ 0 & \text{elsewhere.} \end{cases}$$

Only every fourth autocorrelation coefficient is different from zero; the sequence of these autocorrelation coefficients decreases monotonically like a geometric series. Employing such a model for quarterly data, this AR(4) process captures the correlation between random variables that are distant from each other by a multiplicity of four periods, i.e. the structure of the correlations of all variables which belong to the i-th quarter of a year, i = 1, 2, 3, 4, follows an AR(1) process while the correlations between variables that belong to different quarters are always zero. Such an AR(4) process provides a simple possibility of modelling seasonal effects which typically influence the same quarters of different years. For empirical applications, it is advisable to first eliminate the deterministic component of a seasonal variation by employing seasonal dummies and then to model the remaining seasonal effects by such an AR(4) process.

2.1.4 The Partial Autocorrelation Function

Due to the stability conditions, autocorrelation functions of stationary finite order autoregressive processes are always sequences that converge to zero but do not break off. This makes it difficult to distinguish between processes of different orders when using the autocorrelation function. To cope with this problem, we introduce a new concept, the *partial autocorrelation function*. The partial correlation between two random variables is the correlation that remains if the possible impact of all other random variables has been eliminated. To define the partial autocorrelation coefficient, we use the new notation,

$$X_t = \phi_{k1} X_{t-1} + \phi_{k2} X_{t-2} + \ldots + \phi_{kk} X_{t-k} + u_t,$$

where ϕ_{ki} is the coefficient of the variable with lag i if the process has order k. (According to the former notation it holds that $\alpha_i = \phi_{ki}$ i = 1,2,...,k.) The coefficients ϕ_{kk} are the partial autocorrelation coefficients (of order k), k = 1,2,... . The partial autocorrelation measures the correlation between x_t and x_{t-k} which remains when the influences of x_{t-1}, x_{t-2}, ..., x_{t-k+1} on x_t and x_{t-k} have been eliminated.

Due to the Yule-Walker equations (2.35), we can derive the partial autocorrelation coefficients ϕ_{kk} from the autocorrelation coefficients if we calculate the coefficients ϕ_{kk}, which belong to x_{t-k}, for k = 1, 2, ... from the corresponding linear equation systems

$$\begin{bmatrix} 1 & \rho(1) & \rho(2) & \cdots & \rho(k-1) \\ \rho(1) & 1 & \rho(2) & \cdots & \rho(k-2) \\ \vdots & & & & \vdots \\ \rho(k-1) & \rho(k-2) & \rho(k-3) & \cdots & 1 \end{bmatrix} \begin{bmatrix} \phi_{k1} \\ \phi_{k2} \\ \vdots \\ \phi_{kk} \end{bmatrix} = \begin{bmatrix} \rho(1) \\ \rho(2) \\ \vdots \\ \rho(k) \end{bmatrix}, \quad k = 1, 2, \ldots .$$

With Cramer's rule we get

$$(2.37) \quad \phi_{kk} = \frac{\begin{vmatrix} 1 & \rho(1) & \cdots & \rho(1) \\ \rho(1) & 1 & \cdots & \rho(2) \\ \vdots & \vdots & & \vdots \\ \rho(k-1) & \rho(k-2) & \cdots & \rho(k) \end{vmatrix}}{\begin{vmatrix} 1 & \rho(1) & \cdots & \rho(k-1) \\ \rho(1) & 1 & \cdots & \rho(k-2) \\ \vdots & \vdots & & \vdots \\ \rho(k-1) & \rho(k-2) & \cdots & 1 \end{vmatrix}}, \quad k = 1, 2, \ldots .$$

Thus, if the data generating process (DGP) is an AR(1) process, we get for the partial autocorrelation function:

$$\phi_{11} = \rho(1)$$

$$\phi_{22} = \frac{\begin{vmatrix} 1 & \rho(1) \\ \rho(1) & \rho(2) \end{vmatrix}}{\begin{vmatrix} 1 & \rho(1) \\ \rho(1) & 1 \end{vmatrix}} = \frac{\rho(2) - \rho(1)^2}{1 - \rho(1)^2} = 0,$$

because of $\rho(2) = \rho(1)^2$. Generally, the partial autocorrelation coefficients $\phi_{kk} = 0$ for $k > 1$ in an AR(1) process.

If the DGP is an AR(2) process, we get

$$\phi_{11} = \rho(1), \quad \phi_{22} = \frac{\rho(2) - \rho(1)^2}{1 - \rho(1)^2}, \quad \phi_{kk} = 0 \text{ for } k > 2.$$

The same is true for an AR(p) process: all partial autocorrelation coefficients of order higher than p are zero. Thus, for finite order autoregressive processes, the partial autocorrelation function provides the possibility of identifying the order of the process by the order of the last non-zero partial autocorrelation coefficient. We can estimate the partial autocorrelation coefficients consistently by substituting the theoretical values in (2.37) by their consistent estimates (1.10). For the partial autocorrelation coefficients which have a theoretical value of zero, i.e. the order of which is larger than the order of the process, we get asymptotically that they are normally distributed with $E[\hat{\phi}_{kk}] = 0$ and $V[\hat{\phi}_{kk}] = 1/T$ for $k > p$.

Example 2.8

The AR(1) process of *Example 2.1* has the following theoretical partial autocorrelation function: $\phi_{11} = \rho(1) = \alpha$ and zero elsewhere. In this example, α takes on the values 0.9, 0.5 and -0.9. The estimates of the partial autocorrelation functions for the realisations in *Figures 2.1* and *2.3* are presented in *Figure 2.8*. It is obvious for both processes that these are AR(1) processes. The estimated value for the process with $\alpha = 0.9$ is $\hat{\phi}_{11} = 0.91$, while all other partial autocorrelation coefficients are not significantly different from zero. We get $\hat{\phi}_{11} = -0.91$ for the process with $\alpha = -0.9$, while all estimated higher order partial autocorrelation coefficients do not deviate significantly from zero.

The AR(2) process of *Example 2.4* has the following theoretical partial autocorrelation function: $\phi_{11} = 0.96$, $\phi_{22} = -0.56$ and zero elsewhere. The realisation of this process, which is given in *Figure 2.5*, leads to the empirical partial autocorrelation function in *Figure 2.8*. It corresponds quite closely to the theoretical function; we get $\hat{\phi}_{11} = 0.95$ and $\hat{\phi}_{22} = -0.60$ and all higher order partial autocorrelation coefficients are not significantly different from zero. The same holds for the AR(2) process with the theoretical non-zero partial autocorrelations $\phi_{11} = 0.76$ and $\phi_{22} = -0.85$ given in *Example 2.5*. We get the estimates $\hat{\phi}_{11} = 0.76$ and $\hat{\phi}_{22} = -0.78$, whereas all higher order partial correlation coefficients are not significantly different from zero.

2.1 Autoregressive Processes

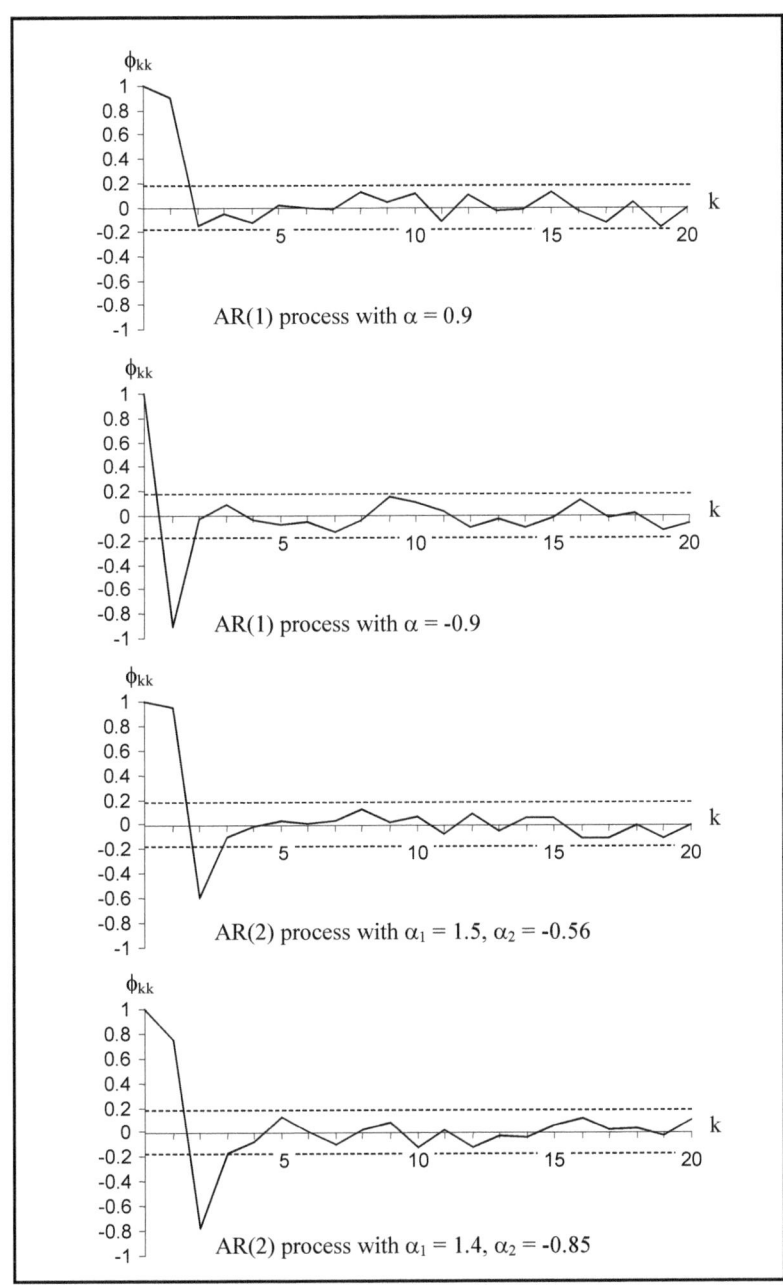

Figure 2.8: Estimated partial autocorrelation functions

2.1.5 Estimating Autoregressive Processes

Under the assumption of a known order p we have different possibilities to estimate the parameters:

(i) If we know the distribution of the white noise process that generates the AR(p) process, the parameters can be estimated by using maximum likelihood (ML) methods.

(ii) The parameters can also be estimated with the method of moments by using the Yule-Walker equations.

(iii) A further possibility is to treat

(2.26) $\quad x_t = \delta + \alpha_1 x_{t-1} + \alpha_2 x_{t-2} + \ldots + \alpha_p x_{t-p} + u_t,$

as a regression equation and apply the ordinary least squares (OLS) method for estimation. OLS provides consistent estimates. Moreover, if (2.26) fulfils the stability conditions, $\sqrt{T}(\hat{\delta} - \delta)$ as well as $\sqrt{T}(\hat{\alpha}_i - \alpha_i)$, $i = 1, 2, \ldots, p$, are asymptotically normally distributed.

If the order of the AR process is unknown, it can be estimated with the help of information criteria. For this purpose, AR processes with successively increasing orders $p = 1, 2, \ldots, p^{max}$ are estimated. Finally, the order p* is chosen which minimises the respective criterion. The following criteria are often used:

(i) The final prediction error which goes back to HIROTUGU AKAIKE (1969)

$$\text{FPE} = \frac{T+m}{T-m} \cdot \frac{1}{T} \sum_{t=1}^{T} (\hat{u}_t^{(p)})^2 \;.$$

(ii) Closely related to this is the Akaike information criterion (HIROTUGU AKAIKE (1974))

$$\text{AIC} = \ln \frac{1}{T} \sum_{t=1}^{T} (\hat{u}_t^{(p)})^2 + m \frac{2}{T} \;.$$

(iii) Alternatives are the Bayesian criterion of GIDEON SCHWARZ (1978)

$$\text{SC} = \ln \frac{1}{T} \sum_{t=1}^{T} (\hat{u}_t^{(p)})^2 + m \frac{\ln T}{T}$$

(iv) as well as the criterion developed by EDWARD J. HANNAN and BARRY G. QUINN (1979)

$$\mathrm{HQ} \;=\; \ln\frac{1}{T}\sum_{t=1}^{T}(\hat{u}_t^{(p)})^2 + m\frac{2\ln(\ln T)}{T} \;.$$

$\hat{u}_t^{(p)}$ are the estimated residuals of the AR(p) process, while m is the number of estimated parameters. If the constant term is estimated, too, m = p + 1 for an AR(p) process. These criteria are always based on the same principle: They consist of one part, the sum of squared residuals (or its logarithm), which decreases when the number of estimated parameters increases, and of a 'penalty term', which increases when the number of estimated parameters increases. Whereas the first two criteria overestimate the true (finite) order asymptotically, the two other criteria estimate the true order of the process consistently. For $T \geq 16$, the penalty term of SC is larger than the one of HQ which itself is larger than the one of AIC. This leads to the following ordering of the estimated AR orders:

$$\text{SC order} \;\leq\; \text{HQ order} \;\leq\; \text{AIC order.}$$

Please note that choosing such an order does not always imply that we have white noise residuals. This has to be checked independently. Many computer programmes like, for example, EViews, do not exactly report the criteria given in (ii) through (iv). Relying on the log-likelihood function instead of on the sum of squared residuals directly, they add $1 + \ln(2\pi) \approx 2.8379$, which does, of course, neither affect the order nor which value of p minimises the information criteria.

Example 2.9

As in *Example 2.6*, we take a look at the development of the three month money market interest rate in Frankfurt am Main. If, for this series, we estimate AR processes up to the order p = 4, we get the following results (for T = 116):

p = 0: AIC = 4.8334, HQ = 4.8430, SC = 4.8571;

p = 1: AIC = 2.7180, HQ = 2.7373, SC = 2.7655;

p = 2: AIC = 2.4457, HQ = 2.4746, SC = 2.5169;

p = 3: AIC = 2.4609, HQ = 2.4995, SC = 2.5559;

p = 4: AIC = 2.4778, HQ = 2.5260, SC = 2.5965.

With all three criteria we get the minimum for p = 2. Thus, the optimal number of lags is p* = 2, as used in *Example 2.6*.

2.2 Moving Average Processes

Moving average processes of an infinite order have already occurred when we presented the Wold decomposition theorem. They are, above all, of theoretical importance as, in practice, only a finite number of (different) parameters can be estimated. In the following, we consider finite order moving average processes. We start with the first order moving average process and then discuss general properties of finite order moving average processes.

2.2.1 First Order Moving Average Processes

The first order moving average process (MA(1)) is given by the following equation:

(2.38) $$x_t = \mu + u_t - \beta u_{t-1},$$

or

(2.38') $$x_t - \mu = (1-\beta L)u_t,$$

with u_t again being a pure random process. The Wold representation of an MA(1) process (as of any finite order MA process) has a finite number of terms. In this special case, the Wold coefficients are $\psi_0 = 1$, $\psi_1 = -\beta$ and $\psi_j = 0$ for $j \geq 2$. Thus, $\sum_j \psi_j^2$ is finite for all finite values of β, i.e. an MA(1) process is always stationary.

Taking expectations of (2.38) leads to

$$E[x_t] = \mu + E[u_t] - \beta E[u_{t-1}] = \mu.$$

The variance can also be calculated directly,

$$\begin{aligned} V[x_t] &= E[(x_t - \mu)^2] \\ &= E[(u_t - \beta u_{t-1})^2] \\ &= E[(u_t^2 - 2\beta u_t u_{t-1} + \beta^2 u_{t-1}^2)] \\ &= (1+\beta^2)\sigma^2 = \gamma(0). \end{aligned}$$

Therefore, the variance is constant at any point of time.

For the covariances of the process we get

$$\begin{aligned} E[(x_t - \mu)(x_{t+\tau} - \mu)] &= E[(u_t - \beta u_{t-1})(u_{t+\tau} - \beta u_{t+\tau-1})] \\ &= E[(u_t u_{t+\tau} - \beta u_t u_{t+\tau-1} - \beta u_{t-1} u_{t+\tau} + \beta^2 u_{t-1} u_{t+\tau-1})]. \end{aligned}$$

The covariances are different from zero only for $\tau = \pm 1$, i.e. for adjoining random variables. In this case

$$\gamma(1) = -\beta \sigma^2.$$

Thus, for an MA(1) process, all autocovariances and therefore all autocorrelations with an order higher than one disappear, i.e. $\gamma(\tau) = \rho(\tau) = 0$ for $\tau \geq 2$.

The correlogram of an MA(1) process is

$$\rho(0) = 1, \quad \rho(1) = \frac{-\beta}{1+\beta^2}, \quad \rho(\tau) = 0 \text{ for } \tau \geq 2.$$

If we consider $\rho(1)$ as a function of β, $\rho(1) = f(\beta)$, it holds that $f(0) = 0$ and $f(\beta) = -f(-\beta)$, i.e. that $f(\beta)$ is point symmetric to the origin, and that $|f(\beta)| \leq 0.5$. $f(\beta)$ has its maximum at $\beta = -1$ and its minimum at $\beta = 1$. Thus, an MA(1) process cannot have a first order autocorrelation above 0.5 or below -0.5.

If we know the autocorrelation coefficient $\rho(1) = \rho_1$, for example, by estimation, we can derive (estimate) the corresponding parameter β by using the equation for the first order autocorrelation coefficient,

$$(1 + \beta^2) \rho_1 + \beta = 0.$$

The quadratic equation can also be written as

(2.39)
$$\beta^2 + \frac{1}{\rho_1} \beta + 1 = 0,$$

and it has the two solutions

$$\beta_{1,2} = -\frac{1}{2\rho_1} \left(1 \pm \sqrt{1 - 4\rho_1^2} \right).$$

Thus, the parameters of the MA(1) process can be estimated non-linearly with the method of moments: the theoretical moments are substituted by their consistent estimates and the resulting equation is used for estimating the parameters consistently.

Because of $|\rho_1| \leq 0.5$, the quadratic equation always results in real roots. They also have the property that $\beta_1 \beta_2 = 1$. This gives us the possibility to model the same autocorrelation structure with two different parameters, where one is the inverse of the other.

In order to get a unique parameterisation, we require a further property of the MA(1) process. We ask under which conditions the MA(1) process (2.38) can have an autoregressive representation. By using the lag operator representation (2.38') we get

$$u_t = -\frac{\mu}{1-\beta} + \frac{1}{1-\beta L} x_t .$$

An expansion of the series $1/(1 - \beta L)$ is only possible for $|\beta| < 1$ and results in the following AR(∞) process

$$u_t = -\frac{\mu}{1-\beta} + x_t + \beta x_{t-1} + \beta^2 x_{t-2} + \ldots$$

or

$$x_t + \beta x_{t-1} + \beta^2 x_{t-2} + \ldots = \frac{\mu}{1-\beta} + u_t .$$

This representation requires the condition of *invertibility* ($|\beta| < 1$). In this case, we get a unique parameterisation of the MA(1) process. Applying the lag polynomial in (2.38'), we can formulate the *invertibility condition* in the following way: An MA(1) process is invertible if and only if the root of the lag polynomial

$$1 - \beta L = 0$$

is larger than one in modulus.

Example 2.10

The following MA(1) process is given:

(E2.5) $\qquad x_t = \varepsilon_t - \beta \varepsilon_{t-1}, \quad \varepsilon_t \sim N(0, 2^2),$

with $\beta = -0.5$. For this process we get

$$E[x_t] = 0,$$
$$V[x_t] = (1 + 0.5^2) \cdot 4 = 5,$$
$$\rho(1) = \frac{0.5}{1+0.5^2} = 0.4,$$
$$\rho(\tau) = 0 \quad \text{for} \quad \tau \geq 2.$$

Solving the corresponding quadratic equation (2.39) for this value of $\rho(1)$ leads to the two roots $\beta_1 = -2.0$ and $\beta_2 = -0.5$. If we now consider the process

(E2.5a) $\qquad y_t = \eta_t + 2 \eta_{t-1}, \quad \eta_t \sim N(0, 1),$

we obtain the following results:

$$E[y_t] = 0,$$
$$V[y_t] = (1 + 2.0^2) \cdot 1 = 5,$$

2.2 Moving Average Processes

$$\rho(1) = \frac{2.0}{1+2.0^2} = 0.4,$$

$$\rho(\tau) = 0 \text{ for } \tau \geq 2,$$

i.e. the variances and the autocorrelogram of the two processes (E2.5) and (E2.5a) are identical. The only difference between them is that (E2.5) is invertible, because the invertibility condition $|\beta| < 1$ holds, whereas (E2.5a) is not invertible. Thus, given the structure of the correlations, we can choose the one of the two processes that fulfils the invertibility condition without imposing any restrictions on the structure of the process.

With equation (2.37), the partial autocorrelation function of the MA(1) process can be calculated in the following way:

$$\phi_{11} = \rho(1),$$

$$\phi_{22} = \frac{\begin{vmatrix} 1 & \rho(1) \\ \rho(1) & 0 \end{vmatrix}}{\begin{vmatrix} 1 & \rho(1) \\ \rho(1) & 1 \end{vmatrix}} = -\frac{\rho(1)^2}{1-\rho(1)^2} < 0,$$

$$\phi_{33} = \frac{\begin{vmatrix} 1 & \rho(1) & \rho(1) \\ \rho(1) & 1 & 0 \\ 0 & \rho(1) & 0 \end{vmatrix}}{\begin{vmatrix} 1 & \rho(1) & 0 \\ \rho(1) & 1 & \rho(1) \\ 0 & \rho(1) & 1 \end{vmatrix}} = \frac{\rho(1)^3}{1-2\rho(1)^2} \gtreqless 0 \text{ for } \beta \lesseqgtr 0,$$

$$\phi_{44} = \frac{\begin{vmatrix} 1 & \rho(1) & 0 & \rho(1) \\ \rho(1) & 1 & \rho(1) & 0 \\ 0 & \rho(1) & 1 & 0 \\ 0 & 0 & \rho(1) & 0 \end{vmatrix}}{\begin{vmatrix} 1 & \rho(1) & 0 & 0 \\ \rho(1) & 1 & \rho(1) & 0 \\ 0 & \rho(1) & 1 & \rho(1) \\ 0 & 0 & \rho(1) & 1 \end{vmatrix}} = \frac{-\rho(1)^4}{(1-\rho(1)^2)^2 - \rho(1)^2} < 0,$$

etc.

If β is positive, ρ(1) is negative and vice versa. This leads to the two possible patterns of partial autocorrelation functions, exemplified by β = ±0.8:

$$\beta = 0.8, \quad \phi_{ii} \in \{-0.49, -0.31, -0.22, -0.17, \ldots\},$$

$$\beta = -0.8, \quad \phi_{ii} \in \{0.49, -0.31, 0.22, -0.17, \ldots\}.$$

Thus, contrary to the AR(1) process, the autocorrelation function of the MA(1) process breaks off, while the partial autocorrelation function does not. These properties hold generally, since invertible finite order MA processes are equivalent to infinite order AR processes.

2.2.2 MA(1) and Temporal Aggregation

The time series which are discussed in this book are measured in discrete time, with intervals of equal length. Exchange rates, for example, are normally quoted at the end of each trading day. For econometric analyses, however, monthly, quarterly, or even annual data are used, rather than these daily values. Usually, averages or end-of-period data are used for temporal aggregation.

Thus, two aggregation schemes have to be distinguished. The first one is skip sampling (or: systematic sampling) where only every m^{th} data point is recorded. If x_t is the basic series at $t = 1, 2, 3, \ldots$, the skip sampled series y_s with new time scale s is end-of-period data,

$$y_1 = x_m, \quad y_2 = x_{2m}, \quad y_3 = x_{3m}, \quad \ldots, \quad y_s = x_{sm}.$$

Such an aggregation is typical for stock variables. However, the second scheme of averaging over m non-overlapping periods is also widely used, in particular for rates or indices:

$$\tilde{y}_1 = \frac{1}{m}(x_m + x_{m-1} + \ldots + x_1)$$

$$\tilde{y}_2 = \frac{1}{m}(x_{2m} + x_{2m-1} + \ldots + x_{m+1})$$

$$\vdots$$

$$\tilde{y}_s = \frac{1}{m}(x_{sm} + x_{sm-1} + \ldots + x_{(s-1)m+1}).$$

In the following, we do not present a general theory of temporal aggregation but just discuss a special case of particular applied interest, the random walk, with

$$x_t = x_{t-1} + u_t,$$

where an artificial MA(1) structure arises due to aggregation by averaging. It is straightforward to see that systematic sampling does not affect the random walk property, since in this case we can write

$$y_s = x_0 + \sum_{t=1}^{sm} u_t.$$

From this representation we get

$$y_s = y_{s-1} + \eta_s,$$

with η_s being white noise:

$$\eta_s = u_{sm} + u_{sm-1} + \ldots + u_{(s-1)m+1},$$

with $E[\eta_s] = 0$ and

$$E(\eta_s \cdot \eta_{s-\tau}) = \begin{cases} m\sigma_u^2 & \text{for } \tau = 0 \\ 0 & \text{elsewhere} \end{cases}.$$

Hence, the random walk property is inherited by y_s, only the variance of the differences $y_s - y_{s-1}$ is inflated in the obvious way. In case of averaging, $\tilde{y}_s$, matters get more complicated. It can, however, be shown that the differences

$$\tilde{y}_s - \tilde{y}_{s-1} = \tilde{\eta}_s$$

follow no longer a white noise process but an MA(1) scheme hidden behind

$$\tilde{\eta}_s = \frac{1}{m}\left(u_{sm} + 2u_{sm-1} + \ldots + mu_{(s-1)m+1} + \ldots + 2u_{(s-2)m+3} + u_{(s-2)m+2}\right).$$

We omit details but refer to HOLBROOK WORKING (1960) who showed that with increasing aggregation level, $m \to \infty$, one obtains the autocorrelation function

$$\rho(\tau) = \frac{E[\tilde{\eta}_s \tilde{\eta}_{s-\tau}]}{V[\tilde{\eta}_s]} \to \begin{cases} 1, & \tau = 0 \\ \frac{1}{4}, & |\tau| = 1 \\ 0, & \text{elsewhere} \end{cases}.$$

Note that the above autocorrelation function corresponds to the following MA(1)-process

$$\tilde{\eta}_s = \tilde{u}_s - \tilde{\beta}\tilde{u}_{s-1}$$

where $\tilde{u}_s$ is white noise, and the limiting value (for $m \to \infty$) of the MA parameter is

$$\tilde{\beta} = \sqrt{3} - 2 \approx -0.268.$$

GEORGE C. TIAO (1972) generalised this result the following way: If $x_t - x_{t-1}$ is not generated by white noise but by an invertible MA(1) process, then $\tilde{y}_s - \tilde{y}_{s-1}$ behaves with growing m like the MA(1) process $\tilde{u}_s - \tilde{\beta}\tilde{u}_{s-1}$, where $\tilde{\beta}$ is independent of the underlying MA(1) structure of $x_t - x_{t-1}$. This result even continues to hold when the assumption that $x_t - x_{t-1}$ is MA(1) is replaced by a more general moving average process of higher order as introduced in subsection 2.2.3.

Example 2.11

Consider averaging over $m = 2$ periods,

$$\tilde{y}_s = \frac{1}{2}(x_{2s} + x_{2s-1}).$$

For the random walk $x_t = x_{t-1} + u_t$, it holds that

$$\tilde{\eta}_s = \tilde{y}_s - \tilde{y}_{s-1}$$

$$= \frac{1}{2}(x_{2s} + x_{2s-1} - x_{2s-2} - x_{2s-3})$$

$$= \frac{1}{2}(u_{2s} + 2u_{2s-1} + u_{2s-2}).$$

This process can be described as

$$\tilde{\eta}_s = \tilde{u}_s - \beta \tilde{u}_{s-1}$$

with $\beta = 2\sqrt{2} - 3 \approx -0.172$, and

$$E(\tilde{\eta}_s \cdot \tilde{\eta}_{s-\tau}) = \begin{cases} \frac{3}{2}\sigma_u^2 & \text{for } \tau = 0 \\ \frac{1}{4}\sigma_u^2 & \text{for } |\tau| = 1, \\ 0 & \text{elsewhere} \end{cases}$$

such that for m = 2 the autocorrelation coefficient at lag one becomes $\rho(1) = 1/6$.

Example 2.12

Example 1.3 as well as *Figure 1.8* present the end-of-month exchange rate between the Swiss Franc and the U.S. Dollar over the period from January 1974 to December 2011. The autocorrelogram of the first differences of the logarithms of this time series indicates that they follow a pure random process. The tests we applied did not reject this null hypothesis.

If we use monthly averages instead of end-of-month data, the following MA(1) process can be estimated for the first difference of the logarithms of this exchange rate:

$$\Delta\ln(e_t) = \underset{(-1.53)}{-0.003} + \hat{u}_t + \underset{(6.91)}{0.308}\,\hat{u}_{t-1},$$

$$\bar{R}^2 = 0.082, \quad SE = 0.028, \quad Q(11) = 8.216\ (p = 0.694),$$
$$JB = 21.194\ (p = 0.000),$$

with the t values again given in parentheses. $\ln(\cdot)$ denotes the natural logarithm. The estimated coefficient of the MA(1) term is highly significantly different from zero. The Ljung-Box Q-statistic indicates that there is no longer any significant autocorrelation in the residuals. As $m \approx 20$ is relatively large (in this context), the estimated values of the MA(1) term should not be too different from the theoretical value given by GEORGE C. TIAO (1972). The theoretical value -0.268 lies in the two-sigma confidence interval of the estimated parameter -0.308.

2.2.3 Higher Order Moving Average Processes

In general, the *moving average process of order q* (MA(q)) can be written as

(2.40) $\qquad x_t = \mu + u_t - \beta_1 u_{t-1} - \beta_2 u_{t-2} - \dots - \beta_q u_{t-q}$

with $\beta_q \neq 0$ and u_t as a pure random process. Using the lag operator we get

(2.40') $\qquad x_t - \mu = (1 - \beta_1 L - \beta_2 L^2 - \dots - \beta_q L^q)u_t$
$$ = \beta(L)u_t.$$

From (2.40) we see that we already have a finite order Wold representation with $\psi_k = 0$ for $k > q$. Thus, there are no problems of convergence, and every finite MA(q) process is stationary, no matter what values are used for β_j, $j = 1, 2, ..., q$.

For the expectation of (2.40) we immediately get $E[x_t] = \mu$. Thus, the variance can be calculated as:

$$\begin{aligned} V[x_t] &= E[(x_t - \mu)^2] \\ &= E[(u_t - \beta_1 u_{t-1} - ... - \beta_q u_{t-q})^2] \\ &= E[(u_t^2 + \beta_1^2 u_{t-1}^2 + ... + \beta_q^2 u_{t-q}^2 - 2\beta_1 u_t u_{t-1} - ... \\ &\quad - 2\beta_{q-1}\beta_q u_{t-q+1} u_{t-q})] . \end{aligned}$$

From this we obtain

$$V[x_t] = (1 + \beta_1^2 + \beta_2^2 + ... + \beta_q^2)\sigma^2 .$$

For the covariances of order τ we can write

$$\begin{aligned} \text{Cov}[x_t, x_{t+\tau}] &= E[(x_t - \mu)(x_{t+\tau} - \mu)] \\ &= E[(u_t - \beta_1 u_{t-1} - ... - \beta_q u_{t-q}) \\ &\quad \cdot (u_{t+\tau} - \beta_1 u_{t+\tau-1} - ... - \beta_q u_{t+\tau-q})] \\ &= E[u_t(u_{t+\tau} - \beta_1 u_{t+\tau-1} - ... - \beta_q u_{t+\tau-q}) \\ &\quad - \beta_1 u_{t-1}(u_{t+\tau} - \beta_1 u_{t+\tau-1} - ... - \beta_q u_{t+\tau-q}) \\ &\quad \vdots \\ &\quad - \beta_q u_{t-q}(u_{t+\tau} - \beta_1 u_{t+\tau-1} - ... - \beta_q u_{t+\tau-q})] . \end{aligned}$$

Thus, for $\tau = 1, 2, ..., q$ we get

(2.41)
$$\begin{aligned} \tau &= 1: \gamma(1) = (-\beta_1 + \beta_1\beta_2 + ... + \beta_{q-1}\beta_q)\sigma^2, \\ \tau &= 2: \gamma(2) = (-\beta_2 + \beta_1\beta_3 + ... + \beta_{q-2}\beta_q)\sigma^2, \\ &\vdots \\ \tau &= q: \gamma(q) = -\beta_q\sigma^2, \end{aligned}$$

while we have $\gamma(\tau) = 0$ for $\tau > q$.

Consequently, all autocovariances and autocorrelations with orders higher than the order of the process are zero. It is – at least theoretically – possible to identify the order of an MA(q) process by using the autocorrelogram.

It can be seen from (2.41) that there exists a system of non-linear equations for given (or estimated) second order moments that determines (makes it possible to estimate) the parameters $\beta_1, ..., \beta_q$. As we have already seen in the case of the MA(1) process, such non-linear equation systems have multiple solutions, i.e. there exist different values for β_1, β_2, ... and β_q that all lead to the same autocorrelation structure. To get a unique parameterisation, the invertibility condition is again required, i.e. it must be possible to represent the MA(q) process as a stationary AR(∞) process. Starting from (2.40'), this implies that the inverse operator $\beta^{-1}(L)$ can be represented as an infinite series in the lag operator, where the sum of the coefficients has to be bounded. Thus, the representation we get is an AR(∞) process

$$u_t = -\frac{\mu}{\beta(1)} + \beta^{-1}(L) x_t$$

$$= -\frac{\mu}{\beta(1)} + \sum_{j=0}^{\infty} c_j x_{t-j} ,$$

where

$$1 = (1 - \beta_1 L - ... - \beta_q L^q)(1 + c_1 L + c_2 L^2 + ...),$$

and the parameters c_i, $i = 1, 2, ...$ are calculated by using again the method of undetermined coefficients. Such a representation exists if all roots of

$$1 - \beta_1 L - ... - \beta_q L^q = 0$$

are larger than one in absolute value.

Example 2.13

Let the following MA(2) process

$$x_t = u_t + 0.6 \, u_{t-1} - 0.1 \, u_{t-2}$$

be given, with a variance of 1 given for the pure random process u. For the variance of x we get

$$V[x_t] = (1 + 0.36 + 0.01) \cdot 1 = 1.37 .$$

Corresponding to (2.41) the covariances are

$$\gamma(1) = +0.6 - 0.06 = 0.54$$
$$\gamma(2) = -0.1$$
$$\gamma(\tau) = 0 \quad \text{for } \tau > 2$$

This leads to the autocorrelation coefficients $\rho(1) = 0.39$ and $\rho(2) = -0.07$. To check whether the process is invertible, the quadratic equation

$$1 + 0.6 \, L - 0.1 \, L^2 = 0$$

has to be solved. As the two roots -1.36 and 7.36 are larger than 1 in absolute value, the invertibility condition is fulfilled, i.e. the MA(2) process can be written as an AR(∞) process

$$x_t = (1 + 0.6 L - 0.1 L^2) u_t,$$

$$u_t = \frac{1}{1 + 0.6L - 0.1L^2} x_t$$

$$= (1 + c_1 L + c_2 L^2 + c_3 L^3 + \ldots) x_t.$$

The unknowns c_i, $i = 1, 2, \ldots$, can be determined by comparing the coefficients of the polynomials in the following way:

$$1 = (1 + 0.6 L - 0.1 L^2)(1 + c_1 L + c_2 L^2 + c_3 L^3 + \ldots)$$

$$\begin{aligned} 1 = 1 &+ c_1 L + c_2 L^2 + c_3 L^3 + \ldots \\ &+ 0.6 L + 0.6 c_1 L^2 + 0.6 c_2 L^3 + \ldots \\ & - 0.1 L^2 - 0.1 c_1 L^3 - \ldots \end{aligned}$$

It holds that

$$\begin{aligned} c_1 + 0.6 &= 0 \Rightarrow c_1 = -0.60, \\ c_2 + 0.6 c_1 - 0.1 &= 0 \Rightarrow c_2 = 0.46, \\ c_3 + 0.6 c_2 - 0.1 c_1 &= 0 \Rightarrow c_3 = -0.34, \\ c_4 + 0.6 c_3 - 0.1 c_2 &= 0 \Rightarrow c_4 = 0.25, \\ &\vdots \end{aligned}$$

Thus, we get the following AR(∞) representation

$$x_t - 0.6 x_{t-1} + 0.46 x_{t-2} - 0.34 x_{t-3} + 0.25 x_{t-4} - \ldots = u_t.$$

Similarly to the MA(1) process, the partial autocorrelation function of the MA(q) process does not break off. As long as the order q is finite, the MA(q) process is stationary whatever its parameters are. If the order tends towards infinity, however, for the process to be stationary the series of the coefficients has to converge just like in the Wold representation.

2.3 Mixed Processes

If we take a look at the two different functions that can be used to identify autoregressive and moving average processes, we see from *Table 2.1* that the situation in which neither of them breaks off can only arise if there is an MA(∞) process that can be inverted to an AR(∞) process, i.e. if the Wold representation of an AR(∞) process corresponds to an MA(∞) process. However, as pure AR or MA representations, these processes cannot be used for empirical modelling because they can only be characterised by means of infinitely many parameters. After all, according to the *principle of parsimony*, the number of estimated parameters should be as small as possible when applying time series methods.

In the following, we introduce processes which contain both an autoregressive (AR) term of finite order p and a moving average (MA) term of finite order q. Hence, these mixed processes are denoted as ARMA(p,q) processes. They enable us to describe processes in which neither the autocorrelation nor the partial autocorrelation function breaks off after a finite number of lags. Again, we start with the simplest case, the ARMA(1,1) process, and consider the general case afterwards.

Table 2.1: Characteristics of the Autocorrelation and the Partial Autocorrelation Functions of AR and MA Processes

	Autocorrelation Function	Partial Autocorrelation Function
MA(q)	breaks off with q	does not break off
AR(p)	does not break off	breaks off with p

2.3.1 ARMA(1,1) Processes

An ARMA(1,1) process can be written as follows,

(2.42) $$x_t = \delta + \alpha x_{t-1} + u_t - \beta u_{t-1},$$

or, by using the lag operator

(2.42') $$(1 - \alpha L) x_t = \delta + (1 - \beta L) u_t,$$

where u_t is a pure random process. To get the Wold representation of an ARMA(1,1) process, we solve (2.42') for x_t,

$$x_t = \frac{\delta}{1-\alpha} + \frac{1-\beta L}{1-\alpha L} u_t.$$

It is obvious that $\alpha \neq \beta$ must hold, because otherwise x_t would be a pure random process fluctuating around the mean $\mu = \delta/(1-\alpha)$. The ψ_j, j = 0, 1, 2, ..., can be determined as follows:

$$\frac{1-\beta L}{1-\alpha L} = \psi_0 + \psi_1 L + \psi_2 L^2 + \psi_3 L^3 + \ldots$$

$$1-\beta L = (1-\alpha L)(\psi_0 + \psi_1 L + \psi_2 L^2 + \psi_3 L^3 + \ldots)$$

$$1-\beta L = \psi_0 + \psi_1 L + \psi_2 L^2 + \psi_3 L^3 + \ldots$$
$$ - \alpha\psi_0 L - \alpha\psi_1 L^2 - \alpha\psi_2 L^3 - \ldots \,.$$

Comparing the coefficients of the two lag polynomials we get

$$L^0: \quad \psi_0 = 1$$
$$L^1: \quad \psi_1 - \alpha\psi_0 = -\beta \;\Rightarrow\; \psi_1 = \alpha - \beta$$
$$L^2: \quad \psi_2 - \alpha\psi_1 = 0 \;\Rightarrow\; \psi_2 = \alpha(\alpha - \beta)$$
$$L^3: \quad \psi_3 - \alpha\psi_2 = 0 \;\Rightarrow\; \psi_3 = \alpha^2(\alpha - \beta)$$
$$\vdots$$
$$L^j: \quad \psi_j - \alpha\psi_{j-1} = 0 \;\Rightarrow\; \psi_j = \alpha^{j-1}(\alpha - \beta)\,.$$

The ψ_j, $j \geq 2$ can be determined from the linear homogeneous difference equation

$$\psi_j - \alpha\psi_{j-1} = 0$$

with $\psi_1 = \alpha - \beta$ as initial condition. The ψ_j converge towards zero if and only if $|\alpha| < 1$. This corresponds to the stability condition of the AR term. Thus, the ARMA(1,1) process is stationary if, with stochastic initial conditions, it has a stable AR(1) term. The Wold representation is

$$(2.43) \quad x_t = \frac{\delta}{1-\alpha} + u_t + (\alpha-\beta)u_{t-1} + \alpha(\alpha-\beta)u_{t-2} + \alpha^2(\alpha-\beta)u_{t-3} + \ldots \,.$$

Thus, the ARMA(1,1) process can be written as an MA(∞) process.

To invert the MA(1) part, $|\beta| < 1$ must hold. Starting from (2.42') leads to

$$u_t = \frac{-\delta}{1-\beta} + \frac{1-\alpha L}{1-\beta L} x_t\,.$$

If $1/(1-\beta L)$ is developed into a geometric series we get

$$u_t = \frac{-\delta}{1-\beta} + (1-\alpha L)(1 + \beta L + \beta^2 L^2 + \ldots)\, x_t$$

$$= \frac{-\delta}{1-\beta} + x_t + (\beta-\alpha)x_{t-1} + \beta(\beta-\alpha)x_{t-2} + \beta^2(\beta-\alpha)x_{t-3} + \ldots \,.$$

2.3 Mixed Processes

This proves to be an AR(∞) representation. It shows that the combination of an AR(1) and an MA(1) term leads to a process with both MA(∞) and AR(∞) representation if the AR term is stable and the MA term invertible.

We obtain the first and second order moments of the stationary process in (2.42) as follows:

$$E[x_t] = E[\delta + \alpha x_{t-1} + u_t - \beta u_{t-1}]$$
$$= \delta + \alpha E[x_{t-1}].$$

Due to $E[x_t] = E[x_{t-1}] = \mu$, we get

$$\mu = \frac{\delta}{1-\alpha},$$

i.e. the expectation is the same as in an AR(1) process.

If we set $\delta = 0$ without loss of generality, the expectation is zero. The autocovariance of order $\tau \geq 0$ can then be written as

(2.44) $\quad E[x_{t-\tau} x_t] = E[x_{t-\tau}(\alpha x_{t-1} + u_t - \beta u_{t-1})],$

which leads to

$$\gamma(0) = \alpha \gamma(1) + E[x_t u_t] - \beta E[x_t u_{t-1}]$$

for $\tau = 0$. Due to (2.43), $E[x_t u_t] = \sigma^2$ and $E[x_t u_{t-1}] = (\alpha - \beta) \sigma^2$. Thus, we can write

(2.45) $\quad \gamma(0) = \alpha \gamma(1) + (1 - \beta(\alpha - \beta)) \sigma^2.$

(2.44) leads to

$$\gamma(1) = \alpha \gamma(0) + E[x_{t-1} u_t] - \beta E[x_{t-1} u_{t-1}]$$

for $\tau = 1$. Because of (2.43) this can be written as

(2.46) $\quad \gamma(1) = \alpha \gamma(0) - \beta \sigma^2.$

If we insert (2.46) in (2.45) and solve for $\gamma(0)$, the resulting variance of the ARMA(1,1) process is

(2.47) $\quad \gamma(0) = \frac{1+\beta^2 - 2\alpha\beta}{1-\alpha^2} \sigma^2.$

Inserting this into (2.46), we get

(2.48) $\quad \gamma(1) = \frac{(\alpha-\beta)(1-\alpha\beta)}{1-\alpha^2} \sigma^2$

for the first order autocovariance. For $\tau \geq 2$, (2.44) results in the autocovariances

(2.49) $$\gamma(\tau) = \alpha\,\gamma(\tau-1)$$

and the autocorrelations

(2.50) $$\rho(\tau) = \alpha\,\rho(\tau-1).$$

This results in the same difference equation as in an AR(1) process but, however, with the different initial condition

$$\rho(1) = \frac{(\alpha-\beta)(1-\alpha\beta)}{1+\beta^2 - 2\alpha\beta}.$$

The first order autocorrelation coefficient is influenced by the MA term, while the higher order autocorrelation coefficients develop in the same way as in an AR(1) process.

If the process is stable and invertible, i.e. for $|\alpha| < 1$ and $|\beta| < 1$, the sign of $\rho(1)$ is determined by the sign of $(\alpha - \beta)$ because of $(1 + \beta^2 - 2\alpha\beta) > 0$ and $(1 - \alpha\beta) > 0$. Moreover, it follows from (2.49) that the autocorrelation function – as in the AR(1) process – is monotonic for $\alpha > 0$ and oscillating for $\alpha < 0$. Due to $|\alpha| < 1$ with τ increasing, the autocorrelation function also decreases in absolute value.

Thus, the following typical autocorrelation structures are possible:

(i) $\alpha > 0$ and $\alpha > \beta$: The autocorrelation function is always positive.

(ii) $\alpha < 0$ and $\alpha < \beta$: The autocorrelation function oscillates; the initial condition $\rho(1)$ is negative.

(iii) $\alpha > 0$ and $\alpha < \beta$: The autocorrelation function is negative from $\rho(1)$ onwards.

(iv) $\alpha < 0$ and $\alpha > \beta$: The autocorrelation function oscillates; the initial condition $\rho(1)$ is positive.

Figure 2.9 shows the development of the corresponding autocorrelation functions up to $\tau = 20$ for the parameter values $\alpha, \beta \in \{0.8, 0.5, -0.5, -0.8\}$ in which, of course, $\alpha \neq \beta$ must always hold, as otherwise the ARMA(1,1) process degenerates to a pure random process.

For the partial autocorrelation function we get

$$\phi_{11} = \rho(1) = \frac{(\alpha-\beta)(1-\alpha\beta)}{1+\beta^2 - 2\alpha\beta},$$

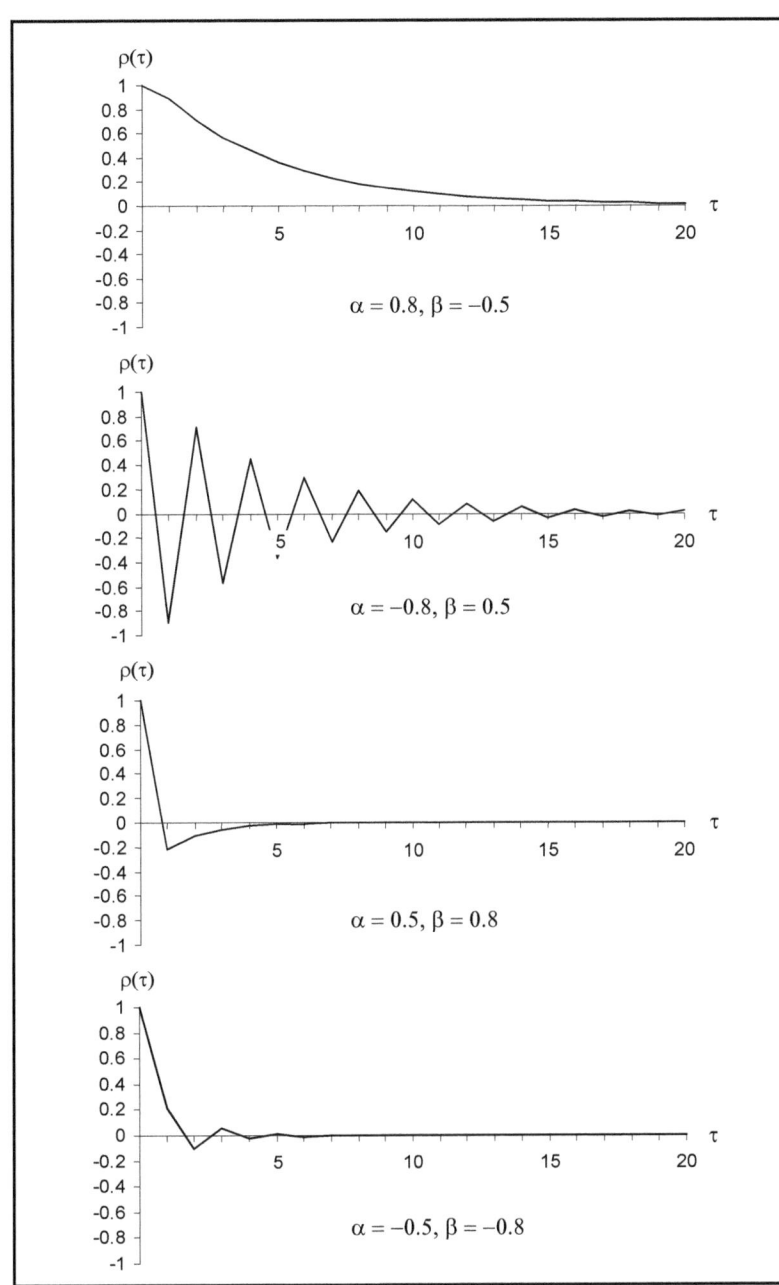

Figure 2.9: Theoretical autocorrelation functions of ARMA(1,1) processes

$$\phi_{22} = \frac{\begin{vmatrix} 1 & \rho(1) \\ \rho(1) & \rho(2) \end{vmatrix}}{\begin{vmatrix} 1 & \rho(1) \\ \rho(1) & 1 \end{vmatrix}} = \frac{\rho(2) - \rho(1)^2}{1 - \rho(1)^2} = \frac{\rho(1)(\alpha - \rho(1))}{1 - \rho(1)^2},$$

because of $\rho(2) = \alpha \rho(1)$,

$$\phi_{33} = \frac{\begin{vmatrix} 1 & \rho(1) & \rho(1) \\ \rho(1) & 1 & \rho(2) \\ \rho(2) & \rho(1) & \rho(3) \end{vmatrix}}{\begin{vmatrix} 1 & \rho(1) & \rho(2) \\ \rho(1) & 1 & \rho(1) \\ \rho(2) & \rho(1) & 1 \end{vmatrix}} = \frac{\begin{vmatrix} 1 & \rho(1) & \rho(1) \\ \rho(1) & 1 & \alpha\rho(1) \\ \alpha\rho(1) & \rho(1) & \alpha^2\rho(1) \end{vmatrix}}{1 + 2\alpha\rho(1)^3 - \rho(1)^2(2 + \alpha^2)}$$

$$= \frac{\rho(1)(\alpha - \rho(1))^2}{1 + 2\alpha\rho(1)^3 - \rho(1)^2(2 + \alpha^2)}, \text{ etc.}$$

Thus, the ARMA(1,1) process is a stationary stochastic process where neither the autocorrelation nor the partial autocorrelation function breaks off.

The following example shows how, due to measurement error, an AR(1)-process becomes an ARMA(1,1) process.

Example 2.14

The 'true' variable $\tilde{x}_t$ is generated by a stationary AR(1) process,

(E2.8) $$\tilde{x}_t = \alpha \tilde{x}_{t-1} + u_t,$$

but it can only be measured with an error v_t, i.e. for the observed variable x_t it holds that

(E2.9) $$x_t = \tilde{x}_t + v_t,$$

where v_t is a pure random process uncorrelated with the random process u_t. (The same model was used in *Example 2.3* but with a different interpretation.) If we transform (E2.8) to

$$\tilde{x}_t = \frac{u_t}{1 - \alpha L}$$

and insert it into (E2.9) we get

$$(1 - \alpha L) x_t = u_t + v_t - \alpha v_{t-1}.$$

For the combined error term $\zeta_t = u_t + v_t - \alpha\, v_{t-1}$ we get

$$\gamma_\zeta(0) = \sigma_u^2 + (1 + \alpha^2)\,\sigma_v^2$$

$$\gamma_\zeta(1) = -\alpha\,\sigma_v^2$$

$$\gamma_\zeta(\tau) = 0 \quad \text{for } \tau \geq 2,$$

or

$$\rho_\zeta(1) = \frac{-\alpha\sigma_v^2}{\sigma_u^2 + (1+\alpha^2)\,\sigma_v^2},\quad \rho_\zeta(\tau) = 0 \text{ for } \tau \geq 2.$$

Thus, the observable variable x_t follows an ARMA(1,1) process,

$$(1 - \alpha L)\,x_t = (1 - \beta L)\,\eta_t,$$

where β can be calculated by means of $\rho_\zeta(1)$ and η_t is a pure random process. (See also the corresponding results in *Section 2.2.1*.)

2.3.2 ARMA(p,q) Processes

The general *autoregressive moving average process* with AR order p and MA order q can be written as

(2.51) $\quad x_t = \delta + \alpha_1 x_{t-1} + \ldots + \alpha_p x_{t-p} + u_t - \beta_1 u_{t-1} - \ldots - \beta_q u_{t-q},$

with u_t being a pure random process and $\alpha_p \neq 0$ and $\beta_q \neq 0$ having to hold. Using the lag operator, we can write

(2.51') $\quad (1 - \alpha_1 L - \ldots - \alpha_p L^p)\,x_t = \delta + (1 - \beta_1 L - \ldots - \beta_q L^q)\,u_t,$

or

(2.51") $\quad\quad\quad \alpha(L)\,x_t = \delta + \beta(L)\,u_t.$

As factors that are common in both polynomials can be reduced, $\alpha(L)$ and $\beta(L)$ cannot have identical roots. The process is stationary if – with stochastic initial conditions – the stability conditions of the AR term are fulfilled, i.e. if $\alpha(L)$ only has roots that are larger than 1 in absolute value. Then we can derive the Wold representation for which

$$\beta(L) = \alpha(L)(1 + \psi_1 L + \psi_2 L^2 + \ldots)$$

must hold. Again, the ψ_j, $j = 1, 2, \ldots$, can be calculated by comparing the coefficients. If, likewise, all roots of $\beta(L)$ are larger than 1 in absolute value, the ARMA(p,q) process is also invertible.

A stationary and invertible ARMA(p,q) process may either be represented as an AR(∞) or as an MA(∞) process. Thus, neither its autocorrelation nor its partial autocorrelation function breaks off. In short, it is possible to generate stationary stochastic processes with infinite AR and MA orders by using only a finite number of parameters.

Under the assumption of stationarity, (2.51) directly results in the constant mean

$$E[x_t] = \mu = \frac{\delta}{1-\alpha_1 - ... - \alpha_p}.$$

If, without loss of generality, we set $\delta = 0$ and thus also $\mu = 0$, we get the following relation for the autocovariances:

$$\gamma(\tau) = E[x_{t-\tau} x_t]$$
$$= E[x_{t-\tau}(\alpha_1 x_{t-1} + ... + \alpha_p x_{t-p} + u_t - \beta_1 u_{t-1} - ... - \beta_q u_{t-q})].$$

This relation can also be written as

$$\gamma(\tau) = \alpha_1 \gamma(\tau-1) + \alpha_2 \gamma(\tau-2) + ... + \alpha_p \gamma(\tau-p)$$
$$+ E[x_{t-\tau} u_t] - \beta_1 E[x_{t-\tau} u_{t-1}] - ... - \beta_q E[x_{t-\tau} u_{t-q}].$$

Due to the Wold representation, the covariances between $x_{t-\tau}$ and u_{t-i}, $i = 0$, ..., q, are zero for $\tau > q$, i.e. the autocovariances for $\tau > q$ and $\tau > p$ are generated by the difference equation of an AR(p) process,

$$\gamma(\tau) - \alpha_1 \gamma(\tau-1) - \alpha_2 \gamma(\tau-2) - ... - \alpha_p \gamma(\tau-p) = 0 \quad \text{for } \tau > q \wedge \tau > p$$

whereas the first q autocovariances are also influenced by the MA part. Normalisation with $\gamma(0)$ leads to exactly the same results for the autocorrelations.

If the orders p and q are given and the distribution of the white noise process u_t is known, the parameters of an ARMA(p,q) process can be estimated consistently by using maximum likelihood methods. These estimates are also asymptotically efficient. If there is no such programme available, it is possible to estimate the parameters consistently with least squares. As every invertible ARMA(p,q) process is equivalent to an AR(∞) process, first of all an AR(k) process is estimated with k sufficiently larger than p. From this, one can get estimates of the non-observable residuals $\hat{u}_t$. By employing these residuals, the ARMA(p,q) process can be estimated with the least squares method,

$$x_t = \delta + \alpha_1 x_{t-1} + ... + \alpha_p x_{t-p} - \beta_1 \hat{u}_{t-1} - ... - \beta_q \hat{u}_{t-q} + v_t.$$

This approach can also be used if p and q are unknown. These orders can, for example, be determined by using the information criteria shown in *Section 2.1.5*.

2.3 Mixed Processes

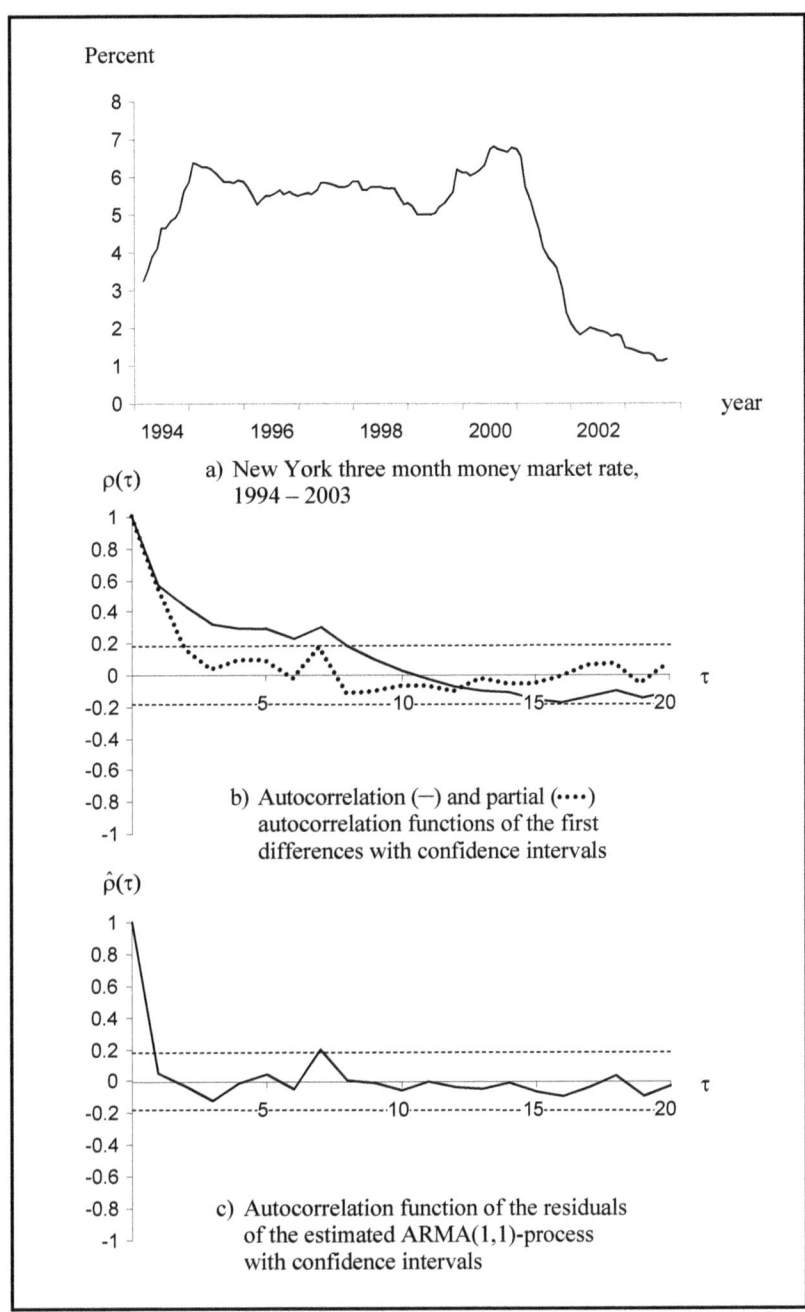

Figure 2.10: Three month money market rate in New York, 1994 – 2003

Example 2.15

Figure 2.10 shows the development of the US three month money market rate (USR) as well as the estimated autocorrelation and partial autocorrelation function of the first differences of this time series for the period from March 1994 to August 2003 (114 observations). Both functions do not show a clear break-off behaviour. Therefore, the following ARMA(1,1) model has been estimated for this time series:

$$\Delta USR_t = \underset{(-0.73)}{-0.006} + \underset{(10.91)}{0.831\ \Delta USR_{t-1}} + \hat{u}_t - \underset{(-3.57)}{0.457\ \hat{u}_{t-1}},$$

$$\bar{R}^2 = 0.351,\ SE = 0.166,\ Q(10) = 7.897\ (p = 0.639).$$

The AR(1) as well as the MA(1) terms are different from zero and from one at any usual significance level. The autocorrelogram of the estimated residuals, which is also given in *Figure 2.10*, as well as the Ljung-Box Q statistic, which is calculated for this model with 12 autocorrelation coefficients (i.e. with 10 degrees of freedom), do not provide any evidence of a higher order process.

2.4 Forecasting

As mentioned in the introduction, in the 1970's, one of the reasons for the broad acceptance of time series analysis using the Box-Jenkins approach was the fact that forecasts with this comparably simple method often outperformed forecasts generated by large econometric models. In the following, we show how ARMA models can be used for making forecasts about the future development of time series. In doing so, we assume that all observations of the time series up to time t are known.

2.4.1 Forecasts with Minimal Mean Squared Errors

We want to solve the problem of making a τ-step ahead forecast for x_t with a *linear prediction function*, given a stationary and/or invertible data generating process.

Let $\hat{x}_t(\tau)$ be such a prediction function for $x_{t+\tau}$. Thus, $\hat{x}_t(\tau)$ is a random variable for given t and τ. As all stationary ARMA processes have a Wold representation, we assume the existence of such a representation without loss of generality. Thus,

$$x_t = \mu + \sum_{j=0}^{\infty} \psi_j u_{t-j},\quad \psi_0 = 1,\quad \sum_{j=0}^{\infty} \psi_j^2 < \infty,$$

2.4 Forecasting

where u_t is a pure random process with the usual properties $E[u_t] = 0$,

$$E[u_t u_s] = \begin{cases} \sigma^2 & \text{for } t = s \\ 0 & \text{for } t \neq s \end{cases}.$$

Therefore, it also holds that

$$(2.52) \qquad x_{t+\tau} = \mu + \sum_{j=0}^{\infty} \psi_j u_{t+\tau-j}, \quad \tau = 1, 2, \ldots.$$

For a linear prediction function with the information given up to time t, we assume the following representation

$$(2.53) \qquad \hat{x}_t(\tau) = \mu + \sum_{k=0}^{\infty} \theta_k^\tau u_{t-k}, \quad \tau = 1, 2, \ldots,$$

where the θ_k^τ, $k = 0, 1, 2, \ldots$, $\tau = 1, 2, \ldots$, are unknown. The forecast error of a τ-step forecast is $f_t(\tau) = x_{t+\tau} - \hat{x}_t(\tau)$, $\tau = 1, 2, \ldots,$. In order to make a good forecast, these errors should be small. The expected quadratic forecast error $E[(x_{t+\tau} - \hat{x}_t(\tau))^2]$, which should be minimised, is used as the criterion to determine the unknowns θ_k^τ. Taking into account (2.52) and (2.53) we can write

$$E[f_t^2(\tau)] = E\left[\left(\sum_{j=0}^{\infty} \psi_j u_{t+\tau-j} - \sum_{k=0}^{\infty} \theta_k^\tau u_{t-k}\right)^2\right]$$

$$= E\left[\left(u_{t+\tau} + \psi_1 u_{t+\tau-1} + \ldots + \psi_{\tau-1} u_{t+1} + \sum_{k=0}^{\infty} (\psi_{\tau+k} - \theta_k^\tau) u_{t-k}\right)^2\right].$$

From this it follows that

$$(2.54) \qquad E[f_t^2(\tau)] = \left(1 + \psi_1^2 + \ldots + \psi_{\tau-1}^2\right)\sigma^2 + \sigma^2 \sum_{k=0}^{\infty} \left(\psi_{\tau+k} - \theta_k^\tau\right)^2.$$

The variance of the forecast error reaches its minimum if we set $\theta_k^\tau = \psi_{\tau+k}$ for $k = 0, 1, 2, \ldots,$. Thus, we get the optimal linear prediction function for a τ-step ahead forecast from (2.53), as

$$(2.55) \qquad \hat{x}_t(\tau) = \mu + \sum_{k=0}^{\infty} \psi_{\tau+k} u_{t-k}, \quad \tau = 1, 2, \ldots.$$

For the conditional expectation of u_{t+s}, given u_t, u_{t-1}, ..., it holds that

$$E[u_{t+s}|u_t, u_{t-1}, ...] = \begin{cases} u_{t+s} & \text{for } s \leq 0 \\ 0 & \text{for } s > 0 \end{cases}.$$

Thus, we get the conditional expectation of $x_{t+\tau}$, because of (2.52), as

$$E[x_{t+\tau}|u_t, u_{t-1}, ...] = \mu + \sum_{k=0}^{\infty} \psi_{\tau+k} u_{t-k}.$$

Due to (2.55), the conditional expectation of $x_{t+\tau}$, with all information available at time t given, is identical to the optimal prediction function. This leads to the following result: The conditional expectation of $x_{t+\tau}$, with all information up to time t given, provides the τ-step forecast with minimal mean squared prediction error.

With (2.52) and (2.55) the τ-step forecast error can be written as

(2.56) $\quad f_t(\tau) = x_{t+\tau} - \hat{x}_t(\tau) = u_{t+\tau} + \psi_1 u_{t+\tau-1} + \psi_2 u_{t+\tau-2} + ... + \psi_{\tau-1} u_{t+1}$

with

$$E[f_t(\tau)|u_t, u_{t-1}, ...] = E[f_t(\tau)] = 0.$$

From these results we can immediately draw some *conclusions*:

1. Best linear unbiased predictions (BLUP) of stationary ARMA processes are given by the conditional expectation for $x_{t+\tau}$, $\tau = 1,2, ...$

$$\hat{x}_t(\tau) = E[x_{t+\tau}|x_t, x_{t-1}, ...] = E_t[x_{t+\tau}].$$

2. For the one-step forecast errors ($\tau = 1$), $f_t(1) = u_{t+1}$, we get

$$E[f_t(1)] = E[u_{t+1}] = 0, \text{ and}$$

$$E[f_t(1)f_s(1)] = E[u_{t+1}u_{s+1}] = \begin{cases} \sigma^2 & \text{for } t = s \\ 0 & \text{for } t \neq s \end{cases}.$$

The one-step forecast errors are a pure random process; they are identical with the residuals of the data generating process. If the one-step prediction errors were correlated, the prediction could be improved by using the information contained in the prediction errors. In such a case, however, $\hat{x}_t(1)$ would not be an optimal forecast.

3. For the τ-step forecast errors ($\tau > 1$) we get

$$f_t(\tau) = u_{t+\tau} + \psi_1 u_{t+\tau-1} + \psi_2 u_{t+\tau-2} + ... + \psi_{\tau-1} u_{t+1},$$

i.e. they follow a MA(τ-1) process with $E[f_t(\tau)] = 0$ and the variance

(2.57) $\qquad V[f_t(\tau)] = \left(1 + \psi_1^2 + \ldots + \psi_{\tau-1}^2\right) \sigma^2 .$

This variance can be used for constructing confidence intervals for τ-step forecasts. However, these intervals are too narrow for practical applications because they do not take into account the uncertainty in the estimation of the parameters ψ_i, $i = 1, 2, \ldots, \tau$-1.

4. It follows from (2.57) that the forecast error variance increases monotonically with increasing forecast horizon τ:

$$V[f_t(\tau)] \geq V[f_t(\tau\text{-}1)] .$$

5. Due to (2.57) we get for the limit

$$\lim_{\tau \to \infty} V[f_t(\tau)] = \lim_{\tau \to \infty}\left(1 + \psi_1^2 + \ldots + \psi_{\tau-1}^2\right)\sigma^2 = \sigma^2 \sum_{j=0}^{\infty} \psi_j^2 = V[x_t] ,$$

i.e. the variance of the τ-step forecast error is not larger than the variance of the underlying process.

6. The following variance decomposition follows from (2.55) and (2.56):

(2.58) $\qquad V[x_{t+\tau}] = V[\hat{x}_t(\tau)] + V[f_t(\tau)] .$

7. Furthermore,

$$\lim_{\tau \to \infty} \hat{x}_t(\tau) = \lim_{\tau \to \infty}\left(\mu + \sum_{k=0}^{\infty} \psi_{\tau+k} u_{t-k}\right) = \mu = E[x_t] ,$$

i.e. for increasing forecast horizons, the forecasts converge to the (unconditional) mean of the series.

The concept of 'weak' rational expectations whose information set is restricted to the current and past values of a variable exactly corresponds to the optimal prediction approach used here.

2.4.2 Forecasts of ARMA(p,q) Processes

The Wold decomposition employed in the previous section has advantages when it comes to the derivation of theoretical results, but it is not practically useful for forecasting. Thus, in the following, we will discuss forecasts directly using AR, MA, or ARMA representations.

Forecasts with a Stationary AR(1) Process

For this process, it holds that

$$x_t = \delta + \alpha\, x_{t-1} + u_t,$$

with $|\alpha| < 1$. The optimal τ-step forecast is the conditional mean of $x_{t+\tau}$, i.e.

$$E_t[x_{t+\tau}] = E_t[\delta + \alpha\, x_{t+\tau-1} + u_{t+\tau}] = \delta + \alpha\, E_t[x_{t+\tau-1}].$$

Due to the *first conclusion*, we get the following first order difference equation for the prediction function

$$\hat{x}_t(\tau) = \delta + \alpha\, \hat{x}_t(\tau-1),$$

which can be solved recursively:

$$\tau = 1: \quad \hat{x}_t(1) = \delta + \alpha\, \hat{x}_t(0) = \delta + \alpha\, x_t$$

$$\tau = 2: \quad \hat{x}_t(2) = \delta + \alpha\, \hat{x}_t(1) = \delta + \alpha\,\delta + \alpha^2 x_t$$

$$\vdots$$

$$\hat{x}_t(\tau) = \delta(1 + \alpha + \ldots + \alpha^{\tau-1}) + \alpha^\tau x_t$$

$$\hat{x}_t(\tau) = \frac{1-\alpha^\tau}{1-\alpha}\delta + \alpha^\tau x_t = \frac{\delta}{1-\alpha} + \alpha^\tau\left(x_t - \frac{\delta}{1-\alpha}\right).$$

As $\mu = \delta/(1-\alpha)$ is the mean of a stationary AR(1) process,

$$\hat{x}_t(\tau) = \mu + \alpha^\tau(x_t - \mu) \quad \text{with} \quad \lim_{\tau\to\infty}\hat{x}_t(\tau) = \mu,$$

i.e., with increasing forecast horizon τ, the predicted values of an AR(1) process converge geometrically to the unconditional mean μ of the process. The convergence is monotonic if α is positive, and oscillating if α is negative.

To calculate the τ-step prediction error, the Wold representation, i.e. the MA(∞) representation of the AR(1) process, can be used,

$$x_t = \mu + u_t + \alpha\, u_{t-1} + \alpha^2 u_{t-2} + \alpha^3 u_{t-3} + \ldots.$$

Due to (2.56) and (2.57) we get the MA(τ-1) process

$$f_t(\tau) = u_{t+\tau} + \alpha\, u_{t+\tau-1} + \alpha^2 u_{t+\tau-2} + \ldots + \alpha^{\tau-1} u_{t+1}$$

for the forecast error with the variance

$$V[f_t(\tau)] = \left(1 + \alpha^2 + \ldots + \alpha^{2(\tau-1)}\right)\sigma^2 = \frac{1-\alpha^{2\tau}}{1-\alpha^2}\sigma^2.$$

With increasing forecast horizons, it follows that

$$\lim_{\tau \to \infty} V[f_t(\tau)] = \frac{\sigma^2}{1-\alpha^2} = V[x_t],$$

i.e. the prediction error variance converges to the variance of the AR(1) process.

Forecasts with Stationary AR(p) Processes

Starting with the representation

$$x_t = \delta + \alpha_1 x_{t-1} + \alpha_2 x_{t-2} + \ldots + \alpha_p x_{t-p} + u_t,$$

the conditional mean of $x_{t+\tau}$ is given by

$$E_t[x_{t+\tau}] = \delta + \alpha_1 E_t[x_{t+\tau-1}] + \ldots + \alpha_p E_t[x_{t+\tau-p}].$$

Here,

$$E_t[x_{t+s}] = \begin{cases} \hat{x}_t(s) & \text{for } s > 0 \\ x_{t+s} & \text{for } s \le 0 \end{cases}.$$

Thus, the above difference equation can be solved recursively:

$$\tau = 1: \quad \hat{x}_t(1) = \delta + \alpha_1 x_t + \alpha_2 x_{t-1} + \ldots + \alpha_p x_{t+1-p}$$

$$\tau = 2: \quad \hat{x}_t(2) = \delta + \alpha_1 \hat{x}_t(1) + \alpha_2 x_t + \ldots + \alpha_p x_{t+2-p}, \text{ etc.}$$

Forecasts with an Invertible MA(1) Process

For this process, it holds that

$$x_t = \mu + u_t - \beta u_{t-1}$$

with $|\beta| < 1$. The conditional mean of $x_{t+\tau}$ is

$$E_t[x_{t+\tau}] = \mu + E_t[u_{t+\tau}] - \beta E_t[u_{t+\tau-1}].$$

For $\tau = 1$, this leads to

(2.59) $$\hat{x}_t(1) = \mu - \beta u_t,$$

and for $\tau \ge 2$, we get

$$\hat{x}_t(\tau) = \mu,$$

i.e. the unconditional mean is the optimal forecast of $x_{t+\tau}$, $\tau = 2, 3, \ldots,$. For the τ-step prediction errors and their variances we get:

$$f_t(1) = u_{t+1}, \qquad V[f_t(1)] = \sigma^2$$
$$f_t(2) = u_{t+2} - \beta u_{t+1}, \qquad V[f_t(2)] = (1+\beta^2)\sigma^2$$
$$\vdots \qquad\qquad \vdots$$
$$f_t(\tau) = u_{t+\tau} - \beta u_{t+\tau-1}, \qquad V[f_t(\tau)] = (1+\beta^2)\sigma^2.$$

To be able to perform the one-step forecasts (2.59), the unobservable variable u has to be expressed as a function of the observable variable x. To do this, it must be taken into account that for $s \leq t$, the one-step forecast errors can be written as

(2.60) $$u_s = x_s - \hat{x}_{s-1}(1).$$

For $t = 0$, we get from (2.59)
$$\hat{x}_0(1) = \mu - \beta u_0$$

with the non-observable but fixed u_0. Taking (2.60) into account, we get for $t = 1$

$$\begin{aligned}\hat{x}_1(1) = \mu - \beta u_1 &= \mu - \beta(x_1 - \hat{x}_0(1)) \\ &= \mu - \beta x_1 + \beta(\mu - \beta u_0) \\ &= \mu(1+\beta) - \beta x_1 - \beta^2 u_0.\end{aligned}$$

Correspondingly, we get for $t = 2$

$$\begin{aligned}\hat{x}_2(1) = \mu - \beta u_2 &= \mu - \beta(x_2 - \hat{x}_1(1)) \\ &= \mu - \beta x_2 + \beta(\mu(1+\beta) - \beta x_1 - \beta^2 u_0) \\ &= \mu(1+\beta+\beta^2) - \beta x_2 - \beta^2 x_1 - \beta^3 u_0.\end{aligned}$$

If we continue this procedure, the so-called *backcasting*, we finally arrive at a representation of the one-step prediction which – except for u_0 – consists only of observable terms,

$$\hat{x}_t(1) = \mu(1+\beta+\ldots+\beta^t) - \beta x_t - \beta^2 x_{t-1} - \ldots - \beta^t x_1 - \beta^{t+1} u_0.$$

Due to the invertibility of the MA(1) process, i.e. for $|\beta| < 1$, the impact of the unknown initial value u_0 finally disappears.

Similarly, one can show that, after q forecast steps, the optimal forecasts of invertible MA(q) processes, $q > 1$ are equal to the unconditional mean of the process and that the variance of the forecast errors is equal to the variance of the underlying process. The forecasts in observable terms are represented similarly to those of the MA(1) process.

Forecasts with ARMA(p,q) Processes

Forecasts for these processes result from combining the approaches of pure AR and MA processes. Thus, the one-step ahead forecast for a stationary and invertible ARMA(1,1) process is given by

$$\hat{x}_t(1) = \delta + \alpha x_t - \beta u_t.$$

Starting with $t = 0$ and taking (2.60) into account, forecasts are successively generated by backcasting. We first get

$$\hat{x}_0(1) = \delta + \alpha x_0 - \beta u_0,$$

where x_0 and u_0 are assumed to be any fixed numbers. For $t = 1$ we get

$$\hat{x}_1(1) = \delta + \alpha x_1 - \beta u_1 = \delta + \alpha x_1 - \beta(x_1 - \hat{x}_0(1))$$
$$= \delta(1 + \beta) + (\alpha - \beta) x_1 + \beta \alpha x_0 - \beta^2 u_0,$$

which finally leads to

$$(2.61) \quad \hat{x}_t(1) = \delta(1 + \beta + \ldots + \beta^t) + (\alpha - \beta) x_t + \beta(\alpha - \beta) x_{t-1} + \ldots$$
$$+ \beta^{t-1}(\alpha - \beta) x_1 + \beta^t \alpha x_0 - \beta^{t+1} u_0.$$

Due to the invertibility condition, i.e. for $|\beta| < 1$, the one-step forecast for large values of t does no longer depend on the unknown initial values x_0 and u_0.

For the τ-step forecast, $\tau = 2, 3, \ldots$, we get

$$\hat{x}_t(2) = \delta + \alpha \hat{x}_t(1)$$
$$\hat{x}_t(3) = \delta + \alpha \hat{x}_t(2)$$
$$\vdots$$

Using (2.61), these forecasts can be calculated recursively.

2.4.3 Evaluation of Forecasts

Forecasts can be evaluated ex post, i.e. when the realised values are available. There are many kinds of measures to do this. Quite often, only graphs and/or scatter diagrams of the predicted values and the corresponding observed values of a time series are plotted. Intuitively, a forecast is 'good' if the predicted values describe the development of the series in the graphs relatively well or if the points in the scatter diagram are concentrated around the angle bisecting line in the first and/or third quadrant. Such intuitive arguments are, however, not founded on the above-mentioned considerations on optimal predictions. For example, as (2.59) shows, the optimal one-step forecast of a MA(1) process is a pure random process. This implies that the graphs compare two quite different processes. *Conclusion 6* given above states that the following decomposition holds for the variances of the data generating processes, the forecasts and the forecast errors,

$$V[x_{t+\tau}] = V[\hat{x}_t(\tau)] + V[f_t(\tau)].$$

Thus, it is obvious that predicted and realised values are generally generated by different processes.

As a result, a measure for the *predictability* of stationary processes can be developed. It is defined as follows,

(2.62) $$P(\tau)^2 = \frac{V[\hat{x}_t(\tau)]}{V[x_{t+\tau}]} = 1 - \frac{V[f_t(\tau)]}{V[x_{t+\tau}]},$$

with $0 \leq P(\tau)^2 \leq 1$. At the same time, $P(\tau)^2$ is the correlation coefficient between the predicted and the realised values of x. The optimal forecast of a pure random process with mean zero is $\hat{x}_t(\tau) = 0$, i.e. $P(\tau)^2 = 0$. Such a process cannot be predicted. On the other hand, for the one-step forecast of a MA(1) process, we can write

$$P(1)^2 = \frac{\beta^2 \sigma^2}{(1+\beta^2)\sigma^2} = \frac{\beta^2}{1+\beta^2} > 0.$$

However, the decomposition (2.58), theoretically valid for optimal forecasts, does not hold for actual (empirical) forecasts, even if they are generated by using (estimated) ARMA processes. This is due to the fact that forecast errors are hardly ever totally uncorrelated with the forecasts. Therefore, the value of $P(\tau)^2$ might even become negative for 'bad' forecasts.

JACOB MINCER and VICTOR ZARNOWITZ (1969) made the following suggestion to check the consistency of forecasts. By using OLS the following regression equation is estimated

(2.63) $$x_{t+\tau} = a_0 + a_1 \hat{x}_t(\tau) + \varepsilon_{t+\tau}.$$

It is tested either individually with t tests or commonly with a F test whether $a_0 = 0$ and $a_1 = 1$. If this is fulfilled, the forecasts are said to be consistent. However, such a regression produces consistent estimates of the parameters if and only if $\hat{x}_t(\tau)$ and $\varepsilon_{t+\tau}$ are asymptotically uncorrelated.

Moreover, to get consistent estimates of the variances, which is necessary for the validity of the test results, the residuals have to be pure random processes. Even under the null hypothesis of optimal forecasts, this only holds for one-step predictions. Thus, the usual F and t tests can only be used for $\tau = 1$. For $\tau > 1$, the MA(τ-1) process of the forecast errors has to be taken into account when the variances are estimated. A procedure for such situations combines Ordinary Least Squares for the estimation of the parameters and Generalised Least Squares for the estimation of the variances, as proposed by BRYAN W. BROWN and SHLOMO MAITAL (1981).

JINOOK JEONG and GANGADHARRAO S. MADDALA (1991) have pointed out another problem which is related to these tests. Even rational forecasts are usually not without errors; they contain measurement errors. This implies, however, that (2.63) cannot be estimated consistently with OLS; an instrumental variables estimator must be used. An alternative to the estimation of (2.63) is therefore to estimate a univariate MA(τ-1) model for the forecast errors of a τ-step prediction,

$$\hat{f}_t(\tau) = a_0 + u_t + a_1 u_{t-1} + a_2 u_{t-2} + \ldots + a_{\tau-1} u_{t-\tau+1},$$

and to check the null hypothesis H_0: $a_0 = 0$ and whether the estimated residuals $\hat{u}_t$ are white noise.

On the other hand, simple descriptive measures, which are often employed to evaluate the performance of forecasts, are based on the average values of the forecast errors over the forecast horizon. The simple arithmetic mean indicates whether the values of the variable are – on average – over- or underestimated. However, the disadvantage of this measure is that large over- and underestimates cancel each other out. The *mean absolute error* is often used to avoid this effect. Starting the forecasts from a fixed point of time, t_0, and assuming that realisations are available up to t_0+m, we get

$$MAE(\tau) = \frac{1}{m+1-\tau} \sum_{j=0}^{m-\tau} \left| f_{t_0+j}(\tau) \right|, \quad \tau = 1, 2, \ldots.$$

Every forecast error gets the same weight in this measure. The *root mean square error* is often used to give particularly large errors a stronger weight:

$$RMSE(\tau) = \sqrt{\frac{1}{m+1-\tau} \sum_{j=0}^{m-\tau} f_{t_0+j}^2(\tau)}, \quad \tau = 1, 2, \ldots.$$

These measures are not normalised, i.e. their size depends on the scale of the data.

The inequality measure proposed by HENRY THEIL (1961) avoids this problem by comparing the actual forecasts with so-called naïve forecasts, i.e. the realised values of the last available observation,

$$U(\tau) = \sqrt{\frac{\sum_{j=0}^{m-\tau} f_{t_0+j}^2(\tau)}{\sum_{j=0}^{m-\tau} (x_{t_0+\tau+j} - x_{t_0+j})^2}}, \quad \tau = 1, 2, \ldots.$$

If $U(\tau) = 1$, the forecast is as good as the naïve forecast, $\hat{x}_t(\tau) = x_t$. For $U(\tau) < 1$ the forecasts perform better than the naïve one. MAE, RMSE and Theil's U all become zero if predicted and realised values are identical over the whole forecast horizon.

Example 2.16

All these measures can also be applied to forecasts which are not generated by ARMA models, as, for example, the forecasts of the Council of Economic Experts or the Association of German Economic Research Institutes. Since the end of the 1960's, both institutions have published forecasts of the German economic development for the following year, the institutes usually in October and the Council at the end of November. HANNS MARTIN HAGEN and GEBHARD KIRCHGÄSSNER (1996) investigated the annual forecasts of the growth rates of GNP for the period from 1970 to 1995 as well as for the sub-periods from 1970 to 1982 and from 1983 to 1995. These periods correspond to the social-liberal government of SPD and FDP and the conservative-liberal government of CDU/CSU and FDP.

The results are given in *Table 2.2*. Besides the criteria given above, the table also indicates the square of the correlation coefficient between realised and predicted values (R^2), the estimated regression coefficient $\hat{a}_1$ of the test equation (2.63) as well as the mean error (ME). According to almost all criteria, the forecasts of the Council outperform those of the institutes. This was to be expected, as the Council's forecasts are produced slightly later, at a time when more information is available. It holds for the forecasts of both institutions that the mean absolute error, the root mean squared error as well as Theil's U are smaller in the second period compared to the first one. This is some evidence that the forecasts might have improved over time. On the other hand, the correlation coefficient between predicted and realised values has also become smaller. This indicates a deterioration of the forecasts. It has to be taken into account that the variance of the variable to be predicted was considerably smaller in the second period as compared to the first one. Thus, the smaller errors do not necessarily indicate improvements of the forecasts. It is also interesting to note that on average the forecast errors of both institutions were negative in the first and positive in the second sub-period. They tended to overestimate the development in the period of the social-liberal coalition and to underestimate it in the period of the conservative-liberal coalition.

Table 2.2: *Forecasts of the Council of Economic Experts and of the Economic Research Institutes*

	Period	R^2	RMSE	MAE	ME	$\hat{a}_1$	U
Institutes	1970 – 1995	0.369	1.838	1.346	-0.250*	1.005*	0.572
	1970 – 1982	0.429	2.291	1.654	-0.731	1.193*	0.625
	1983 – 1995	0.399	1.229	1.038	0.231	1.081	0.457
Council of Economic Experts	1970 – 1995	0.502*	1.647*	1.171*	-0.256	1.114	0.512*
	1970 – 1982	0.599*	2.025*	1.477*	-0.723*	1.354	0.552*
	1983 – 1995	0.472*	1.150*	0.865*	0.212*	1.036*	0.428*

'*' denotes the 'better' of the two forecasts.

2.5 The Relation between Econometric Models and ARMA Processes

The ARMA model-based forecasts discussed in the previous section are *unconditional forecasts*. The only information that is used to generate these forecasts is the information contained in the current and past values of the time series. There is demand for such forecasts, and – as mentioned above – one of the reasons for the development and the popularity of the Box-Jenkins methodology presented in this chapter is that by applying the above-mentioned approaches, these predictions perform – at least partly – much better than forecasts generated by large scale econometric models. Thus, the Box-Jenkins methodology seems to be a (possibly much better) alternative to the traditional econometric methodology.

However, this perspective is rather restricted. On the one hand, conditional rather than unconditional forecasts are required in many cases, for example, in order to evaluate the effect of a tax reform on economic growth. Such forecasts cannot be generated by using (only) univariate models. On the other hand, and more importantly, the separation of the two approaches is much less strict than it seems to be at first glance. As ARNOLD ZELLNER and FRANZ C. PALM (1974) showed, linear dynamic simultaneous equation systems as used in traditional econometrics can be transformed into ARMA models. (Inversely, multivariate time series models as discussed in the next chapters can be transformed into traditional econometric models.) The univariate ARMA models correspond to the *fi-*

nal equations of econometric models in the terminology of JAN TINBERGEN (1940).

Let us consider a very simple model. An exogenous, weakly stationary variable x, as defined in (2.64b), has a current and lagged impact on the dependent variable y, while the error term might be autocorrelated. Thus, we get the model

(2.64a) $$y_t = \eta_1(L) x_t + \eta_2(L) u_{1,t},$$

(2.64b) $$\alpha(L) x_t = \beta(L) u_{2,t},$$

where $\eta_1(L)$ and $\eta_2(L)$ are lag polynomials of finite order. If we insert (2.64b) in (2.64a), we get for y the univariate model

(2.64a') $$\alpha(L) y_t = \zeta(L) v_t$$

with

$$\zeta(L) v_t := \eta_1(L) \beta(L) u_{2,t} + \eta_2(L) \alpha(L) u_{1,t}.$$

As $\zeta(L)v_t$ is an MA process of finite order, we get a finite order ARMA representation for y. It must be pointed out that the univariate representations of the two variables have the same finite order AR term.

References

Since the time when HERMAN WOLD developed the class of ARMA processes in his dissertation and GEORGE E.P. BOX and GWILYM M. JENKINS (1970) popularised and further developed this model class in the textbook mentioned above, there have been quite a lot of **textbooks** dealing with these models at different technical levels. An introduction focusing on empirical applications is, for example, to be found in

ROBERT S. PINDYCK and DANIEL L. RUBINFELD, *Econometric Models and Economic Forecasts*, McGraw-Hill, Boston et al., 4th edition 1998, Chapter 17f. pp. 521 – 578,

PETER J. BROCKWELL and RICHARD A. DAVIS, *Introduction to Time Series and Forecasting*, Springer, New York et al. 1996, as well as

TERENCE C. MILLS, *Time Series Techniques for Economists*, Cambridge University Press, Cambridge (England) 1990. Contrary to this,

PETER J. BROCKWELL and RICHARD A. DAVIS, *Time Series: Theory and Methods*, Springer, New York et al. 1987,

give a rigorous presentation in probability theory. Along with the respective proofs of the theorems, this textbook shows, however, many empirical examples.

Autoregressive processes for the residuals of an estimated regression equation were used for the first time in econometrics by

DONALD COCHRANE and GUY H. ORCUTT, Application of Least Squares Regression to Relationships Containing Autocorrelated Error Terms, *Journal of the American Statistical Association* 44 (1949), pp. 32 – 61.

The different **information criteria** to detect the order of an autoregressive process are presented in

HIROTUGU AKAIKE, Fitting Autoregressive Models for Prediction, *Annals of the Institute of Statistical Mathematics* AC-19 (1974), pp. 364 – 385,

HIROTUGU AKAIKE, A New Look at the Statistical Model Identification, *IEEE Transactions on Automatic Control* 21 (1969), pp. 234 – 237,

GIDEON SCHWARZ, Estimating the Dimensions of a Model, *Annals of Statistics* 6 (1978), pp. 461 – 464, as well as in

EDWARD J. HANNAN and BARRY G. QUINN, The Determination of the Order of an Autoregression, *Journal of the Royal Statistical Society* B 41 (1979), pp. 190 – 195.

The effect of **temporal aggregation** on the first differences of temporal averages have first been investigated by

HOLBROOK WORKING, Note on the Correlation of First Differences of Averages in a Random Chain, *Econometrica* 28 (1960), pp. 916 – 918

and later on, in more detail, by

GEORGE C. TIAO, Asymptotic Behaviour of Temporal Aggregates of Time Series, *Biometrika* 59 (1972), pp. 525 – 531.

The approach to check the **consistency of predictions** was developed by

JACOB MINCER and VICTOR ZARNOWITZ, The Evaluation of Economic Forecasts, in: J. MINCER (ed.), *Economic Forecasts and Expectations*, National Bureau of Economic Research, New York 1969.

The use of MA processes of the forecast errors to estimate the variances of the estimated parameters was presented by

BRYAN W. BROWN and SHLOMO MAITAL, What Do Economists Know? An Empirical Study of Experts' Expectations, *Econometrica* 49 (1981), pp. 491 – 504.

The fact that measurement errors also play a role in rational forecasts and that, therefore, instrumental variable estimators should be used, was indicated by

JINOOK JEONG and GANGADHARRAO S. MADDALA, Measurement Errors and Tests for Rationality, *Journal of Business and Economic Statistics* 9 (1991), pp. 431 – 439.

These procedures have been applied to the common forecasts of the German economic research institutes by

GEBHARD KIRCHGÄSSNER, Testing Weak Rationality of Forecasts with Different Time Horizons, *Journal of Forecasting* 12 (1993), pp. 541 – 558.

Moreover, the forecasts of the German Council of Economic Experts as well as those of the German Economic Research Institutes were investigated in

HANNS MARTIN HAGEN and GEBHARD KIRCHGÄSSNER, Interest Rate Based Forecasts of German Economic Growth: A Note, *Weltwirtschaftliches Archiv* 132 (1996), pp. 763 – 773.

The **measure of inequality** (Theil's U) was proposed by

HENRY THEIL, *Economic Forecasts and Policy*, North-Holland, Amsterdam 1961.

An alternative measure is given in

HENRY THEIL, *Applied Economic Forecasting*, North-Holland, Amsterdam 1966.

Today, both measures are used in computer programmes. Quite generally, **forecasts for time series data** are discussed in

CLIVE W.J. GRANGER, *Forecasting in Business and Economics*, Academic Press, 2nd edition 1989.

On the **evaluation of the predictive accuracy** of forecasts see

FRANCIS X. DIEBOLD and ROBERTO S. MARIANO, Comparing Predictive Accuracy, *Journal of Business and Economic Statistics* 13 (1995), pp. 253 – 263.

The **relationship between time series models and econometric equation systems** is analysed in

ARNOLD ZELLNER and FRANZ C. PALM, Time Series Analysis and Simultaneous Equation Econometric Models, *Journal of Econometrics* 2 (1974), pp. 17 – 54.

See for this also

FRANZ C. PALM, Structural Econometric Modeling and Time Series Analysis: An Integrated Approach, in: A. ZELLNER (ed.), Applied Time Series Analysis of Economic Data, U.S. Department of Commerce, Economic Research Report ER-S, Washington 1983, pp. 199 – 230.

The term **final equation** originates from

JAN TINBERGEN, Econometric Business Cycle Research, *Review of Economic Studies* 7 (1940), pp. 73 – 90.

An introduction into the solution of **difference equations** is given in

WALTER ENDERS, *Applied Econometric Time Series*, 3rd edition, Wiley, Hoboken, N.J. 2010, Chapter 1.

The **permanent income hypothesis** as a determinant of consumption expenditure was developed by

MILTON FRIEDMAN, *A Theory of the Consumption Function*, Princeton University Press, Princeton N.J. 1957.

The example of the **estimated popularity function** is given in

GEBHARD KIRCHGÄSSNER, Causality Testing of the Popularity Function: An Empirical Investigation for the Federal Republic of Germany, 1971 – 1982, *Public Choice* 45 (1985), pp. 155 – 173.

3 Granger Causality

So far, we have only considered single stationary time series. We analysed their (linear) structure, estimated linear models and performed forecasts based on these models. However, the world does not consist of independent stochastic processes. Just the contrary: in accordance with general equilibrium theory, economists usually assume that everything depends on everything else. Therefore, the next question that arises is about (causal) relationships between different time series.

In principle, we can answer this question in two different ways. Following a *bottom up* strategy, one might first assume that the data generating processes of the different time series are independent of each other. In a second step, one might ask whether some specific time series are related to each other. This statistical approach follows the proposals of CLIVE W.J. GRANGER (1969) and is today usually employed when causality tests are performed. The alternative is a *top down* strategy which assumes that the generating processes are not independent and which, in a second step, asks whether some specific time series are generated independently of the other time series considered. This approach is pursued when using *vector autoregressive processes*. The methodology, which goes back to CHRISTOPHER A. SIMS (1980), will be described in the next chapter. Both approaches are employed to investigate the causal relationships which potentially exist between different time series.

However, before we ask these questions we should clarify the meaning of the term *causality*. Ever since GALILEO GALILEI and DAVID HUME, this term is closely related to the terms *cause* and *effect*. Accordingly, a variable x would be causal to a variable y if x could be interpreted as the cause of y and/or y as the effect of x. However, where do we get the necessary information from? In traditional econometrics, when distinguishing endogenous and exogenous (or predetermined) variables, one assumes that such information is a priori available. Problems arise, however, if there are simultaneities between the variables, i.e. if it is possible that x is causal to y and y is causal to x. The usual rank and order conditions for the identification of econometric simultaneous equations systems show that the different relations can only be identified (and estimated) if additional information is available, for example on different impacts of third variables on the dependent variables. It is impossible to determine the direction of causality of instantaneous relations between different variables if there is no such information. In this case, the only possibility is to estimate a reduced form of the system.

As far as possible, modern time series analysis abstains from using exogenous information, so that the way in which the identification problem is treated in traditional econometrics is ruled out. On the other hand, the idea of causality is closely related to the idea of succession in time, at the latest since DAVID HUME who said that cause always precedes effect. Traditional econometrics shared the same view. However, the time periods represented by a single observation are too long to assume that a change in one variable might only influence other variables in later time periods, especially when using annual data. As time series analyses are usually performed with data of higher frequencies, the situation looks different here. Using monthly data, we assume in many cases that changes in one variable only influence other variables in later months. For example, the change in mineral oil prices on the international spot markets might only have a delayed effect on Swiss or German consumer prices for petrol or light heating oil. Thus, it is reasonable to use succession in time as a criterion to find out whether or not there exists a causal relation between two series.

If such a causal relation exists, it should be possible to exploit it when making forecasts. As seen above, it is often possible to make quite good forecasts with univariate models. The precondition for this is that the information contained in the past values of the variable is optimally exploited. Identification and estimation of ARMA models, for example, are attempts in this direction. However, if x is causal to y, current and lagged values of x should contain information that can be used to improve the forecast of y. This implies that the information is not contained in the current and lagged values of y. Otherwise it would be sufficient to consider only the present and past values of y. Accordingly, the definition of causality proposed in 1969 by CLIVE W.J. GRANGER looks at this *incremental predictability*, i.e. it examines whether the forecasts of the future values of y can be improved if – besides all other available information – the current and lagged values of x are also taken into account.

There is, however, another reason why the lagged values of the corresponding variables are taken into account when it comes to the question of causality. Even if they are stationary, economic variables often show a high degree of persistence. This may lead to spurious correlations (regressions) between x_t and y_t, in case x_t has no impact on y_t and y_t depends on y_{t-1} which is not included in the regression equation. CLIVE W. GRANGER and PAUL NEWBOLD (1974) as well as GEBHARD KIRCHGÄSSNER (1981) showed that such spurious regressions can arise even if highly autocorrelated variables are generated independently from each other. If past values of both the dependent and the explanatory variables are included, the risk diminishes as this implies that the time series are filtered. With respect to the causal relation between (two) time series, only the innovations of these series do matter. Correspondingly, G. WILLIAM SCHWERT (1979) also refers to the results of causality tests as "the message in the innovations".

In the following, we present the definition of Granger causality and the different possibilities of causal events resulting from it (*Section 3.1*). This is followed by a characterisation of these causal events within the framework of bivariate autoregressive and moving average models as well as by using the residuals of the univariate models as developed in the preceding chapter (*Section 3.2*). *Section 3.3* presents three test procedures to investigate causal relations between time series: the direct GRANGER procedure, the HAUGH-PIERCE test and the HSIAO procedure. In *Section 3.4*, we ask how these procedures can be applied in situations where more than just two variables are considered. The chapter closes with some remarks on the relation between the concepts of Granger causality and rational expectations if applied to the analysis of economic policy (reaction) functions (*Section 3.5*).

3.1 The Definition of Granger Causality

In the following, we again assume that we have weakly stationary time series. Let I_t be the total information set available at time t. This information set includes, above all, the two time series x and y. Let $\bar{x}_t$ be the set of all current and past values of x, i.e. $\bar{x}_t := \{x_t, x_{t-1}, ..., x_{t-k}, ...\}$ and analogously of y. Let $\sigma^2(\cdot)$ be the variance of the corresponding forecast error. For such a situation, Clive W.J. GRANGER (1969) proposed the following definition of causality between x and y:

(i) *Granger Causality*: x is (simply) Granger causal to y if and only if the application of an optimal linear prediction function leads to

$$\sigma^2(y_{t+1} | I_t) < \sigma^2(y_{t+1} | I_t - \bar{x}_t),$$

i.e. if future values of y can be better predicted, i.e. with a smaller forecast error variance, if current and past values of x are used.

(ii) *Instantaneous Granger Causality*: x is instantaneously Granger causal to y if and only if the application of an optimal linear prediction function leads to

$$\sigma^2(y_{t+1} | \{I_t, x_{t+1}\}) < \sigma^2(y_{t+1} | I_t),$$

i.e. if the future value of y, y_{t+1}, can be better predicted, i.e. with a smaller forecast error variance, if the future value of x, x_{t+1}, is used in addition to the current and past values of x.

(iii) *Feedback*: There is feedback between x and y if x is causal to y and y is causal to x.

Feedback is only defined for the case of simple causal relations. The reason is that the direction of instantaneously causal relations cannot be identified without additional information or assumptions. Thus, the following theorem holds:

Theorem 3.1: x is instantaneously causal to y if and only if y is instantaneously causal to x.

According to Granger's definition of causality there are eight different, exclusive possibilities of causal relations between two time series:

(i)   x and y are independent: (x, y)

(ii)   There is only instantaneous causality: (x–y)

(iii)   x is causal to y, without instantaneous causality: (x→y)

(iv)   y is causal to x, without instantaneous causality: (x←y)

(v)   x is causal to y, with instantaneous causality: (x⇒y)

(vi)   y is causal to x, with instantaneous causality: (x⇐y)

(vii)   There is feedback without instantaneous causality: (x↔y)

(viii) There is feedback with instantaneous causality: (x⇔y)

In the definition given above, I_t includes all information available at time t. Normally, however, only the current and lagged values of the two time series x and y are considered:

$$I_t := \{x_t, x_{t-1}, ..., x_{t-k}, ..., y_t, y_{t-1}, ..., y_{t-k}, ...\}.$$

In many cases, the limitation of the information set does hardly make sense. Thus, when discussing the test procedures, we must also ask how these procedures can be applied if (relevant) 'third variables' play a role.

3.2 Characterisation of Causal Relations in Bivariate Models

In *Chapter 1* we already explained that, according to the Wold decomposition theorem, any weakly stationary process can be represented as an (infinite) moving average of a white noise process. Correspondingly, each pair of time series can be represented by a bivariate MA(∞) process. If this process is invertible, it can also be represented as a bivariate (infinite) AR process. In the following, starting with the above-mentioned definition of causality, causal relations between two time series are first of all characterised by AR representation and then by MA representation. Finally, according to LARRY D. HAUGH (1976), causal relations between two time series can also be characterised by the residuals of their univariate ARMA models. These three characterisations, which are the basis of different testing procedures, are presented in the following.

3.2.1 Characterisation of Causal Relations Using the Autoregressive and Moving Average Representations

Each bivariate system of invertible weakly stationary processes has the following autoregressive representation (deterministic terms are neglected without loss of generality):

$$(3.1) \quad A(L) \begin{bmatrix} y_t \\ x_t \end{bmatrix} = \begin{bmatrix} \alpha_{11}(L) & \alpha_{12}(L) \\ \alpha_{21}(L) & \alpha_{22}(L) \end{bmatrix} \begin{bmatrix} y_t \\ x_t \end{bmatrix} = \begin{bmatrix} u_t \\ v_t \end{bmatrix}.$$

A(L) is a matrix polynomial. Its elements, $\alpha_{ij}(L)$, i, j = 1,2, are one-sided (infinite) polynomials in the lag operator L. These polynomials are identical to zero, ($\alpha_{ij}(L) \equiv 0$), if all their coefficients, which are denoted as α_{ij}^k, are equal to zero. u and v are white noise residuals which might be contemporaneously correlated with each other. In order to normalise the equations, we set

$$\alpha_{11}^0 = \alpha_{22}^0 = 1.$$

As (3.1) is a reduced form, it must hold that

$$(3.2) \quad \alpha_{12}^0 = \alpha_{21}^0 = 0.$$

In this system, instantaneous causality exists if and only if u and v are contemporaneously correlated because then the forecast errors of y and x can be reduced if the current value of x or y is included in the forecast equation along with all lagged values of x and y. Then, however, there always exist representations with either $\alpha_{12}^0 \neq 0$ and $\alpha_{21}^0 = 0$ or $\alpha_{12}^0 = 0$ and $\alpha_{21}^0 \neq 0$. Both representations are observationally equivalent. However, because of these two representations there is also one with $\alpha_{12}^0 \neq 0$ and $\alpha_{21}^0 \neq 0$ which is observationally equivalent to the two other representations.

In the terminology of traditional econometrics, this implies that the structural form (3.1) is not identified. It is well known that a specific structural form of any econometric model can be transformed into another structural form which is observationally equivalent by pre-multiplying it with any quadratic regular matrix P whose rank is equal to the number of endogenous variables. The same happens if we go from one representation to another. Instantaneous causality therefore results in:

$$(3.3) \quad ((x-y) \vee (x \Rightarrow y) \vee (x \Leftarrow y) \vee (x \Leftrightarrow y))$$
$$\approx \rho_{uv}(0) \neq 0 \vee \alpha_{12}^0 \neq 0 \vee \alpha_{21}^0 \neq 0,$$

where '$\approx$' denotes equivalence and $\rho_{uv}(0)$ the contemporaneous correlation between u and v. In the following, we only consider the reduced form, i.e. relation (3.2) holds.

The individual causal events lead to the following representations:

(3.4a) $((x, y) \vee (x - y)) \quad \approx \quad \alpha_{12}(L) \equiv \alpha_{21}(L) \equiv 0,$

(3.4b) $((x \rightarrow y) \vee (x \Rightarrow y)) \quad \approx \quad \neg(\alpha_{12}(L) \equiv 0) \wedge \alpha_{21}(L) \equiv 0,$

(3.4c) $((x \leftarrow y) \vee (x \Leftarrow y)) \quad \approx \quad \alpha_{12}(L) \equiv 0 \wedge \neg(\alpha_{21}(L) \equiv 0),$

(3.4d) $((x \leftrightarrow y) \vee (x \Leftrightarrow y)) \quad \approx \quad \neg(\alpha_{12}(L) \equiv 0) \wedge \neg(\alpha_{21}(L) \equiv 0).$

Thus, a simple causal relation from x to y only exists if all coefficients of the lag polynomial $\alpha_{21}(L)$ are equal to zero, $(\alpha_{21}(L) \equiv 0)$, and if there exists at least one non-zero coefficient of the lag polynomial $\alpha_{12}(L)$, $\neg(\alpha_{12}(L) \equiv 0)$.

Analogous to (3.1) and (3.4), we can also characterise the different causal relations by using the moving average representation

$$(3.5) \quad \begin{bmatrix} y_t \\ x_t \end{bmatrix} = B(L) \begin{bmatrix} u_t \\ v_t \end{bmatrix} = \begin{bmatrix} \beta_{11}(L) & \beta_{12}(L) \\ \beta_{21}(L) & \beta_{22}(L) \end{bmatrix} \begin{bmatrix} u_t \\ v_t \end{bmatrix}.$$

B(L) is also a matrix polynomial, whose elements $\beta_{ij}(L)$, $i,j = 1,2$, are one-sided (infinite) polynomials in the lag operator L. To normalise the system we set

$$\beta_{11}^0 = \beta_{22}^0 = 1.$$

(3.2) also leads to

(3.6) $$\beta_{12}^0 = \beta_{21}^0 = 0.$$

As B(L) results from the inversion of A(L), the following relations between the parameters of the MA and the AR representation hold:

(3.7a) $$\beta_{11}(L) = \alpha_{22}(L) / \delta(L),$$
(3.7b) $$\beta_{12}(L) = -\alpha_{12}(L) / \delta(L),$$
(3.7c) $$\beta_{21}(L) = -\alpha_{21}(L) / \delta(L),$$
(3.7d) $$\beta_{22}(L) = \alpha_{11}(L) / \delta(L),$$

with

$$\delta(L) = \alpha_{11}(L)\alpha_{22}(L) - \alpha_{12}(L)\alpha_{21}(L).$$

This leads to

(3.8a) $$\beta_{12}(L) \equiv 0 \approx \alpha_{12}(L) \equiv 0,$$
(3.8b) $$\beta_{21}(L) \equiv 0 \approx \alpha_{21}(L) \equiv 0.$$

Thus, in analogy to (3.4) the different causal events result in

(3.9a) $((x, y) \vee (x - y)) \approx \beta_{12}(L) \equiv \beta_{21}(L) \equiv 0,$
(3.9b) $((x \rightarrow y) \vee (x \Rightarrow y)) \approx \neg(\beta_{12}(L) \equiv 0) \wedge \beta_{21}(L) \equiv 0,$
(3.9c) $((x \leftarrow y) \vee (x \Leftarrow y)) \approx \beta_{12}(L) \equiv 0 \wedge \neg(\beta_{21}(L) \equiv 0),$
(3.9d) $((x \leftrightarrow y) \vee (x \Leftrightarrow y)) \approx \neg(\beta_{12}(L) \equiv 0) \wedge \neg(\beta_{21}(L) \equiv 0).$

The conditions for the different polynomials hold independently of whether we choose the AR or the MA representation.

3.2.2 Characterisation of Causal Relations Using the Residuals of the Univariate Processes

As an alternative to (3.1) and (3.5), x and y can also be represented by two separate univariate ARMA models. In the Wold representation, this leads to:

(3.10) $$\begin{bmatrix} y_t \\ x_t \end{bmatrix} = \Psi(L) \begin{bmatrix} a_t \\ b_t \end{bmatrix} = \begin{bmatrix} \psi_{11}(L) & 0 \\ 0 & \psi_{22}(L) \end{bmatrix} \begin{bmatrix} a_t \\ b_t \end{bmatrix}.$$

Once again, $\psi_{ii}(L)$, $i = 1, 2$, are one-sided infinite polynomials in the lag operator L normalised by
$$\psi_{11}^0 = \psi_{22}^0 = 1.$$

The residuals a and b are again white noise, and they might also be contemporaneously correlated. We assume that the two MA processes are again invertible. The following representation shows the relation between (3.5) and (3.10):

(3.11) $$\begin{bmatrix} y_t \\ x_t \end{bmatrix} = \Psi(L)\,\Psi(L)^{-1}\,B(L)\begin{bmatrix} u_t \\ v_t \end{bmatrix},$$

or

(3.11a) $$\begin{bmatrix} y_t \\ x_t \end{bmatrix} = \Psi(L)H(L)\begin{bmatrix} u_t \\ v_t \end{bmatrix} = \Psi(L)\begin{bmatrix} \eta_{11}(L) & \eta_{12}(L) \\ \eta_{21}(L) & \eta_{22}(L) \end{bmatrix}\begin{bmatrix} u_t \\ v_t \end{bmatrix}$$

with $H(L) = \Psi(L)^{-1} B(L)$. The different lag polynomials result in

(3.12a) $$\eta_{11}(L) = \beta_{11}(L) / \psi_{11}(L),$$

(3.12b) $$\eta_{12}(L) = \beta_{12}(L) / \psi_{11}(L),$$

(3.12c) $$\eta_{21}(L) = \beta_{21}(L) / \psi_{22}(L),$$

(3.12d) $$\eta_{22}(L) = \beta_{22}(L) / \psi_{22}(L).$$

This leads to the following relation between the residuals u and v and the residuals a and b:

(3.13) $$\begin{bmatrix} a_t \\ b_t \end{bmatrix} = \Psi(L)^{-1}\begin{bmatrix} y_t \\ x_t \end{bmatrix} = \Psi(L)^{-1} B(L)\begin{bmatrix} u_t \\ v_t \end{bmatrix} = H(L)\begin{bmatrix} u_t \\ v_t \end{bmatrix},$$

with the following equivalencies because of (3.7) and (3.12):

(3.14a) $\quad\quad\quad \alpha_{12}(L) \equiv 0 \approx \beta_{12}(L) \equiv 0 \approx \eta_{12}(L) \equiv 0,$

(3.14b) $\quad\quad\quad \alpha_{21}(L) \equiv 0 \approx \beta_{21}(L) \equiv 0 \approx \eta_{21}(L) \equiv 0.$

Analogous to (3.4) and (3.9) the different causal events can be expressed as restrictions on the η_{ij}'s:

(3.15a) $((x, y) \vee (x - y)) \quad\approx\quad \eta_{12}(L) \equiv \eta_{21}(L) \equiv 0,$

(3.15b) $((x \rightarrow y) \vee (x \Rightarrow y)) \quad\approx\quad \neg(\eta_{12}(L) \equiv 0) \wedge \eta_{21}(L) \equiv 0,$

(3.15c) $((x \leftarrow y) \vee (x \Leftarrow y)) \quad\approx\quad \eta_{12}(L) \equiv 0 \wedge \neg(\eta_{21}(L) \equiv 0),$

3.2 Characterisation of Causal Relations in Bivariate Models

(3.15d) $((x \leftrightarrow y) \vee (x \Leftrightarrow y)) \approx \neg(\eta_{12}(L) \equiv 0) \wedge \neg(\eta_{21}(L) \equiv 0)$.

Thus, η_{ij} is subject to the same conditions as α_{ij} and β_{ij}.

For the cross-correlation function between the residuals of the univariate processes, a and b, $\rho_{ab}(k)$, we get:

(3.16) $\qquad \rho_{ab}(k) = \dfrac{E[a_t \cdot b_{t-k}]}{\sqrt{E[a_t^2] \cdot E[b_t^2]}} = \dfrac{\gamma_{ab}(k)}{\sqrt{\gamma_a(0) \cdot \gamma_b(0)}}$,

with:

(3.17) $\quad \gamma_{ab}(k) = E[a_t \, b_{t-k}]$,

$\qquad = E[(\eta_{11}(L) u_t + \eta_{12}(L) v_t) \cdot (\eta_{21}(L) u_{t-k} + \eta_{22}(L) v_{t-k})]$

$\qquad = E[\eta_{11}(L) u_t \cdot \eta_{21}(L) u_{t-k}] + E[\eta_{11}(L) u_t \cdot \eta_{22}(L) v_{t-k}]$

$\qquad + E[\eta_{12}(L) v_t \cdot \eta_{21}(L) u_{t-k}] + E[\eta_{12}(L) v_t \cdot \eta_{22}(L) v_{t-k}]$.

Without instantaneous causality this is reduced to

$\quad \gamma_{ab}(k) = E[\eta_{11}(L) u_t \cdot \eta_{21}(L) u_{t-k}] + E[\eta_{12}(L) v_t \cdot \eta_{22}(L) v_{t-k}]$

because of the orthogonality of u and v.

Thus, if we exclude instantaneous causality, we get:

(i) x is not causal to y:

In this case, $\eta_{12}(L) \equiv 0$ and u_t and a_t are white noise. Because of normalisation it holds $\eta_{11}(L) \equiv 1$, i.e. $a_t = u_t$. This leads to

(3.18a) $\rho_{ab}(k) = 0$ because of $E[u_t \cdot \eta_{21}(L) u_{t-k}] = 0$ for $k \geq 0$.

(ii) y is not causal to x:

In this case, $\eta_{21}(L) \equiv 0$ and v_t and b_t are white noise. Because of normalisation it holds $\eta_{22}(L) \equiv 1$, i.e. $b_t = v_t$. This leads to

(3.18b) $\rho_{ab}(k) = 0$ since $E[\eta_{12}(L) v_t \cdot v_{t-k}] = 0$ for $k \leq 0$.

(iii) y and x are independent:

In this case, $\eta_{12}(L) \equiv \eta_{21}(L) \equiv 0$ and u_t, v_t, a_t and b_t are white noise. It follows $\eta_{11}(L) \equiv \eta_{22}(L) \equiv 1$, i.e. $a_t = u_t$ and $b_t = v_t$. This leads to

(3.18c) $\qquad \rho_{ab}(k) = 0 \; \forall \, k$.

From the above results we get

(3.19a) $(x \rightarrow y) \approx (\exists \, k, k > 0: \rho_{ab}(k) \neq 0) \wedge (\forall \, k, k \leq 0: \rho_{ab}(k) = 0)$.

(3.19b) $(x \leftarrow y) \approx (\exists\, k, k < 0: \rho_{ab}(k) \neq 0) \land (\forall\, k, k \geq 0: \rho_{ab}(k) = 0)$.

(3.19c) $(x \leftrightarrow y) \approx (\exists\, k_1, k_1 < 0: \rho_{ab}(k_1) \neq 0) \land (\exists\, k_2, k_2 > 0: \rho_{ab}(k_2) \neq 0)$.

As far as instantaneous causality between x and y can be excluded, the causal relation may also be characterised by using the cross-correlation function between the residuals a and b of the univariate ARMA processes. If there is instantaneous causality, (3.17) leads to

(3.20) $\qquad\qquad\qquad \rho_{ab}(0) \neq 0$.

However, if there is feedback, this condition is neither necessary nor sufficient for the existence of instantaneous causality.

3.3 Causality Tests

All these characterisations can be used for testing causality. In 1972, CHRISTOPHER A. SIMS was the first to propose a test for simple Granger causal relations. This test was based on the moving average representation. However, some problems occurred with this procedure. Therefore, it is hardly applied today and will not be discussed here. THOMAS J. SARGENT (1976) proposed a procedure which is directly derived from the Granger causality definition. It is usually denoted as the *direct Granger procedure*. LARRY D. HAUGH and DAVID A. PIERCE (1977) proposed a test which uses the estimated residuals of the univariate models for x and y. Finally, CHENG HSIAO (1979) proposed a procedure to identify and estimate bivariate models which – like the direct Granger procedure – is based on autoregressive representation and can also be interpreted (at least implicitly) as causality tests. We will present these three procedures and illustrate them by examples.

3.3.1 The Direct Granger Procedure

As mentioned above, this procedure proposed by THOMAS J. SARGENT (1976) is directly derived from the Granger definition of causality. Similar to the method of CLIVE W.J. GRANGER (1969), a linear prediction function is employed. In the following, let x and y be two stationary variables. To test for simple causality from x to y, it is examined whether the lagged values of x in the regression of y on lagged values of x and y significantly reduce the error variance. By using OLS, the following equation is estimated:

$$(3.21) \quad y_t = \alpha_0 + \sum_{k=1}^{k_1} \alpha_{11}^k y_{t-k} + \sum_{k=k_0}^{k_2} \alpha_{12}^k x_{t-k} + u_{1,t},$$

with $k_0 = 1$. A F test is applied to test the null hypothesis, H_0: $\alpha_{12}^1 = \alpha_{12}^2 = \ldots = \alpha_{12}^{k_2} = 0$. By changing x and y in (3.21), it can be tested whether a simple causal relation from y to x exists. There is a feedback relation if the null hypothesis is rejected in both directions. To test whether there is instantaneous causality we finally set $k_0 = 0$ in relation (3.21) and perform a t or F test for the null hypothesis H_0: $\alpha_{12}^0 = 0$. Accordingly, the corresponding null hypothesis can be tested for x. According to *Theorem 3.1* given above, we expect the same result for testing the equation for y and for x. However, as our data are based on finite samples, we will generally get different numerical values for the test statistics. However, with $k_1 = k_2$, i.e. if we include the same number of lagged variables for the dependent as well as for the explanatory variable in both test equations, we get exactly the same numerical values for the test statistics. The reason for this is that the t or F statistics are functions of the partial correlation coefficient between x and y. Its value does not depend on the direction of the regression; it only depends on the correlation between the two variables and the set of conditioning variables which are included. If $k_1 = k_2$, the same conditioning variables are included, irrespectively of the dependent variable.

One problem with this test is that the results are strongly dependent on the number of lags of the explanatory variable, k_2. There is a trade-off: the more lagged values we include, the better the influence of this variable can be captured. This argues for a high maximal lag. On the other hand, the power of this test is the lower the more lagged values are included.

Two procedures have been developed to solve this problem. In general, different values of k_2 (and possibly also of k_1) are used to inspect the sensitivity of the results to the number of lagged variables. One of the different information criteria presented in *Section 2.1.5* can be used alternatively. As we have included an explanatory variable, the number of estimated parameters, m, has to be adjusted. If, besides the constant term on the right hand side, we include k_1 lagged values of the dependent and k_2 values of additional variables, it holds that $m = k_1 + k_2 + 1$.

Example 3.1

When, in the 1970's, Granger causality tests were applied for the first time, the focus of interest was on the relation between money and income. (See, for example, CHRISTOPHER A. SIMS (1972) as well as EDGAR L. FEIGE and DOUGLAS K. PEARCE (1979).) The simple causal relation from the (real) quantity of money to the real gross national product was interpreted as evidence for the monetarist hypothesis of short-run real effects of monetary policy, whereas the reverse relation was interpreted as evidence for Keynesian doctrines. If such a relation exists, it can be used for predictive purposes.

In the 1980's and 1990's, there was an intensive discussion to what extent the real economic development can be predicted by the term structure of interest, especially by using the difference between long-run and short-run interest rates. *Figure 3.1* demonstrates this possibility by presenting the annual growth rates of the real German GDP and the four quarters lagged interest rate spread for the period from 1970 to 1989. The precondition for using this spread as a predictor is a simple Granger causal relation between this spread and real GDP. The question is which one is 'better' suited to indicate the real effects of monetary policy.

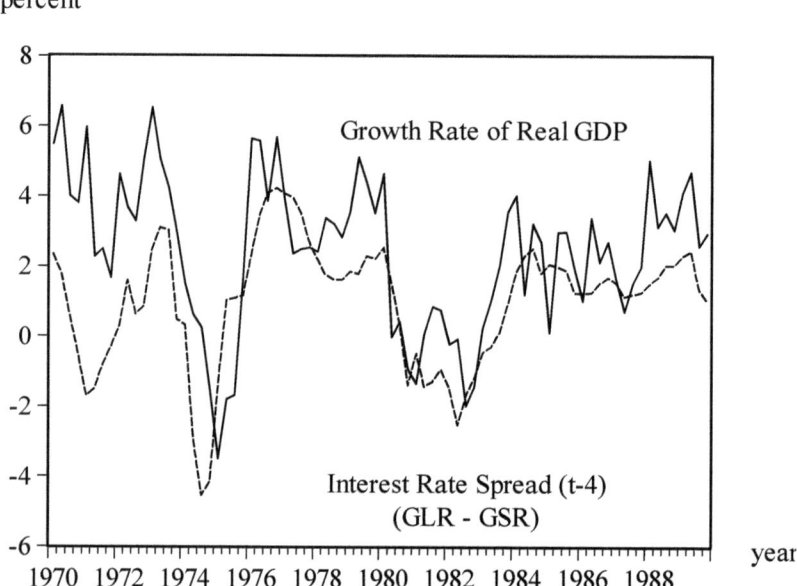

Figure 3.1: Growth rate of real GDP and the four quarters lagged interest rate spread in the Federal Republic of Germany, 1970 – 1989 (in percent)

In the following, we investigate by using quarterly data whether Granger causal relations existed in the Federal Republic of Germany for the period from 1965 to 1989 between the quantity of money M1 or the interest rate differential and the real GDP. (As the German reunification in 1990 is a real structural break we only use data for the period before.) For the dependent as well as for the explanatory variable, we always use four or eight lags, respectively. $\Delta_4\ln(GDP_r)$ denotes the annual growth rate of real GDP, $\Delta_4\ln(M1_r)$ the annual growth rate of the real quantity of money M1, GLR the rate of government bonds (as a long-run interest rate), and GSR the three month money market rate in Frankfurt (as a short-run interest rate).

Table 3.1 Test for Granger Causality (I): Direct Granger Procedure
1/65 – 4/89, 100 Observations

y	x	k_1	k_2	$F(y \leftarrow x)$	$F(y \rightarrow x)$	$F(y-x)$
$\Delta_4 \ln(GDP_r)$	$\Delta_4 \ln(M1_r)$	4	4	6.087***	1.918	0.391
		8	8	3.561**	1.443	0.001
$\Delta_4 \ln(GDP_r)$	GLR – GSR	4	4	3.160*	3.835**	0.111
		8	8	1.927(*)	2.077*	0.279
$\Delta_4 \ln(M1_r)$	GLR – GSR	4	4	5.615***	1.489	10.099**
		8	8	2.521*	1.178	15.125***

'(*)', '*', '**', or '***' denote that the null hypothesis that no causal relation exists can be rejected at the 10, 5, 1 or 0.1 percent significance level, respectively.

The results in *Table 3.1* show that there is only a simple causal relation from money to GDP. The null hypothesis that no such relation exists can be rejected at the 1 percent significance level by using eight lags, and even at the 0.1 percent level by using four lags. By contrast, the null hypothesis that no reversed causal relation exists cannot even be rejected at the 10 percent significance level. The same is true for an instantaneous relation.

The results for the relation between the interest rate differential and the GDP are quite different. There is a simple causal relation from the monetary indicator to GDP, too, but this relation is much less pronounced than the relation between money and income, and, in addition, there is a simple relation in the reverse direction. Thus, there exists feedback between these two variables.

There is, first of all, a very pronounced instantaneous relation between the two monetary indicators. Besides this, there is a simple relation from the interest rate differential to money growth, while no relation seems to exist in the reverse direction. This reflects the fact that the German Bundesbank used the quantity of money as an intermediate target which it tries to influence. It can, however, only do this indirectly via (money market) interest rates. (Before 1987, the Bundesbank had used central bank money as its intermediate target, from then on it used the quantity of money M3.) It takes some time before money growth has fully adjusted to a monetary impulse based on interest rates. This is reflected in the simple Granger causal relation from interest rate differential to money growth as well as in the instantaneous relation between these two variables.

3.3.2 The Haugh-Pierce Test

This procedure which was first proposed by LARRY D. HAUGH (1976) and later on by LARRY D. HAUGH and DAVID A. PIERCE (1977) is based on the cross-correlations $\rho_{ab}(k)$ between the residuals a and b of the univariate ARMA models for x and y. In a first step, these models have to be estimated. By using the Box-Pierce Q* statistic given in (1.11) (or the Ljung-Box Q statistic given in (1.12)) it is checked whether the null hypothesis – that the estimated residuals are white noise – cannot be rejected. Then, analogous to the Q* statistic, the following statistic is calculated:

$$(3.22) \qquad S = T \cdot \sum_{k=k_1}^{k_2} \hat{\rho}_{ab}^2(k).$$

Under the null hypothesis H_0: $\rho_{ab}(k) = 0$ for all k with $k_1 \leq k \leq k_2$, this statistic is asymptotically χ^2 distributed with $k_2 - k_1 + 1$ degrees of freedom. It can be checked for $k_1 < 0 \wedge k_2 > 0$ whether there is any causal relation at all. If this hypothesis can be rejected, it can be checked for $k_1 = 1 \wedge k_2 \geq 1$ whether there is a simple causal relation from x to y. In the reverse direction, for $k_1 \leq -1 \wedge k_2 = -1$, it can be checked whether there is a simple causal relation from y to x. Finally, it can be tested by using $\rho_{ab}(0)$ whether there exists an instantaneous relation. However, the results of the last test are questionable as long as the existence of a feedback relation cannot be excluded.

But this is not the only problem that might arise with this procedure. G. WILLIAM SCHWERT (1979) showed that the power of this procedure, which uses correlations, is smaller than the power of the direct Granger procedure which uses regressions. Thus, following a remark by EDGAR L. FEIGE and DOUGLAS K. PEARCE (1979), this test might only be a first step to analyse causal relations between time series. On the other hand, information on the relations between two time series, which is contained in cross-correlations, can be useful even if no formal test is applied. This information offers a deeper insight into causal relations than just looking at the F and t statistics of the direct Granger procedure.

Example 3.2

To perform the Haugh-Pierce test we estimate univariate models for the three variables of *Example 3.1* and for the period from the first quarter of 1965 to the last quarter of 1989. The results are presented below; the numbers in parentheses are again the corresponding t statistics:

$$\Delta_4\ln(GDP_{r,t}) = \underset{(3.09)}{0.658} + \underset{(12.80)}{0.861}\,\Delta_4\ln(GDP_{r,t-1}) - \underset{(-1.63)}{0.105}\,\Delta_4\ln(GDP_{r,t-4})$$
$$+ \hat{u}_{1,t} - \underset{(-2.58)}{0.266}\,\hat{u}_{1,t-8},$$

$\bar{R}^2 = 0.669$, SE $= 1.395$, AIC $= 3.542$, SC $= 3.646$, $Q(9) = 5.602$ (p $= 0.779$).

$$\Delta_4\ln(M1_{r,t}) = \underset{(1.99)}{0.296} + \underset{(19.44)}{0.908}\,\Delta_4\ln(M1_{r,t-1}) + \hat{u}_{2,t} - \underset{(-12.99)}{0.772}\,\hat{u}_{2,t-4} - \underset{(-2.30)}{0.137}\,\hat{u}_{2,t-5},$$

$\bar{R}^2 = 0.764$, SE $= 1.897$, AIC $= 4.158$, SC $= 4.262$, $Q(9) = 10.845$ (p $= 0.287$).

$$(GLR - GSR)_t = \underset{(2.81)}{0.291} + \underset{(15.95)}{1.039}\,(GLR - GSR)_{t-1} - \underset{(-3.56)}{0.422}\,(GLR - GSR)_{t-3}$$
$$+ \underset{(3.00)}{0.426}\,(GLR - GSR)_{t-4} - \underset{(-3.17)}{0.297}(GLR - GSR)_{t-5} + \hat{u}_{3,t},$$

$\bar{R}^2 = 0.796$, SE $= 0.771$, AIC $= 2.368$, SC $= 2.498$, $Q(8) = 11.390$ (p $= 0.181$).

In all three cases, the Ljung-Box Q statistic calculated for 12 lags does not indicate any autocorrelation of the estimated residuals.

The next step was to calculate the cross-correlation functions presented in *Figure 3.2*. (The dotted lines are the approximate 95 percent confidence intervals.) It

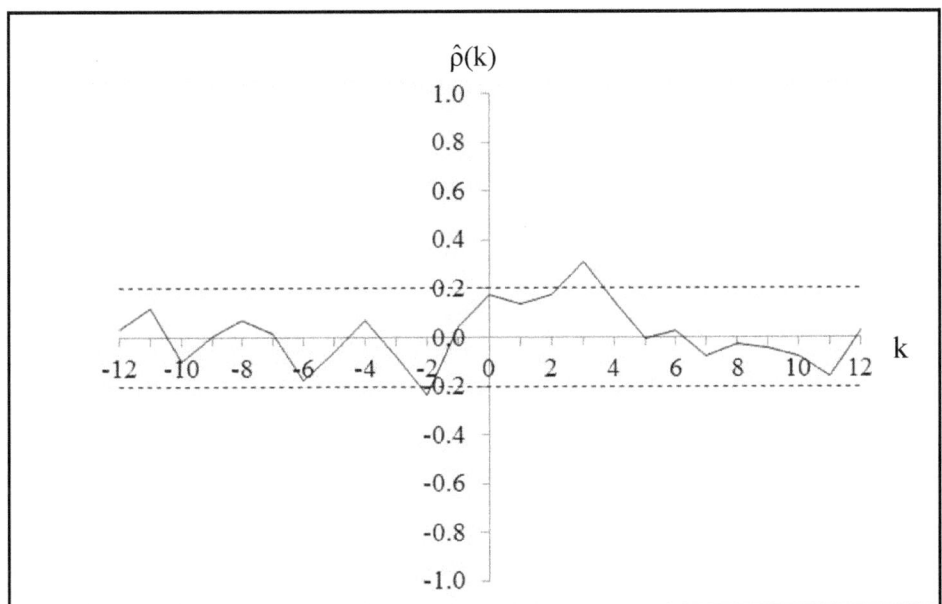

Figure 3.2a: Cross-correlations between the residuals of the univariate models of GDP and the quantity of money M1

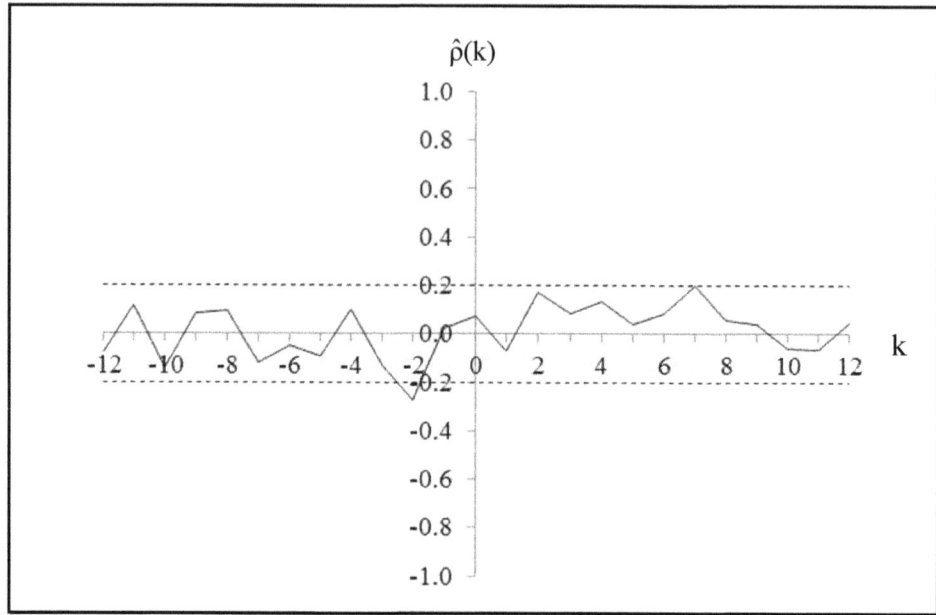

Figure 3.2b: Cross-correlations between the residuals of the univariate models of GDP and the interest rate spread

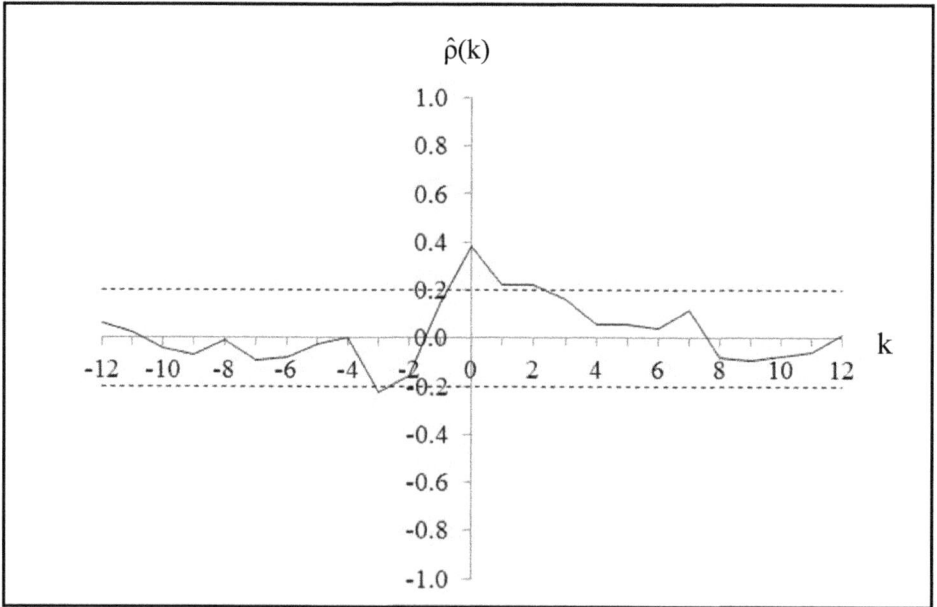

Figure 3.2c: Cross-correlations between the residuals of the univariate models of the quantity of money M1 and the interest rate differential

is quite obvious that this procedure leads to less pronounced (possible) causal relations. Only in a few cases the estimated cross-correlation coefficients exceed the 5 percent bounds. In particular, the causal relation between interest rate differential and GDP cannot be detected.

The impression received by the graphs is confirmed by the test statistic S, see equation (3.22). Again, we use four or eight lags, respectively. The results are quite similar to those of the direct Granger procedure. As *Table 3.2* shows, there is a simple causal relation from the quantity of money to GDP and, in addition, an instantaneous relation which is, however, only significant at the 10 percent level. We find a relation from real GDP growth to the interest rate differential, but nothing in the reverse direction. According to this result, it should be impossible to make better forecasts on real economic development by using the interest rate as predictor. Between the two monetary indicators, we find a strong instantaneous relation and also a feedback relation, but only for four lags. Thus, the only two differences to the results of the direct Granger procedure are that we do not find a direct relation from the interest rate differential to real GDP growth but find a reverse relation from the quantity of money to the interest rate differential.

Table 3.2: Test for Granger Causality (II): Haugh-Pierce Test
1/65 – 4/89, 100 Observations

y	x	$\hat{\rho}(0)$	k	S(y←x)	S(y→x)	S(y<=>x)
$\Delta_4\ln(GDP_r)$	$\Delta_4\ln(M1_r)$	0.179(*)	4	16.547**	7.036	26.771**
			8	17.234*	11.005	31.426*
$\Delta_4\ln(GDP_r)$	GLR – GSR	0.076	4	6.031	10.218*	16.826(*)
			8	11.270	13.718(*)	25.565(*)
$\Delta_4\ln(M1_r)$	GLR – GSR	0.383***	4	11.967*	9.660*	36.295***
			8	14.424(*)	11.270	40.362**

'(*)', '*', '**', or '***' denote that the null hypothesis that no causal relation exists can be rejected at the 10, 5, 1 or 0.1 percent significance level, respectively.

Such results are not untypical for this procedure. Firstly, the application of different test procedures might produce different results: one procedure might detect a causal relation, the other one might not. Reviewing different papers on the relation between money and income, EDGAR L. FEIGE and DOUGLAS K. PEARCE (1979), therefore, referred to the "casual causal relation between money and income". Secondly, 'non-results' are to be expected in particular if the Haugh-Pierce test is applied. DAVID A. PIERCE (1977), for example, was unable to find statistically significant relations between various macroeconomic variables whereas economists are convinced that such relations do exist.

3.3.3 The Hsiao Procedure

The procedure for identifying and estimating bivariate time series models proposed by CHENG HSIAO (1979) initially corresponds to the application of the direct Granger procedure. However, the lag lengths are determined with an information criterion. CHENG HSIAO proposed the use of the final prediction error. Any other criterion presented in *Section 2.1.5* might of course also be used.

Again, the precondition is that the two variables are weakly stationary. The procedure is divided into six steps:

(i) First, the optimal lag length k_1^* of the univariate autoregressive process of y is determined.

(ii) In a second step, by fixing k_1^*, the optimal lag length k_2^* of the explanatory variable x in the equation of y is determined.

(iii) Then k_2^* is fixed and the optimal lag length of the dependent variable y is again determined: $\overline{k}_1^*$.

(iv) If the value of the information criterion applied in the third step is smaller than that of the first step, x has a significant impact on y. Otherwise, the univariate representation of y is used. Thus, we get a (preliminary) model of y.

(v) Steps (i) to (iv) are repeated by exchanging the variables x and y Thus, we get a (preliminary) model for x.

(vi) The last step is to estimate the two models specified in steps (i) to (v) simultaneously to take into account the possible correlation between their residuals. Usually, the procedure to estimate *seemingly unrelated regressions* (SUR) developed by ARNOLD ZELLNER (1962) is applied.

The Hsiao procedure only captures the simple causal relations between the two variables. The possible instantaneous relation is reflected by the correlation between the residuals. However, by making theoretical assumptions about the direction of the instantaneous relation, it is possible to take into account the instantaneous relation in the model for y or in the model for x.

Example 3.3

As explained above, the first steps of the Hsiao procedure are different from the usual application of the direct Granger procedure, where the number of lags is fixed (and might be varied), insofar as an information criterion is used to determine the optimal lag length. In our example, we used a maximal length of eight lags for the dependent as well as for the explanatory variable, and we calculated the values of the Akaike and the Schwarz criterion. In doing so, we did not take into account a possible instantaneous relation.

Table 3.3: Optimal Lag Length for the Hsiao Procedure

Relation	Akaike Criterion			Schwarz Criterion		
	k_1^*	k_2^*	$\bar{k}_1^*$	k_1^*	k_2^*	$\bar{k}_1^*$
$\Delta_4 \ln(M1_r) \rightarrow \Delta_4 \ln(GDP_r)$	4	1	1	1	1	1
$\Delta_4 \ln(GDP_r) \rightarrow \Delta_4 \ln(M1_r)$	5	3	8	4	0	4
$(GLR - GSR) \rightarrow \Delta_4 \ln(GDP_r)$	4	2	1	1	2	1
$\Delta_4 \ln(GDP_r) \rightarrow (GLR - GSR)$	5	5	5	5	0	5

Table 3.3 shows quite different results for the two criteria. As expected, the optimal lag length is sometimes smaller when using the Schwarz criterion as compared to the Akaike criterion. In our example, this leads to economic implications. Both criteria reveal simple causal relations from the quantity of money as well as the interest rate differential to real GDP. Reverse causation, however, can only be found with the Akaike criterion. While we find one-sided relations only with the Schwarz criterion, we get feedback relations with the Akaike criterion.

The models which were estimated using these lags are given in *Table 3.4* for the relation between money and income and in *Table 3.5* for the relation between the interest rate spread and income. In all cases, the simple causal relation from the monetary indicator to GDP is significant. This also holds when – using Wald tests – we check the common null hypotheses that all coefficients as well as the sum of the coefficients of the interest rate differential in the GDP equations are (jointly) zero. In all cases, the null hypothesis can be rejected at the 0.1 percent significance level. The reverse causal relations detected by the Akaike criterion are significant at the 5 percent level in the money equation and at the one percent level in the interest rate equation. On the other hand, none of the models detects an instantaneous relation: in both cases, the values of the correlation coefficient between the residuals of the two equations are below any conventional critical value.

Table 3.4: Models Estimated with the Hsiao Procedure
1/65 – 4/89, 100 Observations

Criterion	Akaike Criterion		Schwarz Criterion	
	Dependent Variable			
Explanatory Variable	$\Delta_4\ln(GDP_{r,t})$	$\Delta_4\ln(M1_{r,t})$	$\Delta_4\ln(GDP_{r,t})$	$\Delta_4\ln(M1_{r,t})$
Constant term	0.146 (0.67)	1.263*** (3.42)	0.136 (0.62)	1.139*** (3.94)
$\Delta_4\ln(GDP_{r,t-1})$	0.751*** (13.59)	-0.195 (1.32)	0.756*** (13.68)	
$\Delta_4\ln(GDP_{r,t-2})$		-0.283 (1.65)		
$\Delta_4\ln(GDP_{r,t-3})$		0.369* (2.54)		
$\Delta_4\ln(M1_{r,t-1})$	0.159*** (4.62)	1.027*** (10.73)	0.159*** (4.61)	0.972*** (10.12)
$\Delta_4\ln(M1_{r,t-2})$		-0.173 (1.29)		-0.135 (0.99)
$\Delta_4\ln(M1_{r,t-3})$		0.185 (1.36)		0.083 (0.61)
$\Delta_4\ln(M1_{r,t-4})$		-0.478*** (3.53)		-0.265** (2.72)
$\Delta_4\ln(M1_{r,t-5})$		0.340* (2.50)		
$\Delta_4\ln(M1_{r,t-6})$		-0.188 (1.36)		
$\Delta_4\ln(M1_{r,t-7})$		0.192 (1.41)		
$\Delta_4\ln(M1_{r,t-8})$		-0.203* (2.08)		
$\hat{\rho}(\hat{u}_1, \hat{u}_2)$	0.012		0.077	
$\bar{R}^2$	0.694	0.750	0.694	0.726
SE	1.340	1.952	1.340	2.041
Q(m)	23.084*	11.226*	23.344*	16.548*
m	11	4	11	8

The numbers in parentheses are the absolute values of the estimated t statistics. '(*)', '*', '**', or '***' denote that the corresponding null hypothesis can be rejected at the 10, 5, 1 or 0.1 percent significance level, respectively. m denotes the number of degrees of freedom of the Q statistic.

Table 3.5: Models Estimated with the Hsiao Procedure
1/65 – 4/89, 100 Observations

Criterion	Akaike Criterion		Schwarz Criterion	
Explanatory Variable	Dependent Variable			
	$\Delta_4 \ln(GDP_{r,t})$	$(GLR - GSR)_t$	$\Delta_4 \ln(GDP_{r,t})$	$(GLR - GSR)_t$
Constant term	0.327 (1.47)	0.404** (2.80)	0.320 (1.43)	0.293** (2.93)
$\Delta_4 \ln(GDP_{r,t-1})$	0.730*** (12.22)	-0.034 (0.65)	0.733*** (12.27)	
$\Delta_4 \ln(GDP_{r,t-2})$		-0.132* (2.10)		
$\Delta_4 \ln(GDP_{r,t-3})$		0.021 (0.32)		
$\Delta_4 \ln(GDP_{r,t-4})$		0.154* (2.58)		
$\Delta_4 \ln(GDP_{r,t-5})$		-0.083(*) (1.72)		
$(GLR - GSR)_{t-1}$	-0.105 (0.64)	1.128*** (11.91)	-0.103 (0.63)	1.138*** (12.13)
$(GLR - GSR)_{t-2}$	0.441** (2.62)	-0.168 (1.27)	0.438* (2.60)	-0.198 (1.42)
$(GLR - GSR)_{t-3}$		-0.347** (2.69)		-0.316* (2.32)
$(GLR - GSR)_{t-4}$		0.481*** (3.70)		0.448** (3.25)
$(GLR - GSR)_{t-5}$		-0.274** (2.95)		-0.327*** (3.53)
$\hat{\rho}(\hat{u}_1, \hat{u}_2)$	0.053		0.031	
$\bar{R}^2$	0.684	0.816	0.684	0.798
SE	1.362	0.732	1.362	0.768
Q(m)	16.513	4.824	16.648	7.118
m	11	7	11	7

The numbers in parentheses are the absolute values of the estimated t statistics. '(*)', '*', '**', or '***' denote that the corresponding null hypothesis can be rejected at the 10, 5, 1 or 0.1 percent significance level, respectively. m denotes the number of degrees of freedom of the Q statistic.

3.4 Applying Causality Tests in a Multivariate Setting

Whenever such a test is applied, one can hardly assume that there are no other variables with an impact on the relation between the two variables under consideration. The definition of Granger causality given above does not imply such a limitation despite the fact that the relation between just two variables is investigated: besides $\bar{y}_t$ and $\bar{x}_t$, the relevant information set I_t can include the values of any other variables $\bar{z}_{j,t}$, $j = 1, ..., m$. To distinguish between (real) causal and spurious relations, this enlargement of the relevant information set is crucial.

However, the above presented test procedures only take into account the past values of x and y as the relevant information set. In order to apply these models in a multivariate framework, two questions have to be answered: (i) How can the procedures be generalised so that they can be applied in a model with more than two variables? (ii) Which conclusions can be drawn if the procedure considers only two variables, but, nevertheless, relations to additional variables do exist?

3.4.1 The Direct Granger Procedure with More Than Two Variables

As the Haugh-Pierce test uses the cross-correlation function between the residuals of the univariate ARMA models, it is obvious that only two time series can be considered. Thus, it is not possible to generalise as to situations with more than two variables. However, the direct Granger procedure is a different case. Let $z_1, ..., z_m$ be additional variables. According to the definition of Granger causality, the estimation equation (3.21) can be extended to

$$(3.23) \quad y_t = \alpha_0 + \sum_{k=1}^{k_1} \alpha_{11}^k y_{t-k} + \sum_{k=1}^{k_2} \alpha_{12}^k x_{t-k} + \sum_{j=1}^{m} \sum_{k=1}^{k_{j+2}} \beta_j^k z_{j,t-k} + u_t,$$

if we test for simple Granger causal relations, with β_j^k, $k = 1, ..., k_{j+2}$, $j = 1, ..., m$, being the coefficients of the additional variables. It does not matter whether the additional variables are endogenous or exogenous since only lagged values are considered. After determining the numbers of lags k_1, k_2, k_3, ..., (3.23) can be estimated using OLS. As in the bivariate case, it can be checked via an F test whether the coefficients of the lagged values of x are jointly significantly different from zero. By interchanging x and y in

(3.23), it can be tested whether there exists a simple Granger causal relation from y to x and/or feedback.

However, problems arise again if there are instantaneous relations. It is, of course, possible to extend the test equation (3.23) by including the current value of x analogous to (3.21) in order to test for instantaneous causality as per the definition given in *Section 3.1*. Again, it holds that it is impossible to discriminate between whether x is instantaneously causal to y and/or y is instantaneously causal to x without additional information. It also holds that if all conditioning variables have the same maximal lag, i.e. for $k_1 = k_2 = k_3 = ... = k_{m+2}$, the values of the test statistics are identical irrespectively of which equation is used to check for instantaneous causality between x and y. However, as long as the other contemporaneous values of the additional variables z_j are not included, the resulting relations might be spurious instantaneous relations.

Example 3.4

The results of the direct Granger procedure as well as those of the Hsiao procedure given above indicate that both monetary indicators are Granger causal to the real economic development and can therefore be used for predictive purposes. The question that arises is not only whether one of the indicators is 'better', but also whether forecasts can be improved by the use of both indicators. This can be investigated by using the trivariate Granger procedure.

Table 3.6: *Test for Granger Causality:*
Direct Granger Procedure with Three Variables
1/65 – 4/89, 100 Observations

y	x	z	k	F(y←x)	F(y→x)	F(y–x)
$\Delta_4 \ln(GDP_r)$	$\Delta_4 \ln(M1r)$	GLR – GSR	4	2.747*	3.788**	0.577
			8	2.866**	2.362*	0.127
$\Delta_4 \ln(GDP_r)$	GLR – GSR	$\Delta_4 \ln(M1_r)$	4	0.260	2.426(*)	0.247
			8	1.430	1.817(*)	0.229
$\Delta_4 \ln(M1_r)$	GLR – GSR	$\Delta_4 \ln(GDP_r)$	4	7.615***	0.293	7.273***
			8	3.432**	1.009	8.150***

'(*)', '*', '**', or '***' denote that the null hypothesis that no causal relation exists can be rejected at the 10, 5, 1 or 0.1 percent significance level, respectively.

Again, we use four and eight lags. The results are presented in *Table 3.6*. Here, z denotes the respective conditioning (third) variable. The results for M1 and for the interest rate spread are quite different. While we still find a significant simple causal relation from the quantity of money to real GDP as well as a reverse relation, the interest rate differential seems to have no impact at all on real GDP as soon as M1 is considered as a third variable. This indicates that the quantity of money is sufficient for predictive purposes; the interest rate spread does not contain any information which is not already contained in M1 but which is relevant for the prediction of real GDP. This holds despite the fact that (as with the bivariate tests) we find a highly significant simple causal relation from the interest rate differential to M1.

Analogous to this procedure, third variables can also be considered using the Hsiao procedure. In this case, first the optimal lag length of the dependent variable y and the conditioning variables z_1 to z_m must be determined before the optimal lag length k_2^* of the variable of interest x is fixed.

Example 3.5

Applying the trivariate Hsiao procedure, we start with the equation of interest, i.e. the equation for real GDP. Let us first consider the equations of *Table 3.4* with the lagged quantity of money as explanatory variable. If we add the interest rate differential with the Akaike criterion we get the optimal lag length of two compared to the one lag indicated by the Schwarz criterion. In both cases, however, the values of the criterion are higher than when this variable is not included. Thus, the interest rate differential, along with real M1, does not significantly contribute to the explanation of real GDP, and we can stick to the bivariate model of *Table 3.4*.

We get the same results if we add the quantity of money as additional variable to the equations including the lagged interest rate spread in *Table 3.5*. We get the optimal lag one by using both criteria. In both cases, however, the value of the criterion is below the value that results without considering this variable. If, once again, we vary the maximal lag of the interest rate differential we end up with the equation including M1 as explanatory variable. However, we have just found out that the interest rate spread does not have a significant impact. Thus, we stick to the estimated equations of *Table 3.4*.

A quite different procedure is to apply the definition of Granger causality not to single variables but to groups of variables: a vector Y of dependent variables and a vector X of explanatory variables. We can ask for the relations between these two groups of variables. In *Section 4.2* this will be discussed within the framework of vector autoregressive models.

3.4.2 Interpreting the Results of Bivariate Tests in Systems With More Than Two Variables

To what extent do the results of bivariate tests apply for systems with more than two variables? Let us first consider instantaneous relations. Such relations can be detected with the direct Granger procedure as well as with the Haugh-Pierce test. However, definite evidence whether these relations are real or only spurious can only be found in a complete model and by using additional information. Insofar, the results of bivariate tests are only preliminary with respect to instantaneous relations.

What are the consequences for simple causal relations if third variables are not considered? GEBHARD KIRCHGÄSSNER (1981) shows that it usually implies that an existing simple causal relation appears as a feedback relation. In the reverse case it holds that if the relation between x and y is only one-sided in the bivariate model, there are no third variables which are Granger causal to x and y. Thus, whereas the measured feedback relation might be spurious and the inclusion of other variables might reduce it to a one sided relation, the reverse does not hold.

Which are the effects of spurious correlations on the results of Granger causality tests if there is no direct causal relation between x and y but if both depend on a third variable z? CHRISTOPHER A. SIMS (1977) showed that rather extreme assumptions are necessary to avoid such a spurious relation as feedback relation in the data.

With respect to non-considered third relevant variables as well as to spurious correlations as a result of the common dependence on third variables, the following holds: If it is found that, in a bivariate model, only a one-sided causal relation from x to y (or from y to x) without feedback exists, this should also hold when additional variables are included in the model. On the other hand, spurious feedback might occur due to several reasons, without the 'true' relation being a feedback relation. Thus, the fact whether feedback exists or not can only be verified within a full model.

However, it has to be taken into account that spurious feedback relations arising, for example, from omitted variables or from measurement errors are, in most cases, rather weakly pronounced compared to the 'real' causal relations. Thus, they might often not be detected with causality tests. Moreover, as shown above, spurious independence arises quite often when these test procedures are applied. If, however, the (relatively strongly pronounced) direct causal relations cannot be detected in many cases, it is even more unlikely that feedback relations which result from measurement errors or omitted third variables are detected by causality tests. Thus, the interpretation of detected unidirectional causal relations should also be treated cautiously. Finally, it should not be ignored that in case a specific

3.5 Concluding Remarks

The definition of causality proposed by CLIVE W.J. GRANGER (1969) has been heavily criticised in the first years after the publication of his paper as it reduces causality to incremental predictability. ‚Post hoc, ergo propter hoc?' was the question. It is correct that causality implies predictability, but the reverse is not generally correct. In time series analysis, this concept of causality is nevertheless widely accepted today.

Partly, the criticism was definitely exaggerated. Succession in time is a principal element of the classical causality definition of DAVID HUME, and exactly this idea was taken up by the definition of CLIVE W.J. GRANGER. Insofar, the latter is in the classical tradition. However, even if a 'true' causal relation exists, its structure does not have to coincide with the structure represented in the data. Even if the true model contains a temporal asymmetry, the same asymmetry does not have to be reflected in the data. The technical problem how the data can be measured and actually are measured plays a crucial role here. Firstly, as explained above, due to the long periods covered by one observation, simple causal relations may appear to be instantaneous relations. Of course, this holds especially when annual data are used. Secondly, when different variables are measured with different time delays it might even occur that the measured relation is in the reverse direction of the true one. When x is causal to y, the tests might indicate that y is causal to x. Finally, different methods of temporal aggregation might disguise the true relation if, for example, monthly averages are used for one time series and end of month data for another one.

If economic policy follows a given (contingent) rule, there will generally be a feedback relation even if the 'true' relation is a unidirectional one. If the rule is deterministic, it might even be the case that only the reverse causation can be detected. Let x be the economic policy instrument and y the objective variable, which are connected by the simple linear relation

$$(3.24) \qquad y_t = \alpha_0 + \alpha_1 y_{t-1} + \alpha_2 x_t + u_t.$$

Let u be white noise. The coefficients of this relation are assumed to be constant and known to the government. It strives for a constant (optimal) value y*. In this situation, the optimal (deterministic) rule is given by

(3.25) $$x_t = \frac{1}{\alpha_2}[y^* - \alpha_0 - \alpha_1 y_{t-1}].$$

For the objective variable, it holds that

(3.26) $$y_t = y^* + u_t,$$

i.e. it follows a white noise process with mean y^* and variance σ_u^2. In this case, neither past nor current values of x can improve the forecasts of y. By inserting (3.26) into (3.25) we get

(3.27) $$x_t = \frac{1}{\alpha_2}[y^*(1-\alpha_1) - \alpha_0 - \alpha_1 u_{t-1}].$$

As u_{t-1} is contained in y_{t-1}, but not in x_{t-1}, forecasts of x can be improved using past values of y (besides past values of x), i.e. there is a simple Granger causal relation from y to x: the measured causal relation goes into the opposite direction of the true relation.

If however, one assumes that the government is not able to steer exactly the economy as, for example, it does not exactly know the coefficients of the 'true' model, it might, instead of (3.25), follow the stochastic rule

(3.25') $$x_t = \frac{1}{\alpha_2}[y^* - \alpha_0 - \alpha_1 y_{t-1}] + v_t, \quad E[v_t] = 0,$$

where v is independent of u. In such a situation there is also an instantaneous relation between x and y because v, the stochastic part of x, has an impact on y but is independent of the lagged values of y. If, in addition to that, it is assumed that there is a delay in the effect of x on y, we also get a simple causal relation from x to y.

Thus, as soon as the government reacts systematically to past developments we expect reverse causal relations. However, under realistic assumptions, we can also expect that there is a simple Granger causal relation in the 'true' direction. This also holds under the conditions of the New Classical Macroeconomics where people have rational expectations if unexpected changes, for example in monetary policy, affect real and/or nominal economic development with some delay. Insofar, Granger causality tests can be used to investigate the effectiveness of economic policy. On the other hand, we only get distinctive evidence for the true model if we make additional, sometimes rather restrictive assumptions.

There is also an interesting relation between the efficiency of (financial) markets and (instantaneous) Granger causality. If the price in an efficient market really contains all (publicly) available information and can, therefore, be modelled as a random walk or a martingale, there is no simple Granger causal relation from any other variable on this price. Only instantaneous relations are possible, because any simple causal relation would indicate that information is available which has not been used efficiently.

Thus, the existence or non-existence of Granger causal relations between economic variables has substantial implications. But one should not forget that Granger causality is a *statistical* concept: given a specific set of information, it asks for the (incremental) predictability of y using x. The power of these tests, especially of the Haugh-Pierce test, is often rather low and spurious independence might occur, sometimes caused by omitted variables. But, nevertheless, it is not sensible in this context to speak of misspecifications as this always presupposes the existence of a 'true' model. A concept that allows results only according to a specific information set has no room for the idea of a 'true' model. As shown above, this does not preclude that (stochastic) economic models imply Granger causal relations for the variables included in these models.

References

The **definition of Granger causality** was presented in

CLIVE W.J. GRANGER, Investigating Causal Relations by Econometric Models and Cross-Spectral Methods, *Econometrica* 37 (1969), pp. 424 – 438.

The **Sims test** was proposed by

CHRISTOPHER A. SIMS, Money, Income, and Causality, *American Economic Review* 62 (1972), pp. 540 – 552.

This was also the first paper to investigate the relation between money and income by using causality tests. The **direct Granger procedure** was applied for the first time in

THOMAS J. SARGENT, A Classical Macroeconomic Model for the United States, *Journal of Political Economy* 84 (1976), pp. 207 – 237.

This was one of the papers for which he got the Nobel Prize in 2011.

The **Haugh-Pierce test** was proposed by

LARRY D. HAUGH, Checking the Independence of Two Covariance Stationary Time Series: A Univariate Residual Cross-Correlation Approach, *Journal of the American Statistical Association* 71 (1976), pp. 378 – 385,

and was made popular by

DAVID A. PIERCE and LARRY D. HAUGH, Causality in Temporal Systems: Characterizations and a Survey, *Journal of Econometrics* 5 (1977), pp. 265 – 293.

The **Hsiao procedure** was applied for the first time in

CHENG HSIAO, Autoregressive Modeling of Canadian Money and Income Data, *Journal of the American Statistical Association* 74 (1979), pp. 553 – 560.

To estimate the bivariate models this approach employs the **SUR procedure** proposed by

ARNOLD ZELLNER, An Efficient Method of Estimating Seemingly Unrelated Regressions and Tests for Aggregation Bias, *Journal of the American Statistical Association* 57 (1962), pp. 348 – 368.

Possible **impacts of third variables** on the test results are discussed in

CHRISTOPHER A. SIMS, Exogeneity and Causal Ordering in Macroeconomic Models, in: FEDERAL RESERVE BANK OF MINNEAPOLIS (ed.), *New Methods in Business Cycle Research: Proceedings from a Conference*, Minneapolis 1977, pp. 23 – 44.

The problems that can arise with the detection of instantaneous relations when applying the Haugh-Pierce test in situations with feedback relation between the two variables have first been mentioned by

J. MICHAEL PRICE, Causality in Temporal Systems: A Correction, *Journal of Econometrics* 10 (1979), pp. 253 – 256.

An **introduction** to the different testing procedures is given in

GEBHARD KIRCHGÄSSNER, *Einige neuere statistische Verfahren zur Erfassung kausaler Beziehungen zwischen Zeitreihen, Darstellung und Kritik*, Vandenhoeck und Ruprecht, Göttingen 1981.

This book also gives a proof of *Theorem 3.1*. **Critical discussions** of these procedures are given in

G. WILLIAM SCHWERT, Tests of Causality: The Message in the Innovations, in: K. BRUNNER and A.H. MELTZER (eds.), *Three Aspects of Policy and Policymaking: Knowledge, Data, and Institutions*, Carnegie-Rochester Conference Series on Public Policy, Band 10, North-Holland, Amsterdam 1979, pp. 55 – 96;

ARNOLD ZELLNER, Causality and Econometrics, in: K. BRUNNER and A. H. MELTZER (eds.), *Three Aspects of Policy and Policymaking: Knowledge, Data, and Institutions*, Carnegie-Rochester Conference Series on Public Policy, Vol. 10, North-Holland, Amsterdam 1979, pp. 9 – 54; as well as in

EDGAR L. FEIGE and DOUGLAS K. PEARCE, The Casual Causal Relationship Between Money and Income: Some Caveats for Time Series Analysis, *Review of Economics and Statistics* 61(1979), pp. 521 – 533.

The latter paper especially discusses the different results on the **relation between money and income** when different testing procedures are applied.

'Non-results' quite often occur when these tests are applied on economic time series, i.e. it is not possible to detect statistically significant relations between variables where theoretical considerations suggest that there must be causal relations. Thus, one can assume that **spurious independence** occurs. This holds especially when the Haugh-Pierce test is applied as

DAVID A. PIERCE, Relationships – and the Lack Thereof – Between Economic Time Series, with Special Reference to Money and Interest Rates, *Journal of the American Statistical Association* 72 (1977), pp. 11 – 26,

shows. Such spurious independence might result from omitting third variables, as

HELMUT LÜTKEPOHL, Non-Causality due to Omitted Variables, *Journal of Econometrics* 19 (1982), pp. 367 – 378,

shows. On the other hand, it was demonstrated by

CLIVE W.J. GRANGER and PAUL NEWBOLD, Spurious Regressions in Econometrics, *Journal of Econometrics* 2 (1974), pp. 111 – 120,

that the use of (highly autocorrelated) time series in traditional econometrics might show spurious relations.

How far the **temporal aggregation** of variables affects the results of causality tests was investigated by

GEBHARD KIRCHGÄSSNER and JÜRGEN WOLTERS, Implications of Temporal Aggregation on the Relation Between Two Time Series, *Statistische Hefte/Statistical Papers* 33 (1992), pp. 1 – 19.

The implications of the results of Granger causality tests for the evaluation of different economic theories and especially for the effectiveness (or ineffectiveness) of economic policy were discussed in

THOMAS J. SARGENT, The Observational Equivalence of Natural and Unnatural Rate Theories of Macroeconomics, *Journal of Political Economy* 84 (1976), pp. 631 – 640, as well as

WILLEM H. BUITER, Granger Causality and Policy Effectiveness, *Economica* 51 (1984), pp. 151 – 162.

It was first shown by

JÖRG W. KRÄMER and ENNO LANGFELD, Die Zinsdifferenz als Frühindikator für die westdeutsche Konjunktur, *Die Weltwirtschaft*, Issue 1/1993, pp. 34 – 43,

that the **interest rate spread** might be a good predictor for the real economic development in the Federal Republic of Germany, see also

FREDERIC S. MISHKIN, What Does the Term Structure Tell Us about Future Inflation?, *Journal of Monetary Economics* 20 (1990), pp. 77 – 95.

However,

GEBHARD KIRCHGÄSSNER and MARCEL R. SAVIOZ, Monetary Policy and Forecasts for Real GDP Growth: An Empirical Investigation for the Federal Republic of Germany, *German Economic Review* 2 (2001), pp. 339 – 365,

show that the quantity of money M1 is a better predictor. This also holds for the time after the German Unification until the end of 1998. For the time since 1999, i.e. since the European Central Bank has been responsible for monetary policy in the Euro-area, no results on this topic are available.

4 Vector Autoregressive Processes

The previous chapter presented a statistical approach to analyse the relations between time series: starting with univariate models, we asked for relations that might exist between two time series. Subsequently, the approach was extended to situations with more than two time series. Such a procedure where models are developed bottom up to describe relations is hardly compatible with the economic approach of theorising where – at least in principle – all relevant variables of a system are treated jointly. For example, starting out from the general equilibrium theory as the core of economic theory, all quantities and prices in a market are simultaneously determined. This implies that, apart from the starting conditions, everything depends on everything, i.e. there are only endogenous variables. For example, if we consider a single market, supply and demand functions simultaneously determine the equilibrium quantity and price.

In such a system where each variable depends on all the other ones, the structural form of an econometric model is no longer identifiable. We need additional information to identify it. In traditional econometrics, it is usually assumed that such information is available. One might, for example, plausibly assume that some variables are not included in some equations. In a market for agricultural products, for example, there should be no (direct) impact of consumer income on the supply nor of the weather on the demand of such products.

However, CHRISTOPHER A. SIMS (1980) exemplified that such exclusion restrictions are no longer justified as soon as we assume that individuals have rational expectations. For example, the world market prices of coffee largely depend on the Brazilian production, which is put on the market in autumn. If a hard frost in spring destroys a significant part of the Brazilian coffee harvest, supply will be smaller in autumn. This should lead to higher prices. At first glance, this should have no impact on the demand function. However, if American consumers know about the frost, they might try to buy additional (still cheap) coffee in order to stock up. Thus, the weather in Brazil becomes a determinant of the coffee demand in the United States; a variable which was thought to be excludable from the demand function is now included. According to CHRISTOPHER A. SIMS, nearly all exclusion restrictions are incredible.

He developed the approach of *Vector Autoregressive Systems* (VAR) as an alternative to the traditional simultaneous equations system approach. Starting from the autoregressive representation of weakly stationary processes, all included variables are assumed to be jointly endogenous. Thus, in a VAR of order p (VAR(p)), each component of the vector X depends linearly on its own lagged values up to p periods as well as on the lagged values of all other variables up to order p. With the concept of the VAR a method is proposed which allows to identify and interpret economic shocks and to assess their influence on macroeconomic variables.

Starting point is the reduced form of a dynamic econometric model. With such a model we can find out, for example, whether specific Granger causal relations exist in this system. In doing so, we follow a top-down approach based on an econom(etr)ic philosophy contrary to the statistical bottom-up approach of CLIVE W.J. GRANGER. However, it has to be mentioned that the number of variables that can jointly be analysed in such a system is quite small; at least in the usual econometric applications, this is limited by the number of observations which are available. Nevertheless, vector autoregressive systems play a crucial role in modern approaches to analyse economic time series. This holds, for example, for the *LSE-Approach* which was originally developed by J. DENIS SARGAN (1964) at the London School of Economics (LSE) and today is most prominently represented by DAVID F. HENDRY.

This chapter will show the conclusions about the relation between stationary time series that can be drawn from such a system. Essentially, we ask how new information that appears at a certain point in time in one variable is processed in the system and which impact it has over time not only for this particular variable but also for the other variables of the system. In this context, we will introduce two new instruments: the impulse response function and the variance decomposition. The latter depends on the possibility shown in *Section 2.4* that the variance of a weakly stationary variable can be reconstructed as the variance of the forecast error if the prediction horizon goes to infinity.

In the following, the autoregressive and the moving average representations of the system as well as its error correction representation are presented (*Section 4.1*). Furthermore, we will see how forecasts can be generated in such a system. *Section 4.2* asks for possible Granger causal relations between sub-vectors in this system. *Section 4.3* presents the impulse response analysis and *Section 4.4* the variance decomposition. We close with some remarks on the status of the economic theory in such a system (*Section 4.5*).

4.1 Representation of the System

We start with the k-dimensional stochastic process X. The reduced form of the general linear dynamic model of this process, a vector autoregression of order p, VAR(p), can be described as

(4.1) $\quad X_t = \delta + A_1 X_{t-1} + A_2 X_{t-2} + \ldots + A_p X_{t-p} + U_t$.

The A_i, $i = 1, \ldots, p$, are k-dimensional quadratic matrices, and U represents the k-dimensional vector of residuals at time t. The vector of constant terms is denoted as δ. This system can compactly be written as

(4.1') $\quad A(L) X_t = \delta + U_t$,

with

$$A(L) = I_k - A_1 L - A_2 L^2 - \ldots - A_p L^p,$$
$$E[U_t] = 0, \quad E[U_t U_t'] = \Sigma_{uu}, \quad E[U_t U_s'] = 0 \text{ for } t \neq s.$$

The residuals U might be contemporaneously correlated which indicates instantaneous relations between the endogenous variables in relation (4.1).

This system is stable if and only if all roots of the characteristic equation of the lag polynomial are outside the unit circle, i.e.

(4.2) $\quad \det(I_k - A_1 z - A_2 z^2 - \ldots - A_p z^p) \neq 0 \text{ for } |z| \leq 1$.

Under this condition, system (4.1') has the MA representation

(4.3) $\quad X_t = A^{-1}(L) \delta + A^{-1}(L) U_t$
$\quad\quad\quad = \mu + U_t - B_1 U_{t-1} - B_2 U_{t-2} - B_3 U_{t-3} - \ldots$
$\quad\quad\quad = \mu + B(L) U_t, \quad B_0 = I_k$,

with

$$B(L) := I_k - \sum_{j=1}^{\infty} B_j L^j \equiv A^{-1}(L), \quad \mu = A^{-1}(1) \delta = B(1) \delta.$$

The *autocovariance matrices* are defined as:

(4.4) $\quad \Gamma_X(\tau) = E[(X_t - \mu)(X_{t-\tau} - \mu)']$.

Without loss of generality, we set $\delta = 0$ and, therefore, $\mu = 0$. Due to (4.1), it holds that

$$E[X_t X_{t-\tau}'] = A_1 E[X_{t-1} X_{t-\tau}'] + A_2 E[X_{t-2} X_{t-\tau}'] + \ldots$$
$$+ A_p E[X_{t-p} X_{t-\tau}'] + E[U_t X_{t-\tau}'].$$

This leads to the equations determining the autocovariance matrices for $\tau \geq 0$:

(4.5a) $\quad \Gamma_X(\tau) = A_1 \Gamma_X(\tau-1) + A_2 \Gamma_X(\tau-2) + \ldots + A_p \Gamma_X(\tau-p),$

(4.5b) $\quad \Gamma_X(0) = A_1 \Gamma_X(-1) + A_2 \Gamma_X(-2) + \ldots + A_p \Gamma_X(-p) + \Sigma_{uu}$

$\quad\quad\quad\quad = A_1 \Gamma_X(1)' + A_2 \Gamma_X(2)' + \ldots + A_p \Gamma_X(p)' + \Sigma_{uu}.$

The last equation is due to the fact that $\gamma_{ij}(\tau) = \gamma_{ji}(-\tau)$ holds for the ij-element of $\Gamma_X(\tau)$, $\gamma_{ij}(\tau)$. Thus, $\Gamma_X(\tau) = \Gamma_X(-\tau)'$.

The individual correlation coefficients are defined as

$$\rho_{ij}(\tau) = \frac{\gamma_{ij}(\tau)}{\sqrt{\gamma_{ii}(0) \cdot \gamma_{jj}(0)}}, \quad i, j = 1, 2, \ldots, k.$$

Thus, we get the autocorrelation matrices as

(4.6) $\quad\quad\quad\quad\quad R_X(\tau) = D^{-1} \Gamma_X(\tau) D^{-1}$

with

$$D^{-1} = \begin{bmatrix} 1/\sqrt{\gamma_{11}(0)} & 0 & \cdots & 0 \\ 0 & 1/\sqrt{\gamma_{22}(0)} & \cdots & 0 \\ \vdots & \vdots & \ddots & \vdots \\ 0 & 0 & \cdots & 1/\sqrt{\gamma_{kk}(0)} \end{bmatrix}.$$

Example 4.1

Let the following VAR(1) model be given:

$$\begin{bmatrix} x_{1,t} \\ x_{2,t} \end{bmatrix} = \begin{bmatrix} 0.6 & -0.3 \\ -0.3 & 0.6 \end{bmatrix} \begin{bmatrix} x_{1,t-1} \\ x_{2,t-1} \end{bmatrix} + \begin{bmatrix} u_{1,t} \\ u_{2,t} \end{bmatrix}$$

with

$$\Sigma_{uu} = \begin{bmatrix} 1.00 & 0.70 \\ 0.70 & 1.49 \end{bmatrix},$$

or, in the compact representation

(E4.1) $\quad\quad\quad\quad (I_2 - A_1 L) X_t = U_t.$

To check whether the system is stable, the roots of $|I_2 - A_1 z| = 0$ have to be calculated according to (4.2), i.e. we have to solve the system

$$\begin{vmatrix} 1-0.6z & 0.3z \\ 0.3z & 1-0.6z \end{vmatrix} = 0.$$

This results in

$$z_1 = \frac{10}{9}, \quad z_2 = \frac{10}{3},$$

which both are larger than one in modulus. Thus, the system is stable. The MA representation of (E4.1) is given as

$$X_t = (I_2 - A_1 L)^{-1} U_t = (I_2 + A_1 L + A_1^2 L^2 + ...) U_t,$$

or, explicitly written as,

$$\begin{bmatrix} x_{1,t} \\ x_{2,t} \end{bmatrix} = \begin{bmatrix} u_{1,t} \\ u_{2,t} \end{bmatrix} + \begin{bmatrix} 0.6 & -0.3 \\ -0.3 & 0.6 \end{bmatrix} \begin{bmatrix} u_{1,t-1} \\ u_{2,t-1} \end{bmatrix} + \begin{bmatrix} 0.45 & -0.36 \\ -0.36 & 0.45 \end{bmatrix} \begin{bmatrix} u_{1,t-2} \\ u_{2,t-2} \end{bmatrix}$$

$$+ \begin{bmatrix} 0.378 & -0.351 \\ -0.351 & 0.378 \end{bmatrix} \begin{bmatrix} u_{1,t-3} \\ u_{2,t-3} \end{bmatrix} + \quad ... \; .$$

For the variance-covariance matrix we get, because of (4.5),

$$\Gamma_x(0) = A_1 \Gamma_x(1)' + \Sigma_{uu},$$
$$\Gamma_x(1) = A_1 \Gamma_x(0).$$

This leads to

(E4.2) $\qquad \Gamma_x(0) = A_1 \Gamma_x(0) A_1' + \Sigma_{uu}.$

To get the variances $\gamma_{11}(0)$ and $\gamma_{22}(0)$ for x_1 and x_2 as well as their covariance $\gamma_{12}(0)$, we have to solve the following linear equation system because of (E4.2):

$$0.64 \, \gamma_{11}(0) + 0.36 \, \gamma_{12}(0) - 0.09 \, \gamma_{22}(0) = 1.00$$
$$0.18 \, \gamma_{11}(0) + 0.55 \, \gamma_{12}(0) + 0.18 \, \gamma_{22}(0) = 0.70$$
$$-0.09 \, \gamma_{11}(0) + 0.36 \, \gamma_{12}(0) + 0.64 \, \gamma_{22}(0) = 1.49 \,.$$

This leads to

$$\gamma_{11}(0) = 2.17, \quad \gamma_{12}(0) = -0.37, \quad \gamma_{22}(0) = 2.84.$$

Thus, the instantaneous correlation between x_1 and x_2 is -0.15.

VAR(p) models are often used for forecasting. According to the considerations in *Section 2.4* that the best linear unbiased predictor (BLUP) is given by the conditional expectation, the following holds for the autoregressive representation (4.1):

(4.7) $\hat{X}_t(1) = E_t[X_{t+1}]$
$= \delta + A_1 X_t + A_2 X_{t-1} + \ldots + A_p X_{t-p+1}$
$\hat{X}_t(2) = \delta + A_1 \hat{X}_t(1) + A_2 X_t + A_3 X_{t-1} + \ldots + A_p X_{t-p+2}$.

Alternatively, we get

(4.8) $\hat{X}_t(1) = \mu - B_1 U_t - B_2 U_{t-1} - B_3 U_{t-2} - \ldots$

for the MA representation (4.3).

While the autoregressive representation is mainly relevant to generate forecasts, the MA representation is used for calculating the corresponding forecast errors as well as for additional methods to analyse the dynamic properties of the system.

As an alternative to the AR and MA representations (4.1') and (4.3), there is an *error correction representation* for every stationary VAR of order p:

(4.9) $A^*_{p-1}(L) \Delta X_t = \delta - A(1) X_{t-1} + U_t$,

with

$$A^*_{p-1}(L) = I_k - A^*_1 L - \ldots - A^*_{p-1} L^{p-1}$$

and

$$A^*_i = -\sum_{j=i+1}^{p} A_j, \quad i = 1, 2, \ldots, p-1.$$

As the vectors ΔX_{t-i}, $i = 1, \ldots, p-1$, together with X_{t-1}, generate the same vector space as the vectors X_{t-i}, $i = 1, \ldots, p$, the (finite order) autoregressive representation and the error correction representation are observationally equivalent. The advantage of the latter is that $A(1)$, the matrix of the long-run equilibrium relations, can be estimated directly in the framework of a linear model.

Example 4.2

We start with the general dynamic model of a single equation, but (for reasons of simplicity) we consider only one explanatory variable which is assumed to be exogenous:

(E4.3) $\alpha_p(L) y_t = \delta + \beta_q(L) x_t + u_t$.

In the long-run equilibrium it holds that

$$y_t = y_{t-1} = \ldots = y_{t-p} = \ldots = \bar{y},$$

$$x_t = x_{t-1} = \ldots = x_{t-q} = \ldots = \bar{x},$$

$$u_t = 0.$$

From this we get for the long-run equilibrium:

(E4.4) $$\alpha_p(1)\bar{y} = \delta + \beta_q(1)\bar{x},$$

$$\bar{y} = \frac{\delta}{\alpha_p(1)} + \frac{\beta_q(1)}{\alpha_p(1)}\bar{x}$$

$$= \mu + \beta\bar{x}$$

with

$$\mu = \delta/\alpha_p(1), \quad \beta = \beta_q(1)/\alpha_p(1).$$

According to (4.9), if y and x are weakly stationary (or, as discussed in *Chapter 6*, nonstationary but cointegrated), the following representation of the general dynamic linear model is an alternative to (E4.3). Here, the short- and long-run effects are separated and can be directly estimated:

(E4.5) $$\alpha^*_{p-1}(L)(1-L)y_t = \delta + \beta^*_{q-1}(L)(1-L)x_t - \gamma_0 y_{t-1} + \gamma_1 x_{t-1} + u_t$$

with

$$\alpha^*_{p-1}(L) = 1 - \alpha^*_1 L - \ldots - \alpha^*_{p-1} L^{p-1},$$

$$\alpha^*_i = -\sum_{j=i+1}^{p} \alpha_j, \quad i = 1, 2, \ldots, p-1,$$

$$\beta^*_{q-1}(L) = \beta^*_0 - \beta^*_1 L - \ldots - \beta^*_{q-1} L^{q-1},$$

$$\beta^*_i = -\sum_{j=i+1}^{q} \beta_j, \quad j = 1, 2, \ldots, q-1, \quad \beta^*_0 = \beta_0.$$

$$\gamma_0 = \alpha_p(1), \quad \gamma_1 = \beta_q(1).$$

In equilibrium $\Delta y_t = \Delta x_t = 0$ and $u_t = 0$ hold and, therefore, $y_t = \bar{y}$ as well as $x_t = \bar{x}$ for all t. From this it follows that

$$-\gamma_0 \bar{y} + \delta + \gamma_1 \bar{x} = 0$$

or

$$-\alpha_p(1)\bar{y} + \delta + \beta_q(1)\bar{x} = 0,$$

and again we get (E4.4) as representation of the long-run equilibrium.

Example 4.3

We consider the relation between the German (GER) and the Swiss (SER) three month money market rates. We use monthly data for the period from January 1975 to November 1998. Preliminary Granger causality tests (the results of which are not given here) have indicated that, along with an instantaneous relation, there is a simple causal relation from German to Swiss interest rates: The null hypothesis that there is no simple relation in the reverse direction can neither be rejected by using first differences nor by using levels at any conventional significance level. Assuming that the instantaneous causation runs from German to Swiss interest rates, using levels we get the following equation for the Swiss rates:

$$SER_t = \underset{(-1.60)}{-0.121} + \underset{(9.10)}{0.717\,GER_t} + \underset{(18.68)}{0.994\,SER_{t-1}} - \underset{(-1.57)}{0.080\,SER_{t-2}}$$

$$- \underset{(-7.66)}{0.636\,GER_{t-1}} + \hat{u}_t,$$

$\bar{R}^2 = 0.965$, SE $= 0.466$, $Q(10) = 8.810$ (p $= 0.550$).

(The numbers in parentheses are again the estimated t statistics). If we estimate the error correction representation directly, we get the following result:

$$\Delta SER_t = \underset{(-1.60)}{-0.121} + \underset{(9.10)}{0.717\,\Delta GER_t} + \underset{(1.57)}{0.080\,\Delta SER_{t-1}} - \underset{(-4.00)}{0.086\,SER_{t-1}}$$

$$+ \underset{(3.66)}{0.081\,GER_{t-1}} + \hat{u}_t.$$

Both relations are observationally equivalent. Aside from the multiple correlation coefficient, all test statistics for the equation as well as the residual error variance take the same values. On the other hand, as the variance of the dependent variable is reduced by taking first differences, the $\bar{R}^2$ necessarily decreases; its value is now 0.286.

Moreover, the linear estimate of the error correction model is equivalent to the following non-linear estimation:

$$\Delta SER_t = \underset{(9.10)}{0.717\,\Delta GER_t} + \underset{(1.57)}{0.080\,\Delta SER_{t-1}} - \underset{(-4.00)}{0.086\,(SER_{t-1}} + \underset{(1.63)}{1.419}$$

$$- \underset{(-6.98)}{0.946\,GER_{t-1}}) + \hat{u}_t.$$

The estimate shows that during this period Swiss short-run interest rates developed parallelly with the German rates, but on a level lower by about 1.5 percentage points, i.e. the so-called 'Swiss interest rate bonus' was about 1.5 percentage points. As the estimated coefficient of GER_{t-1} is not significantly different from one, this relation is consistent with a relative version of uncovered interest parity.

Relation (4.1), the starting point of the entire approach, is the reduced form of a dynamic linear econometric system where each equation includes the same explanatory variables. Therefore, the different equations of this system can be estimated using OLS. This leads to consistent estimates of the slope coefficients with the same properties as a generalised least squares estimator. However, if there are zero restrictions, the individual equations of the system are considered as *seemingly unrelated* and are therefore simultaneously estimated as a system. Here, the SUR method is applied to get efficient estimates. (Further details of this method are given in *Section 7.1.2*)

To estimate the system, the order p, i.e. the maximal lag of the system, has to be determined. As (4.1) shows, the same maximal lag is used for all variables. In order to fix p, the information criteria described in *Section 2.1.5* can be used again. HELMUT LÜTKEPOHL (2005, pp. 146ff.), for example, showed that in the multivariate case with k variables, T observations, a constant term and a maximal lag of p, these criteria are as follows:

(i) The final prediction error (FPE):

(4.10a) $$FPE(p) = \left[\frac{T + kp + 1}{T - kp - 1}\right]^k |\Sigma_{\hat{u}\hat{u}}(p)|.$$

(ii) The Akaike criterion (AIC):

(4.10b) $$AIC(p) = \ln|\Sigma_{\hat{u}\hat{u}}(p)| + (k + pk^2)\frac{2}{T}.$$

(iii) The Hannan-Quinn criterion (HQ):

(4.10c) $$HQ(p) = \ln|\Sigma_{\hat{u}\hat{u}}(p)| + (k + pk^2)\frac{2\ln(\ln(T))}{T}.$$

(iv) The Schwarz criterion (SC):

(4.10d) $$SC(p) = \ln|\Sigma_{\hat{u}\hat{u}}(p)| + (k + pk^2)\frac{\ln(T)}{T}.$$

$|\Sigma_{\hat{u}\hat{u}}(p)|$ is the determinant of the variance-covariance matrix of the estimated residuals. Again it holds that the Hannan-Quinn criterion as well as the Schwarz criterion consistently determine the (finite) order of the true maximal lag, while the final prediction error and the Akaike criterion tend to overestimate it. This is also reflected in the following relations which, because of the different penalty terms, hold for these criteria:

(i) $\qquad \hat{p}(SC) \leq \hat{p}(HQ)$,

(ii) $\qquad \hat{p}(SC) \leq \hat{p}(AIC)$ for $T \geq 8$,

(iii) $\qquad \hat{p}(HQ) \leq \hat{p}(AIC)$ for $T \geq 16$,

where $\hat{p}(\cdot)$ gives the estimated lag length according to the respective criterion.

Example 4.4

We use the same quarterly data and the same period from 1965 to 1989 as in *Examples 3.1* to *3.5*: the annual growth rate of real GDP ($\Delta_4\ln(GDP_r)$), the annual growth rate of the real quantity of money M1 ($\Delta_4\ln(M1_r)$), and the interest rate differential (GLR − GSR). Considering the whole system, we get the following values for the information criteria:

$p = 1$: AIC = 10.003, HQ = 9.779, SC = 9.315,

$p = 2$: AIC = 9.992, HQ = 9.603, SC = 10.539

$p = 3$: AIC = 10.028, HQ = 10.342, SC = 10.807,

$p = 4$: AIC = 9.991, HQ = 10.402, SC = 11.007.

Thus, according to the Akaike criterion, we get an optimal lag length of four periods while the Hannan-Quinn criterion suggests an optimal lag length of two and the Schwarz criterion an optimal lag length of one. Accepting the Hannan-Quinn criterion leads to the following estimates:

$$\begin{bmatrix} \Delta_4 \ln(GDP_{r,t}) \\ \Delta_4 \ln(M1_{r,t}) \\ (GLR - GSR)_t \end{bmatrix} = \begin{bmatrix} 0.142 \\ 1.094 \\ 0.510 \end{bmatrix} + \begin{bmatrix} 0.611 & 0.078 & -0.133 \\ -0.183 & 0.761 & 0.981 \\ -0.015 & 0.036 & 0.995 \end{bmatrix} \begin{bmatrix} \Delta_4 \ln(GDP_{r,t-1}) \\ \Delta_4 \ln(M1_{r,t-1}) \\ (GLR - GSR)_{t-1} \end{bmatrix}$$

$$+ \begin{bmatrix} 0.096 & 0.091 & 0.205 \\ -0.024 & -0.108 & -0.438 \\ -0.077 & -0.070 & -0.128 \end{bmatrix} \begin{bmatrix} \Delta_4 \ln(GDP_{r,t-2}) \\ \Delta_4 \ln(M1_{r,t-2}) \\ (GLR - GSR)_{t-2} \end{bmatrix} + \begin{bmatrix} \hat{u}_{1,t} \\ \hat{u}_{2,t} \\ \hat{u}_{3,t} \end{bmatrix}.$$

For the individual equations we get the following test statistics:

(i) $\quad \Delta_4\ln(GDP_r)$: SE = 1.327, AIC = 3.472, HQ = 3.545, SC = 3.654,
$\qquad\qquad\qquad$ Q(10) = 16.406 (p = 0.089),

(ii) $\quad \Delta_4\ln(M1_r)$: SE = 1.905, AIC = 4.194, HQ = 4.268, SC = 4.376,
$\qquad\qquad\qquad$ Q(10) = 20.024 (p = 0.029),

(iii)   GLR – GSR: SE = 0.786, AIC = 2.423, HQ = 2.496, SC = 2.605,
         Q(10) = 17.296 (p = 0.068).

Between the residuals the following correlations exist:

$$\hat{\rho}_{12} = 0.102, \quad \hat{\rho}_{13} = 0.045, \quad \hat{\rho}_{23} = 0.286.$$

Again, we see the instantaneous relation between the growth rate of real M1 and the interest rate differential.

The values of the Ljung-Box Q statistic indicate that the residuals of all three equations are still autocorrelated. Thus, the dynamics of the system is not fully captured. However, when specifying vector autoregressive models, in order to guarantee consistency of the estimates, it is important that the residuals are really white noise. Following the Akaike criterion and estimating a VAR(4) model, we get the following values for the test statistics of the different equations:

(i) $\Delta_4 \ln(GDP_r)$: SE = 1.333, AIC = 3.533, HQ = 3.670, SC = 3.872,
        Q(8) = 9.340 (p = 0.314),

(ii) $\Delta_4 \ln(M1_r)$: SE = 1.763, AIC = 4.092, HQ = 4.229, SC = 4.431,
        Q(8) = 11.390 (p = 0.181),

(iii)   GLR – GSR: SE = 0.777, AIC = 2.454, HQ = 2.592, SC = 2.793,
        Q(8) = 9.661 (p = 0.290).

For the instantaneous correlations we get:

$$\hat{\rho}_{12} = 0.082, \quad \hat{\rho}_{13} = 0.053, \quad \hat{\rho}_{23} = 0.279.$$

The values of these criteria change considerably. The standard error of regression slightly improves in the M1 equation and hardly changes in the other equations. The Akaike criterion also improves in the M1 equation, but deteriorates slightly in the other equation. The Hannan-Quinn criterion slightly improves in the M1 but deteriorates in the other two equations, while the Schwarz criterion always deteriorates. On the other hand, the values of the Ljung-Box Q statistic improve considerably in all three equations; now the null hypothesis that there is no autocorrelation left in the residuals can never be rejected. The lowest p value is 0.18. Thus we will use the VAR(4) model for all further calculations in this chapter.

Contrary to the *parsimony principle* applied in the univariate analysis, the VAR(p) models are over-parameterised systems. The individual parameters can hardly be interpreted meaningfully. For this reason, other methods, like Granger causality tests, impulse response analyses and variance decompositions, are employed. These methods are presented in the following.

4.2 Granger Causality

Now we will consider the Granger causal relations between the two subvectors X_1 and X_2 of the vector X. X_1 has the dimension k_1 and X_2 the dimension k_2 with $k_1 + k_2 = k$. For the MA representation we get

$$(4.11a) \quad X_t = \begin{bmatrix} X_{1,t} \\ X_{2,t} \end{bmatrix} = \begin{bmatrix} \mu_1 \\ \mu_2 \end{bmatrix} + \begin{bmatrix} B_{11}(L) & B_{12}(L) \\ B_{21}(L) & B_{22}(L) \end{bmatrix} \begin{bmatrix} U_{1,t} \\ U_{2,t} \end{bmatrix}.$$

The corresponding AR representation is:

$$(4.11b) \quad \begin{bmatrix} A_{11}(L) & A_{12}(L) \\ A_{21}(L) & A_{22}(L) \end{bmatrix} \begin{bmatrix} X_{1,t} \\ X_{2,t} \end{bmatrix} = \begin{bmatrix} \delta_1 \\ \delta_2 \end{bmatrix} + \begin{bmatrix} U_{1,t} \\ U_{2,t} \end{bmatrix}.$$

Irrespective of instantaneous causality; the following is true for (4.11):

(i) X_2 is not Granger causal to X_1 if and only if $B_{12}(L) \equiv 0$. Analogous to Section 3.2.1 it holds that $B_{12}(L) \equiv 0$ is equivalent to $A_{12}(L) \equiv 0$. Thus, it also holds that X_2 is not Granger causal to X_1 if and only if $A_{12}(L) \equiv 0$ in the corresponding AR representation.

(ii) X_1 is not Granger causal to X_2 if and only if $B_{21}(L) \equiv 0$. Analogous to Section 3.2.1 it holds that $B_{21}(L) \equiv 0$ is equivalent to $A_{21}(L) \equiv 0$. Thus, it also holds that X_1 is not Granger causal to X_2 if and only if $A_{21}(L) \equiv 0$ in the corresponding AR representation.

As in the bivariate case, instantaneous relations involve some complications. The variance-covariance matrix of the system (4.1) can be decomposed into:

$$(4.12) \quad \Sigma_{uu} = P P',$$

where P is a regular lower triangular matrix. Such a (Choleski) decomposition exists for each regular variance-covariance matrix. Using this triangular matrix, the MA representation (4.3) can be transformed in the following way:

$$(4.13) \quad X_t = \mu + U_t - \sum_{j=1}^{\infty} \left(B_j U_{t-j} \right)$$

$$X_t = \mu + P P^{-1} U_t - \sum_{j=1}^{\infty} \left(B_j P P^{-1} U_{t-j} \right)$$

$$= \mu + P W_t - \sum_{j=1}^{\infty}\left(\Theta_j W_{t-j}\right)$$

$$= \mu + \Theta(L) W_t,$$

with

$$\Theta_j = B_j P, \quad \Theta_0 = P, \quad W_t = P^{-1} U_t,$$

$$\Sigma_{ww} = P^{-1} \Sigma_{uu} P^{-1\prime} = P^{-1} P P' P^{-1\prime}$$

$$= I_k.$$

Thus, the following decomposition exists for the subvectors:

$$X_t = \begin{bmatrix} X_{1,t} \\ X_{2,t} \end{bmatrix} = \begin{bmatrix} \mu_1 \\ \mu_2 \end{bmatrix} + \begin{bmatrix} \Theta_{11}^0 & 0 \\ \Theta_{21}^0 & \Theta_{22}^0 \end{bmatrix} \begin{bmatrix} W_{1,t} \\ W_{2,t} \end{bmatrix}$$

$$- \begin{bmatrix} \Theta_{11}^1 & \Theta_{12}^1 \\ \Theta_{21}^1 & \Theta_{22}^1 \end{bmatrix} \begin{bmatrix} W_{1,t-1} \\ W_{2,t-1} \end{bmatrix} - \ldots .$$

W is a vector of innovations whose elements – contrary to the elements of U – are also instantaneously uncorrelated. Moreover, the variances of these elements are 1.

The transformation with matrix P implies an ordering of the variables; causal directions are assumed for the instantaneous relations. The variable x_i has an impact on the variable x_j, $j > i$, while the instantaneous relation in the reverse direction is excluded. In terms of traditional econometrics, this implies that the model is exactly identified and, correspondingly, the parameters of the structural form can be consistently estimated using OLS. This method to identify the model is one possibility to proceed from the reduced to the structural form of a simultaneous system of equations and to give the innovations an economic interpretation. This structural form is called *structural VAR*. Due to the exact identification, the residuals of the different equations are not cross-correlated with each other.

The following holds for this system: There is no instantaneous causality between X_1 and X_2 if and only if $\Theta_{21}^0 = 0$. In this situation Σ_{uu} is block diagonal, i.e. it holds that

$$E[U_{1,t} U_{2,t}'] = 0.$$

The fact that X_2 is not causal to X_1 and that there is no instantaneous causality leads to

$$\Theta_{21}^0 = 0 \wedge \Theta_{12}^1 = \Theta_{12}^2 = \ldots = 0.$$

Such results depend, of course, on the sequence of the different variables, i.e. on the kind of causal order assumed for the instantaneous relations.

Example 4.5

If we divide the three variables of the vector of *Example 4.4* in the following way:

$$X_1 = [\Delta_4 \ln(GDP_r)], \quad X_2' = [\Delta_4 \ln(M1_r) \ \ GLR - GSR],$$

we get the following results by using Wald tests:

(i) $H_0: \neg (X_2 \rightarrow X_1): \hat{\chi}^2 = 24.597 \ (p = 0.002),$

(ii) $H_0: \neg (X_1 \rightarrow X_2): \hat{\chi}^2 = 24.115 \ (p = 0.000),$

(iii) $H_0: \neg (X_1 - X_2): \hat{\chi}^2 = 0.658 \ (p = 0.720).$

Thus, there is feedback but no instantaneous relation between the sub-vectors. To test for an instantaneous relation, we included the current values of the growth rate of real money and the interest rate differential in the equation for real GDP growth.

4.3 Impulse Response Analysis

In the following, we show how, at a specific point of time t_0, an impulse that originates from one variable proceeds through the system: How does a change in the residuals u_{i,t_0} or in the innovations w_{i,t_0}, $i = 1, ..., k$, influence the components of the vector X? In system (4.3), the use of the multivariate Wold representation instead of the MA representation

$$X_t = \mu + \Psi_0 U_t + \Psi_1 U_{t-1} + \Psi_2 U_{t-2} + \Psi_3 U_{t-3} + ... ,$$

$$\Psi_0 = I, \quad \Psi_i = -B_i, \quad i = 1, 2, ...,$$

with ψ_{ji}^τ, $\tau = 0, 1, 2, ...$, results in the so-called impulse response sequences. They measure the effect of a unit impulse, i.e. of a shock with the size of one standard deviation of the error term u_i of the variable i at time t_0 on the variable j in later periods. As U_t are the residuals of the reduced form, they are in general cross-correlated and, therefore, have no direct economic interpretation. Thus, it makes sense not to investigate shocks with respect to the residuals U but with respect to the innovations W which are not cross-correlated. Because of the Choleski decomposition of the variance-covariance matrix of the residuals U, in accordance with the considerations in *Section 4.2*, the innovations can be calculated as

4.3 Impulse Response Analysis

$$W_t = P^{-1} U_t,$$

with a lower triangular matrix P. Due to (4.13), the MA representation of X can – analogously to the Wold decomposition – be written as

(4.14) $\quad X_t = \mu + \Phi_0 W_t + \Phi_1 W_{t-1} + \Phi_2 W_{t-2} + \ldots$

with $\Phi_0 = P$ and $\Phi_i = -B_i P = \Psi_i P$, $i = 1, 2, \ldots$. Here, ϕ_{ji}^0 are *impact multipliers* that measure the immediate impact of a unit shock in variable i on variable j. The lagged effects are described by the k^2 impulse response sequences ϕ_{ji}^τ, $i,j = 1, \ldots, k$, $\tau = 1, 2, \ldots,$; they show how each of the k variables are influenced by each of the k innovations. The reaction of the vector X at time t_0+m on the innovations at time t_0 leads to

$$\Psi_m U_{t_0} = \Phi_m W_{t_0},$$

or, if we only consider non cross-correlated unit shocks, we get

$$\Phi_m.$$

If we set $m = 1, 2, \ldots$, we can observe (and graphically represent) the time path. If it is a stationary system, the effect expires over time, i.e. the values of the impulse response function (at least asymptotically) approach zero. This implies that after a unique shock the variables return to their mean.

The cumulative impulse response function describes the effects of a permanent shock on the system. The cumulative effects of a unit shock up to period t_0+m are given by

$$\sum_{j=0}^{m} \Phi_j.$$

If, in a stationary system, m tends to infinity, we get

(4.15) $\quad \lim_{m \to \infty} \left[\sum_{j=0}^{m} \Phi_j \right] = \Phi(1) = B(1) P = A(1)^{-1} P$

for the long-run effect.

Example 4.6

Again, we consider the model of *Example 4.1*. To calculate the innovations of this VAR(1) process, it is assumed that x_2 does not have an instantaneous impact on x_1. For the decomposition $\Sigma_{uu} = P P'$, we denote the elements of the lower triangular matrix as:

142 4 Vector Autoregressive Processes

$$P = \begin{bmatrix} p_{11} & 0 \\ p_{21} & p_{22} \end{bmatrix}.$$

Due to (4.12) we get

$$\begin{bmatrix} 1.00 & 0.70 \\ 0.70 & 1.49 \end{bmatrix} = \begin{bmatrix} p_{11}^2 & p_{11} p_{21} \\ p_{11} p_{21} & p_{21}^2 + p_{22}^2 \end{bmatrix}.$$

From this we choose the positive solution $p_{11} = p_{22} = 1$ and $p_{21} = 0.7$. The innovations W can be calculated as

$$\begin{bmatrix} w_{1,t} \\ w_{2,t} \end{bmatrix} = P^{-1} \begin{bmatrix} u_{1,t} \\ u_{2,t} \end{bmatrix} = \begin{bmatrix} 1.0 & 0.0 \\ -0.7 & 1.0 \end{bmatrix} \begin{bmatrix} u_{1,t} \\ u_{2,t} \end{bmatrix},$$

or

$$w_{1,t} = u_{1,t},$$
$$w_{2,t} = u_{2,t} - 0.7\, u_{1,t}.$$

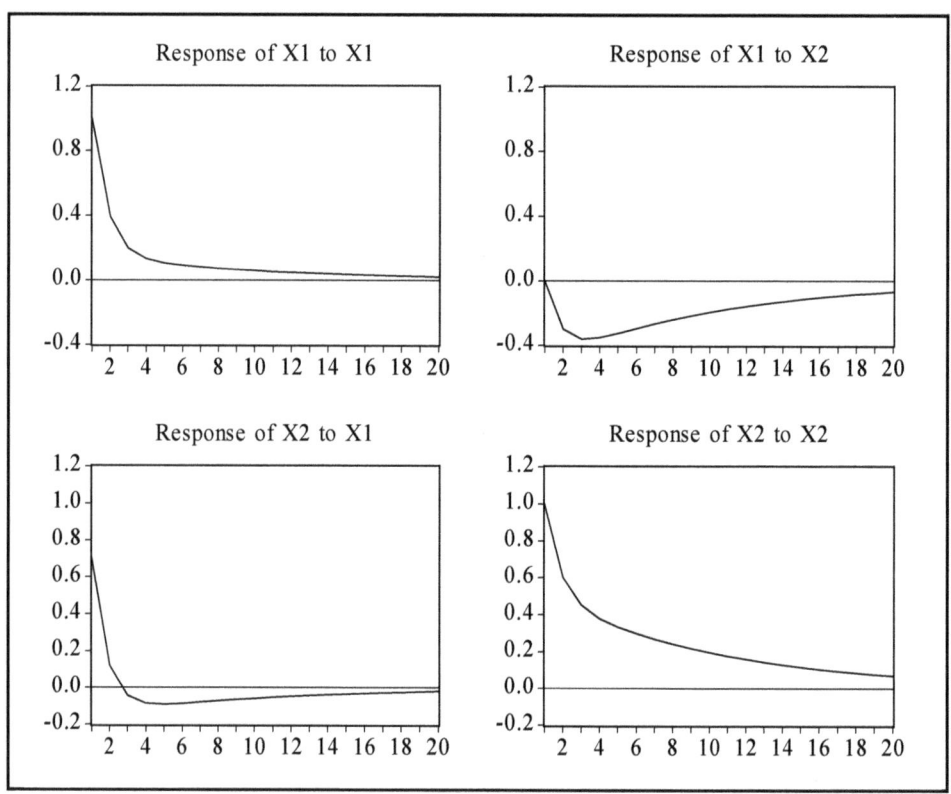

Figure 4.1: Impulse response functions

4.3 Impulse Response Analysis

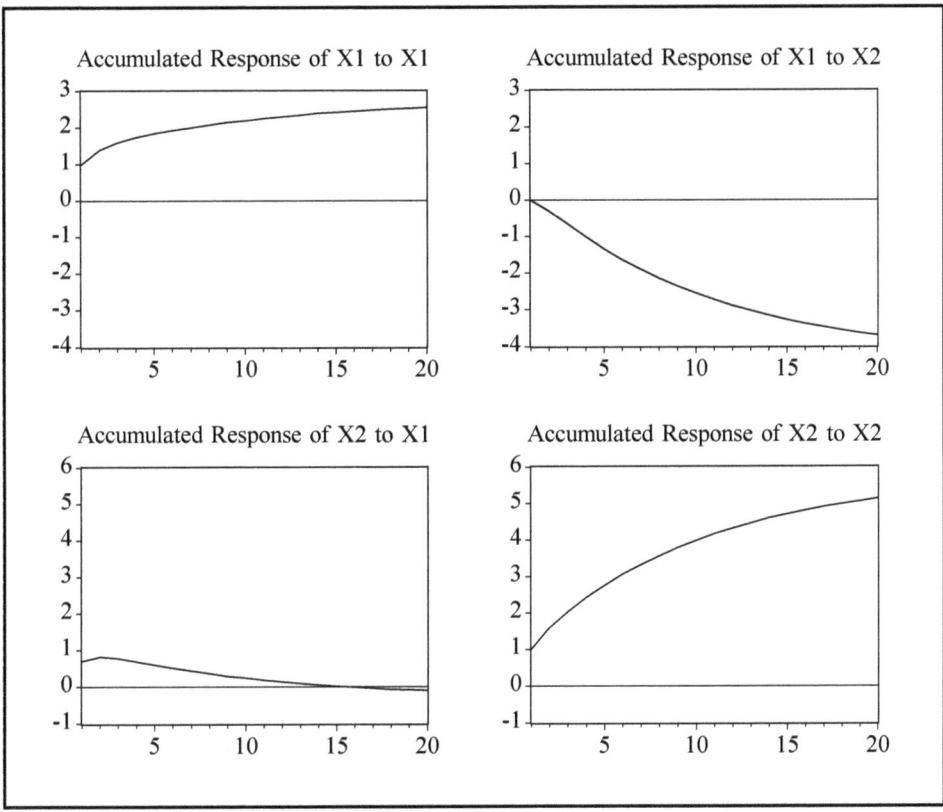

Figure 4.2: Cumulative impulse response functions

For the impulse response analysis, we need representation (4.14). This leads to

$$X_t = A_1 X_{t-1} + U_t,$$

$$= \sum_{i=0}^{\infty} A_1^i U_{t-i}$$

$$= \sum_{i=0}^{\infty} A_1^i P P^{-1} U_{t-i} = \sum_{i=0}^{\infty} \Phi_i W_{t-i},$$

with

$$\Phi_i = A_1^i P, \quad i = 1, 2, \ldots .$$

Thus, we get the following matrices:

$$\Phi_0 = \begin{bmatrix} 1.0 & 0.0 \\ 0.7 & 1.0 \end{bmatrix}, \quad \Phi_1 = \begin{bmatrix} 0.39 & -0.30 \\ 0.12 & 0.60 \end{bmatrix}, \quad \Phi_2 = \begin{bmatrix} 0.20 & -0.36 \\ -0.05 & 0.45 \end{bmatrix},$$

$$\Phi_3 = \begin{bmatrix} 0.13 & -0.35 \\ -0.09 & 0.38 \end{bmatrix}, \ldots$$

The numerical results as well as the graphical representations of the impulse response functions in *Figure 4.1* show that an innovation in x_1 does not have a permanent effect on the system. The impact on the variable itself as well as on x_2 is dying away relatively fast. For the latter, we get a positive impact for the first and second period, and, subsequently, very small negative impacts.

By contrast, a shock in x_2 has a longer lasting impact on the variable itself as well as on x_1. As, in order to identify the system, we assumed that P is a lower triangular matrix, i.e. that x_2 has no instantaneous impact on x_1, the first value in the impulse response function of x_1 on x_2 is zero.

The cumulative impulse response functions in *Figure 4.2* show that after the initial effect of the reaction of x_2 on x_1 the system converges monotonically to its long-run limiting values (multipliers). Because of (4.15) and (E4.1) we get

$$A(1)^{-1} P = (I - A_1)^{-1} P = \begin{bmatrix} 2.714 & -4.286 \\ -0.286 & 5.714 \end{bmatrix}.$$

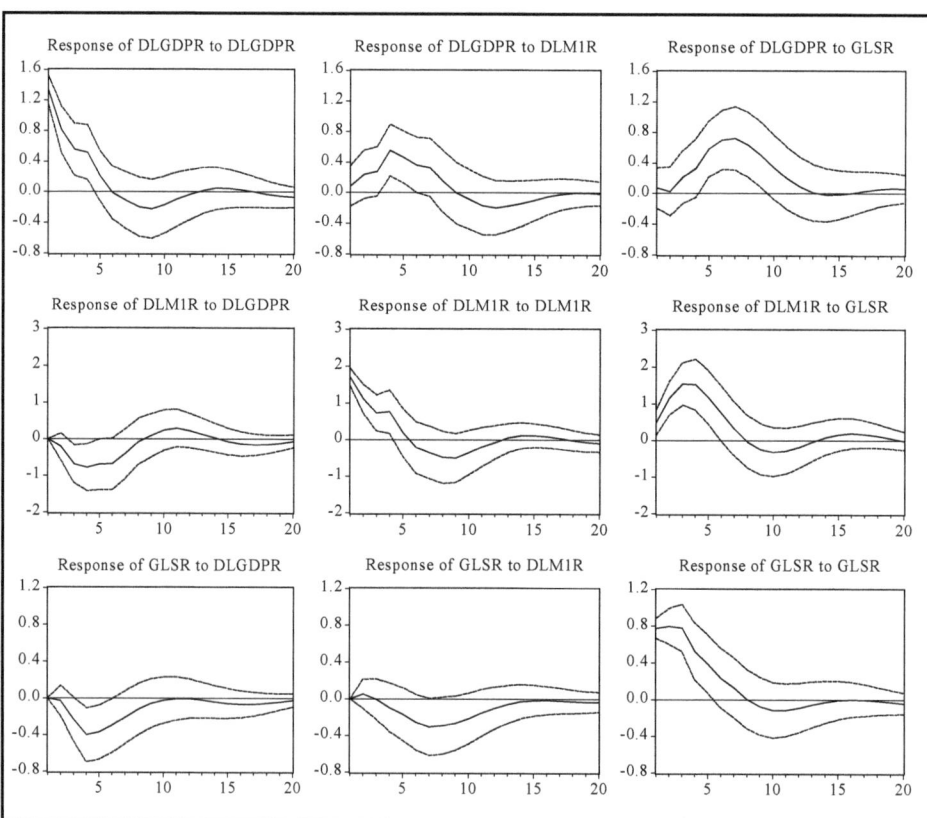

Figure 4.3: Impulse response functions

Example 4.7

For the system given in *Examples 4.4* and *4.5*, ordinary and cumulative impulse response functions are estimated. We assumed for the instantaneous relations that the interest rate differential has an impact on the quantity of money as well as on GDP, while the instantaneous impact of real M1 is restricted to GDP. Thus, we assume the following ordering of the variables: (GLR − GSR) → $\Delta_4\ln(M1_r)$ → $\Delta_4\ln(GDP_r)$.

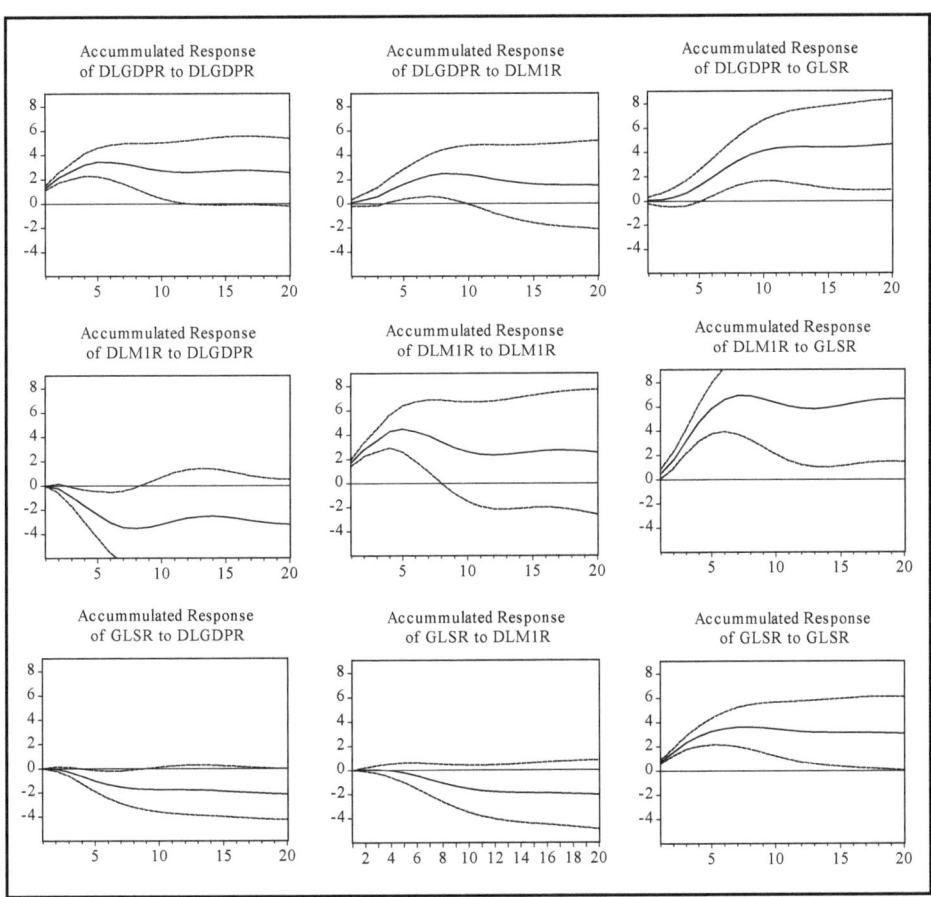

Figure 4.4: Cumulative impulse response functions

The results are presented in *Figures 4.3* and *4.4*. Furthermore, the analytically derived 95 percent confidence intervals are indicated. *Figure 4.3* shows that in the short run, the increase of the interest rate differential has a positive impact on real money as well as – with some delay – on GDP. (GEBHARD KIRCHGÄSSNER and MARCEL R. SAVIOZ (2001) showed that this effect results from the reduction of the short-run interest rate and not from an increase of the long-run interest rate.) Additionally, there is a short-run positive impact of real M1 on GDP. This impact lasts two years at the most; after nine quarters the impulse response function is no longer significantly different from zero. As *Figure 4.4* shows, the only long-run impact is that of the interest rate differential on the two other variables; all other cumulative impulse response functions are no longer significantly different from zero after three years at the latest.

4.4 Variance Decomposition

The following analysis allows to decompose the forecast error variance of a variable into those parts which are generated by the innovations of the different variables in the system. Starting point is the transformed Wold representation (4.14)

$$X_t = \mu + \sum_{j=0}^{\infty}\left(\Phi_j W_{t-j}\right), \quad \Sigma_{ww} = I_k.$$

Taking conditional expectations, we get

$$E_t[X_{t+\tau}] = \mu + \sum_{j=0}^{\tau-1}\left(\Phi_j E_t[W_{t+\tau-j}]\right) + \sum_{j=\tau}^{\infty}\left(\Phi_j W_{t+\tau-j}\right).$$

Due to $E_t[W_{t+s}] = 0$ for $s > 0$ the terms for $j = 0, 1, ..., \tau-1$ can be omitted, while the values for $j \geq \tau$ are already realised and therefore observable. Thus, following the considerations in *Section 2.4*, we get for the optimal forecasts

(4.16) $$\hat{X}_t(\tau) = \mu + \sum_{j=\tau}^{\infty}\left(\Phi_j W_{t+\tau-j}\right).$$

For the data generating process we can write

$$X_{t+\tau} = \mu + \sum_{j=0}^{\infty}\left(\Phi_j W_{t+\tau-j}\right).$$

The forecast error is given by

(4.17) $$F_t(X_{t+\tau}) = X_{t+\tau} - \hat{X}_t(\tau)$$

$$F_t(X_{t+\tau}) = \sum_{j=0}^{\tau-1}\left(\Phi_j W_{t+\tau-j}\right).$$

With an increasing forecast horizon, i.e. for $\tau \to \infty$, the forecast error converges to the stochastic part of the process.

4.4 Variance Decomposition

The forecast error can be decomposed in the following way: For its j-th component, $j \in \{1, ..., k\}$, it holds that

$$x_{j,t+\tau} - \hat{x}_{j,t}(\tau) = \sum_{i=0}^{\tau-1}\left(\phi_{j1}^i\, w_{1,t+\tau-i}\right) + ... + \sum_{i=0}^{\tau-1}\left(\phi_{jk}^i\, w_{k,t+\tau-i}\right)$$

$$= \sum_{m=1}^{k}\left[\sum_{i=0}^{\tau-1}\left(\phi_{jm}^i\, w_{m,t+\tau-i}\right)\right],$$

i.e. we have a summation not only over the different time periods, $i = 0, ..., \tau-1$, but also over the contributions of the different innovations w_m, $m = 1, ..., k$.

As $\Sigma_{ww} = I_k$, i.e. because the individual elements of W are not only white noise and uncorrelated with each other but also have variance of one, it holds for the variance of the components of this forecast error that

$$(4.18) \quad E[(x_{j,t+\tau} - \hat{x}_{j,t}(\tau))^2] = E\left[\left(\sum_{m=1}^{k}\sum_{i=0}^{\tau-1}\phi_{jm}^i\, w_{m,t+\tau-i}\right)^2\right]$$

$$= \sum_{m=1}^{k}\sum_{i=0}^{\tau-1}\left(\phi_{jm}^i\right)^2,$$

i.e. because of $E[w_{m,t+\tau-i}\, w_{r,t+\tau-s}] = 0$ except for $m = r$ and $i = s$, all cross terms are omitted, and because of $E[(w_{m,t+\tau-i})^2] = 1$ only the squares of the coefficients are left.

On the other hand, the variance can be decomposed into those parts that are generated by the impact of the individual innovations w_m, $m = 1, ..., k$, on the variable j when a forecast over τ periods is performed. In this case, we get

$$(4.19) \quad \omega_{jm}^\tau = \frac{\sum_{i=0}^{\tau-1}\left(\phi_{jm}^i\right)^2}{\sum_{s=1}^{k}\sum_{i=0}^{\tau-1}\left(\phi_{js}^i\right)^2}, \quad m = 1, ..., k, \quad \tau = 1, 2, ...$$

for the respective shares.

With an increasing time horizon, i.e. for $\tau \to \infty$, it is not only the variance of the forecast error but also the variance of the variable itself that can be decomposed into those fractions that are generated by the different innovations w_m. As these fractions are, by construction, orthogonal to each other, they add up to one. Thus, the analysis of the forecast errors leads to a decomposition of the variances of the system's variables.

Example 4.8

The variance decomposition of the VAR(1) process described in *Examples 4.1* and *4.6* is presented in *Table 4.1*. Here, the immediate effects in the first period are presented, the effects after 4, 8, and 20 periods as well as the long-run effects. According to the identifying restriction that there is no instantaneous effect from x_2 to x_1; in the first period the variance of x_1 is exclusively generated by its own innovations. The impact of x_2 on x_1 increases monotonically and in the long-run generates about 42 percent of the variance of this variable. Contrary to this, the impact of x_1 on x_2, rather strong with 33 percent in the first period, decreases over time, and in the long-run generates only about 20 percent of the variance of x_2. Thus, 80 percent of the variance of x_2 are generated by its own innovations and only 20 percent by those of x_1, while only 58 percent of the variance of x_1 are generated by its own innovations, but 42 percent by the innovations of x_2.

Table 4.1: Variance Decomposition

Forecast horizon		x_1	x_2
immediate	x_1	100.000	0.000
	x_2	32.834	67.166
4 periods	x_1	77.866	22.134
	x_2	23.089	76.911
8 periods	x_1	65.085	34.915
	x_2	20.957	79.043
20 periods	x_1	58.527	41.473
	x_2	19.838	80.162
infinity	x_1	58.020	41.980
	x_2	19.748	80.252

Example 4.9

The variance decomposition for the vector autoregressive process of *Example 4.4* is given in *Table 4.2a*. First, we again suppose the causal direction (GLR – GSR) $\rightarrow \Delta_4 \ln(M1_r) \rightarrow \Delta_4 \ln(GDP_r)$. We consider the immediate reaction, i.e. the reaction in the same quarter in which the innovation occurs, as well as forecast horizons of one, two, and five years, as well as an infinite forecast horizon in order to capture the decomposition of the total variance.

Table 4.2a: Variance Decomposition
1/65 – 4/89, 100 Observations

Forecast horizon		$\Delta_4\ln(GDP_r)$	$\Delta_4\ln(M1_r)$	GLR – GSR
immediate	$\Delta_4\ln(GDP_r)$	99.231	0.483	0.286
	$\Delta_4\ln(M1_r)$	0.000	92.202	7.798
	GLR – GSR	0.000	0.000	100.000
1 year	$\Delta_4\ln(GDP_r)$	82.899	12.479	4.622
	$\Delta_4\ln(M1_r)$	8.994	41.336	49.670
	GLR – GSR	9.223	0.487	90.289
2 years	$\Delta_4\ln(GDP_r)$	51.948	15.604	32.448
	$\Delta_4\ln(M1_r)$	13.896	34.910	51.194
	GLR – GSR	16.124	8.998	74.878
5 years	$\Delta_4\ln(GDP_r)$	48.235	16.049	35.716
	$\Delta_4\ln(M1_r)$	14.738	35.244	50.018
	GLR – GSR	15.719	13.062	71.219
infinity	$\Delta_4\ln(GDP_r)$	48.187	16.132	35.681
	$\Delta_4\ln(M1_r)$	14.733	35.258	50.009
	GLR – GSR	15.677	13.079	71.244

In the first quarter, the variances of all variables are mainly driven by their own innovations. This also holds for the growth rate of real GDP. Again, this indicates that there is hardly any instantaneous relation between the two monetary variables on the one hand and the real variable on the other hand. During the first year it is mainly the quantity of money that has an impact on GDP, while the interest rate spread, which has already had a considerable impact on the quantity of money in the first year, only fully affects real GDP in the second year. After about two years, the process of monetary policy influencing real developments is almost complete. Altogether, about half of the variance of the growth rate of real GDP is caused by its own innovations, while the other half results from monetary innovations. About two thirds of them are generated by the interest rate differential and less than one third by the quantity of money. Moreover, there is a clear hierarchy between the two monetary variables: while the interest rate has a strong impact on the quantity of money, also in the long-run, the reverse impact is quite weak. In addition, the feedback from real development to monetary variables is also rather weak.

Table 4.2b: Variance Decomposition
1/65 – 4/89, 100 Observations

Forecast horizon		$\Delta_4\ln(GDP_r)$	$\Delta_4\ln(M1_r)$	GLR – GSR
immediate	$\Delta_4\ln(GDP_r)$	99.231	0.667	0.102
	$\Delta_4\ln(M1_r)$	0.000	100.000	0.000
	GLR – GSR	0.000	7.798	92.292
1 year	$\Delta_4\ln(GDP_r)$	82.899	15.740	1.361
	$\Delta_4\ln(M1_r)$	8.994	60.685	30.321
	GLR – GSR	9.223	7.326	83.450
2 years	$\Delta_4\ln(GDP_r)$	51.948	26.995	21.057
	$\Delta_4\ln(M1_r)$	13.896	50.669	35.435
	GLR – GSR	16.124	11.184	72.692
5 years	$\Delta_4\ln(GDP_r)$	48.235	25.978	25.787
	$\Delta_4\ln(M1_r)$	14.738	50.970	34.292
	GLR – GSR	15.719	16.065	68.216
infinity	$\Delta_4\ln(GDP_r)$	48.187	26.033	25.780
	$\Delta_4\ln(M1_r)$	14.733	50.999	34.269
	GLR – GSR	15.677	16.136	68.188

As we have shown repeatedly, there is a well pronounced instantaneous relation between the two monetary variables. Insofar, the order of the variables in the system has a considerable impact on the results. To show this, we have changed the order between these two variables in *Table 4.2b*, i.e. we now suppose the causal ordering $\Delta_4\ln(M1_r) \rightarrow$ (GLR – GSR) $\rightarrow \Delta_4\ln(GDP_r)$. The result is that the two monetary variables have the same impact on the variance of real GDP. On the other hand, the hierarchy between the two monetary variables mentioned above is hardly influenced by this.

4.5 Concluding Remarks

The concept of vector autoregressive processes which was originally proposed by CHRISTOPHER A. SIMS (1980) has become an indispensable instrument for data description, forecasting, structural inference and policy analysis. One reason is that two new methods of analysis were developed, impulse response analysis and variance decomposition, which provided new insights into the dynamic relations between the variables of a system. However, *Chapter 6* will show that this approach is today mainly employed in the analysis of systems with nonstationary variables.

The new procedures are mainly based on the MA representation of the system. First, the AR representation is used, and a finite order AR process is estimated. However, to analyse the effects, a transformation to the MA representation is unavoidable. This shows that the MA representation introduced in *Chapter 2* is not only an analytical device but also crucial to the substantive interpretation of the relations between the different variables of a system.

Considering vector autoregressions, it becomes obvious that – compared to traditional econometrics – the relevance of the residuals has drastically changed. In traditional econometrics, they were merely regarded as unexplained effects 'disturbing' the true relationship between the variables. In vector autoregressions they are the channel through which new information flows into the system. For this reason they require special consideration. As the variance decomposition shows, all stochastic variables are finally generated by such innovations. Statistical analysis has to ask at what time such an innovation first appears in the system and how it 'moves along' the system. All other substantive questions can be traced back to these questions.

Finally, there is the same problem as when we discussed the concept of Granger causality: data analysis alone is not sufficient to make meaningful statements about the relations between (economic) variables. First, we need information on which variables are to be jointly investigated in such a system. When considering vector autoregressions, this question is of special relevance as only rather few variables (with a finite number of lags) can be included, given the large number of parameters to be estimated.

Furthermore, the problem of how to handle instantaneous relations is more severe than when testing for Granger causality. If such relations exist, and they nearly always exist, we need external information, i.e. information not included in the data, to order the variables. Even if, at first glance, the VAR approach seems to get along without theoretical considerations, we need considerable theoretical (pre-)information to apply it correctly to economic data and to be able to interpret it in a meaningful way. Here, 'theory-free' data analysis is as impossible as in other contexts. A further development taking this into account is the approach of *structural vector autoregressions* where identifying restrictions are used to generate the innovations W and to give intuitive meaning to them. CHRISTOPHER A. SIMS proposed recursive ordering for identification, but this is only one possibility. Later on, other approaches were suggested. OLIVER J. BLANCHARD and DANNY QUAH (1989), for example, assumed that demand shocks have no effect on output in the long-run whereas supply shocks do have such effects. HARALD UHLIG (2005) introduced so-called sign restrictions based on the assumption that some short-run effects have a theoretically expected sign whereas others do not.

References

The methodology of **vector autoregressive processes** was first proposed by

CHRISTOPHER A. SIMS, Macroeconomics and Reality, *Econometrica* 48 (1980), pp. 1 – 48.

This is the main paper for which he got the Nobel Prize in 2011.

KATARINA JUSELIUS, *The Cointegrated VAR Model: Methodology and Applications*, Oxford University Press, Oxford 2006, chapter 3

with reference on

DAVID HENDRY and JEAN-FRANÇOIS RICHARD, The Econometric Analysis of Economic Time Series (with discussion), *International Statistical Review* 51 (1983), pp. 111 – 163,

showed that, assuming multivariate normality and time independent first and second moments, the vector autoregressive model is the result of the sequentially decomposition of the joint distribution of the k-dimensional stochastic process X into T conditional distribution functions.

Applications can be found, for example, in

CHRISTOPHER A. SIMS, Comparing Interwar and Postwar Business Cycles: Monetarism Reconsidered, *American Economic Review*, Papers and Proceedings, 70.2 (1981), pp. 250 – 257; or

CHRISTOPHER A. SIMS, Policy Analysis with Econometric Models, *Brookings Papers on Economic Activity* 1/1982, pp. 107 – 164.

The presentation in this chapter is mainly based on

HELMUT LÜTKEPOHL, *New Introduction to Multiple Time Series Analysis*, Springer, Berlin 2005, pp. 13 – 82, 135 – 157.

This textbook offers a comprehensive presentation of this concept and its possibilities. It also shows how confidence intervals can be calculated for impulse response functions (pp. 109ff.). In addition, it compares different criteria to determine the optimal lag length of the VAR (pp. 135ff.). Proficient introductions are given in

GEORGE G. JUDGE, R. CARTER HILL, WILLIAM E. GRIFFITHS, HELMUT LÜTKEPOHL and TSOUNG-CHAO. LEE, *Introduction to the Theory and Practice of Econometrics*, Wiley, New York 1988, Chapter 18;

WALTER ENDERS, *Applied Econometric Time Series*, Wiley, Hoboken NJ, 3rd edition 2010, Chapter 5, as well as in

JAMES H. STOCK and MARK W. WATSON, Vector Autoregressions, *Journal of Economic Perspectives* 15/4 (2001), pp. 101 – 115.

In this article it is assessed how well VAR models have addressed the four macroeconomic tasks: data description, forecasting, structural inference, and policy analysis. A short introduction is also given in

DONALD ROBERTSON and MICHAEL WICKENS, VAR Modeling, in: STEVEN G. HALL (ed.), *Applied Economic Forecasting Techniques*, Harvester Wheatsheaf, New York 1994, pp. 29 – 47.

Error correction models were first used in an investigation on wages and prices in the United Kingdom carried out by

J. DENIS SARGAN, Wages and Prices in the United Kingdom: A Study in Econometric Methodology, in: P.E. HART, G. MILLS and J.K. WHITAKER (eds.), *Econometric Analysis for National Economic Planning*, Butterworth, London 1962, pp. 25 – 54.

This concept became popular by a paper about the consumption function in the United Kingdom,

JAMES E.H. DAVIDSON, DAVID F. HENDRY, FRANK SRBA and Y. STEPHEN YEO, Econometric Modelling of the Aggregate Time Series Relationship between Consumers' Expenditure and Income in the United Kingdom, *Economic Journal* 88 (1978), pp. 661 – 692.

The **LSE approach,** that goes back to J. DENIS SARGAN and DAVID F. HENDRY, is described and confronted with other approaches in

ADRIAN PAGAN, Three Econometric Methodologies: A Critical Appraisal, *Journal of Economic Surveys* 1 (1987), pp. 3 – 24.

A comprehensive introduction to this approach is presented in a textbook by

DAVID F. HENDRY, *Dynamic Econometrics*, Oxford University Press, Oxford et al. 1995.

The difference between statistical and econometric approaches to empirically analyse economic problems is discussed, for example, in

CLIVE W.J. GRANGER, Comparing the Methodologies Used by Statisticians and Economists for Research and Modeling, *Journal of Socio-Economics* 30 (2001), pp. 7 – 14.

For the **structural VAR** see, for example

GIANNI AMISANO and CARLO GIANNINI, *Topics in Structural VAR Econometrics*, Springer, Berlin et al., 2nd edition 1997,

JÖRG BREITUNG, RALF BRÜGGEMANN and HELMUT LÜTKEPOHL, Structural Vector Autoregressive Modeling and Impulse Responses, in: H. LÜTKEPOHL and M. KRÄTZIG *Applied Time Series Econometrics,* Cambridge University Press, Cambridge 2004, pp. 159 – 196.

Different identification schemes for VARs are developed by

OLIVER J. BLANCHARD and DANNY QUAH, The Dynamic Effects of Aggregate Demand and Supply Disturbances, *American Economic Review* 79 (1989), pp. 655 – 673, and

HARALD UHLIG, What are the Effects of Monetary Policy on Output? Results from an Agnostic Identification Procedure," *Journal of Monetary Economics* 52 (2005), pp. 381 – 419.

5 Nonstationary Processes

So far, we have only considered stationary time series. As a matter of fact, however, most economic time series are trending, like, for example, the GDP series investigated in *Chapter 1*. We tried to eliminate the trend by using first differences or growth rates. These filtered series can be investigated by employing the concepts that were developed for the analysis of stationary time series.

There are, however, two basic problems with this procedure. Firstly, if we employ these transformations, information is lost about the trends which have been eliminated. However, if there exist relations between the long-run components of economic time series, this lost information might be of special interest to economists. Secondly, we exclusively used visual inspection to determine whether a series is stationary or nonstationary. This procedure might raise problems whenever the roots of the lag polynomial in the autoregressive part of a possible stationary process are close to one. In this case, it is appropriate to use test procedures in order to decide by means of statistical criteria whether we will consider the time series as a realisation of a stationary or a nonstationary process.

In the following, we first present two different concepts of trending behaviour, the concepts of deterministic and of stochastic trends (*Section 5.1*). Then we discuss the elimination of such trends (*Section 5.2*). In *Section 5.3* we present tests for unit roots (stationarity) and in *Section 5.4* possible decompositions of time series in a stationary and a nonstationary component. In *Section 5.5* we present some generalisations before we finally discuss economic implications of models with either deterministic or stochastic trends (*Section 5.6*).

5.1 Forms of Nonstationarity

Due to the fact that a time series represents only one realisation of a stochastic process, only some special forms of nonstationarity can be handled. One possibility is that the expectations are time dependent, i.e. that the mean is determined by a *deterministic trend*. Such a trend might usually be modelled or at least approximated by a polynomial in t, possibly after having performed logarithmic transformations. Such a process is no longer mean stationary but still covariance stationary. Such *trend stationary* processes can be written as

(5.1) $$y_t = \sum_{j=0}^{m} \delta_j t^j + x_t,$$

where x is a stationary and invertible ARMA(p,q) process with mean zero. Thus, we have

(5.2) $$\alpha(L) x_t = \beta(L) u_t.$$

It is easy to see that

$$E[y_t] = \sum_{j=0}^{m} \delta_j t^j = \mu_t$$

and that

$$E[(y_t - \mu_t)(y_{t+\tau} - \mu_{t+\tau})] = E[x_t x_{t+\tau}] = \gamma_x(\tau).$$

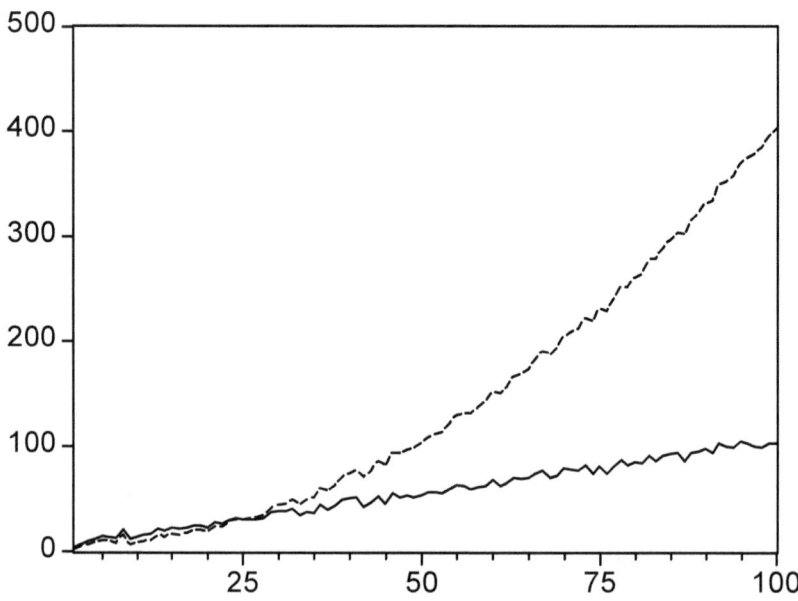

Figure 5.1: Linear and quadratic trend, superimposed by a pure random process

Because of the constant variance of the process, its realisations fluctuate with limited amplitude around the deterministic trend. Refer to *Figure 5.1*, where a linear and quadratic trend is superimposed by a pure random process. The deviations from the trend are always transitory. If long-run forecasts are performed for such a process, these follow the mean function, and the forecast errors stay finite, no matter how long the forecast horizon might be. This is essentially a deterministic approach. Despite the fact that such deterministic trends are quite often used in popular analyses, they are in most cases no appropriate instrument for long-run forecasts.

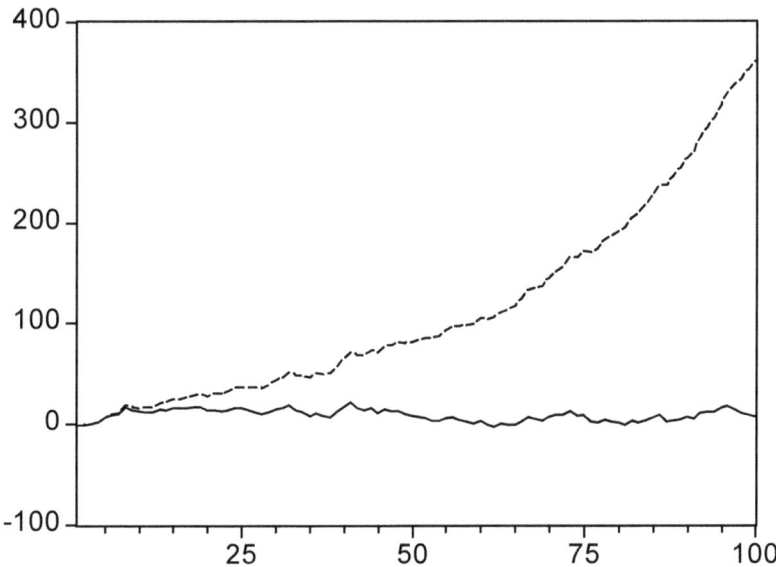

Figure 5.2: Realisations of AR(1) processes
$\alpha = 1.03$ (------), $\alpha = 0.97$ (———)

Another possibility to generate nonstationary processes is to use autoregressive processes which violate the stability conditions. If we consider, for example, an AR(1) process with $\alpha > 1$ and the given initial condition y_0,

$$y_t = \alpha\, y_{t-1} + u_t, \quad \alpha > 1$$

we immediately get

$$y_t = y_0\, \alpha^t + \sum_{j=0}^{t-1} \alpha^j u_{t-j}.$$

Therefore, we get

$$E[y_t] = y_0 \alpha^t = \mu_t .$$

Thus, the mean of this process grows exponentially for $\alpha > 1$.

The variance of this process can be calculated as follows,

$$V[y_t] = (1 + \alpha^2 + \alpha^4 + ... + \alpha^{2(t-1)}) \sigma_u^2 ,$$

$$= \frac{\alpha^{2t} - 1}{\alpha^2 - 1} \sigma_u^2 ,$$

i.e. the variance also grows exponentially with t. Thus, the process is explosive.

We get a stationary development for AR(1) processes if $-1 < \alpha < 1$, but explosive solutions if $|\alpha| > 1$. The realisations of such processes with $\alpha = 1.03$ and $\alpha = 0.97$ are shown in *Figure 5.2*. If $\alpha < -1$ the variance increases in t as for $\alpha > 1$, whereas the mean alternates with an explosive amplitude.

The special case of $\alpha = 1$ results in a *random walk*:

(5.3) $$y_t = y_{t-1} + u_t ,$$

where u is again a pure random process. Adding a constant term leads to a *random walk with drift*,

(5.4) $$y_t = \delta + y_{t-1} + u_t .$$

For a given initial condition y_0 we get the representation

(5.5) $$y_t = y_0 + \delta t + \sum_{i=1}^{t} u_i .$$

All first and second order moments are time dependent. In particular for $0 < \tau < t$ we get

$$E[y_t] = y_0 + \delta t = \mu_t ,$$
$$V[y_t] = t \sigma^2 = \gamma(0,t) ,$$
$$Cov[y_t, y_{t-\tau}] = (t - \tau) \sigma^2 = \gamma(\tau,t) .$$

Thus, the autocorrelation function is also time dependent:

$$\rho(\tau,t) = \frac{t - \tau}{\sqrt{t(t - \tau)}} = \sqrt{\frac{t - \tau}{t}} = \sqrt{1 - \frac{\tau}{t}} .$$

The autocorrelation coefficients converge to one for given τ and increasing t. Thus, we get a relatively smooth development of the realisations, despite the fact that the variance increases with t. Moreover, the random walk is mean stationary for $\delta = 0$. The nonstationarity results from the time dependence of the variance and the covariances. Contrary to the situation of stationary processes which fluctuate around their mean with a limited amplitude, the reversion to a fixed value (*mean reverting behaviour*) rarely occurs for nonstationary processes. *Figure 5.3* shows the behaviour of a random walk with and without drift. The linear trend generated by the positive drift parameter can clearly be recognised.

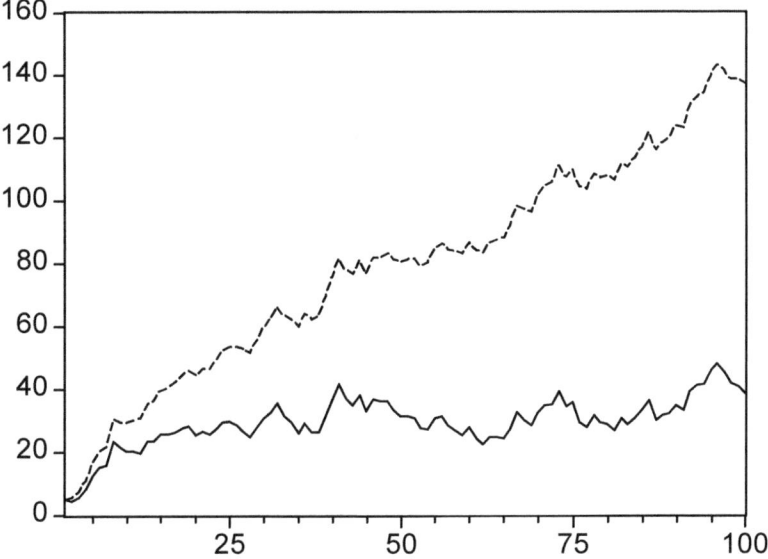

Figure 5.3: Random walk with (-----) and without (——) drift

It results in an obvious generalisation if the pure random process u in (5.3) or (5.4), respectively, is substituted by a general, weakly stationary ARMA(p,q) process, denoted as x:

(5.6) $$y_t = \delta + y_{t-1} + x_t.$$

Transforming (5.6) by using $w_t := y_t - y_{t-1}$ eliminates the nonstationarity, as $w_t = \delta + x_t$ is stationary. Such processes are called *difference stationary* or *integrated* processes, as the data generating process in levels recurs by inverting the process of taking differences, i.e. by summation (integration). Thus, the following definition generally holds:

- A stochastic process y is *integrated of order d* (I(d)), if it can be transformed to a stationary, invertible stochastic process by differencing d times, i.e.

$$(1-L)^d y_t = \delta + x_t,$$

where x is an ARMA(p,q) process. The original process y is then denoted as an ARIMA(p,d,q) process. It contains d roots of 1.0 (unit roots).

Such processes are characterised by *stochastic trends*. For a linear stochastic trend, the expectation of the change in the process is constant, whereas for a linear deterministic trend the change in the process itself is constant.

Let m = 1 and $x_t = u_t$ in relation (5.1). We thus get the trend stationary process

(5.7) $$y_t = \delta_0 + \delta_1 t + u_t,$$

whereas relation (5.5) holds for the random walk with drift:

(5.5) $$y_t = y_0 + \delta t + \sum_{i=1}^{t} u_i.$$

Both processes contain a linear deterministic trend and a stochastic part. The latter is stationary in relation (5.7), but nonstationary in relation (5.5). This implies that shocks only have a transitory effect in (5.7) because they disappear after one period, whereas they have a permanent impact in (5.5).

Let $w_t := y_t - y_{t-1}$ in equation (5.6) and substitute (5.2). We thus have

$$w_t = \delta + \frac{\beta(L)}{\alpha(L)} u_t,$$

or

(5.8) $$\alpha(L) w_t = \alpha(1) \delta + \beta(L) u_t,$$

or

(5.8') $$\alpha(L)(1 - L) y_t = \overline{\delta} + \beta(L) u_t.$$

We thus get an AR part of order p+1 with one root of 1.0, while all other roots are larger than 1.0 (in modulus). This is an ARIMA(p,1,q) process. If first differences are not sufficient to get a weakly stationary process, we have to difference the series d-times. In this case, equation (5.8') can be generalised to the ARIMA(p,d,q) process

(5.9) $$\alpha(L)(1 - L)^d y_t = \overline{\delta} + \beta(L) u_t,$$

as was already done in the definition above.

5.2 Trend Elimination

To transform the nonstationary processes (5.1) and (5.9) into stationary processes, the deterministic or the stochastic trend have to be eliminated, respectively. Let us assume that m = 1 in relation (5.1) and d = 1 in relation (5.9). In this case, we have a linear deterministic or stochastic trend. According to their definition, the nonstationarity of I(1) processes can be eliminated by forming first differences. The same procedure might be applied to models with a linear deterministic trend. Taking first differences on both sides of relation (5.1) we get (for m = 1)

$$y_t - y_{t-1} = \delta_1 + x_t - x_{t-1} .$$

Because of (5.2) this can also be written as

$$\alpha(L)w_t = \alpha(1)\delta_1 + (1 - L)\beta(L)u_t .$$

We get a stationary ARMA(p,q+1) process for w which, however, is not invertible because of the unit root in the MA part. Using first differences does not lead back to the original stationary process x but to a new stationary process which exhibits artificial short-run cycles due to overdifferencing. (In case of a quadratic deterministic trend, we get similar results by differencing the series twice.)

In *Figure 5.4*, the scatter diagrams between the differences of the nonstationary series and the original white noise processes, which have generated the trend stationary and difference stationary series, show clear differences. Whereas differencing the random walk reproduces exactly the realisation of the white noise process, the first differences of the trend stationary process do not correspond to the realisations of the generating white noise process.

One might also try to eliminate the linear trend by a regression on a time trend. The scatter diagrams in *Figure 5.5* show that this method is appropriate for trend stationary processes. The regression residuals largely correspond to the realisations of the generating white noise process. On the other hand, there is no relation between the regression residuals and the realisations of the white noise process for integrated processes.

The results in *Table 5.1* further clarify this situation. For the realisation of a trend stationary process with a constant term of 5.0 and a slope coefficient of 1.0 we get, as expected, estimates of the regression on time which are quite close to the true parameters. The adjusted coefficient of determination is high and the Durbin-Watson statistic gives no indication of first order autocorrelation. Taking the usual t statistic in case of the realisation of the random walk, we also get a highly significant regression coefficient

162   5 Nonstationary Processes

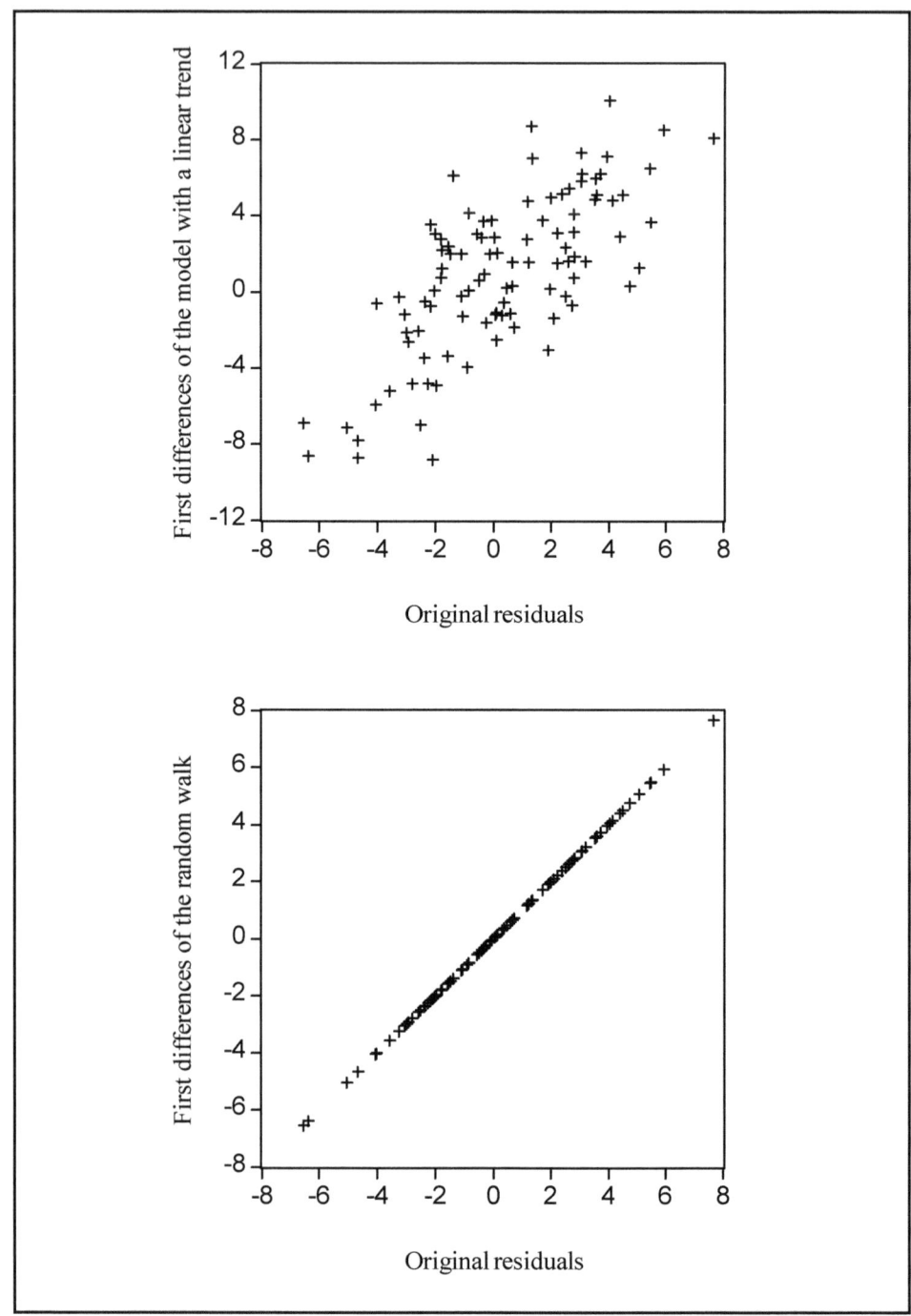

Figure 5.4: Scatter diagrams of the first differences against the original residuals of nonstationary processes

5.2 Trend Elimination 163

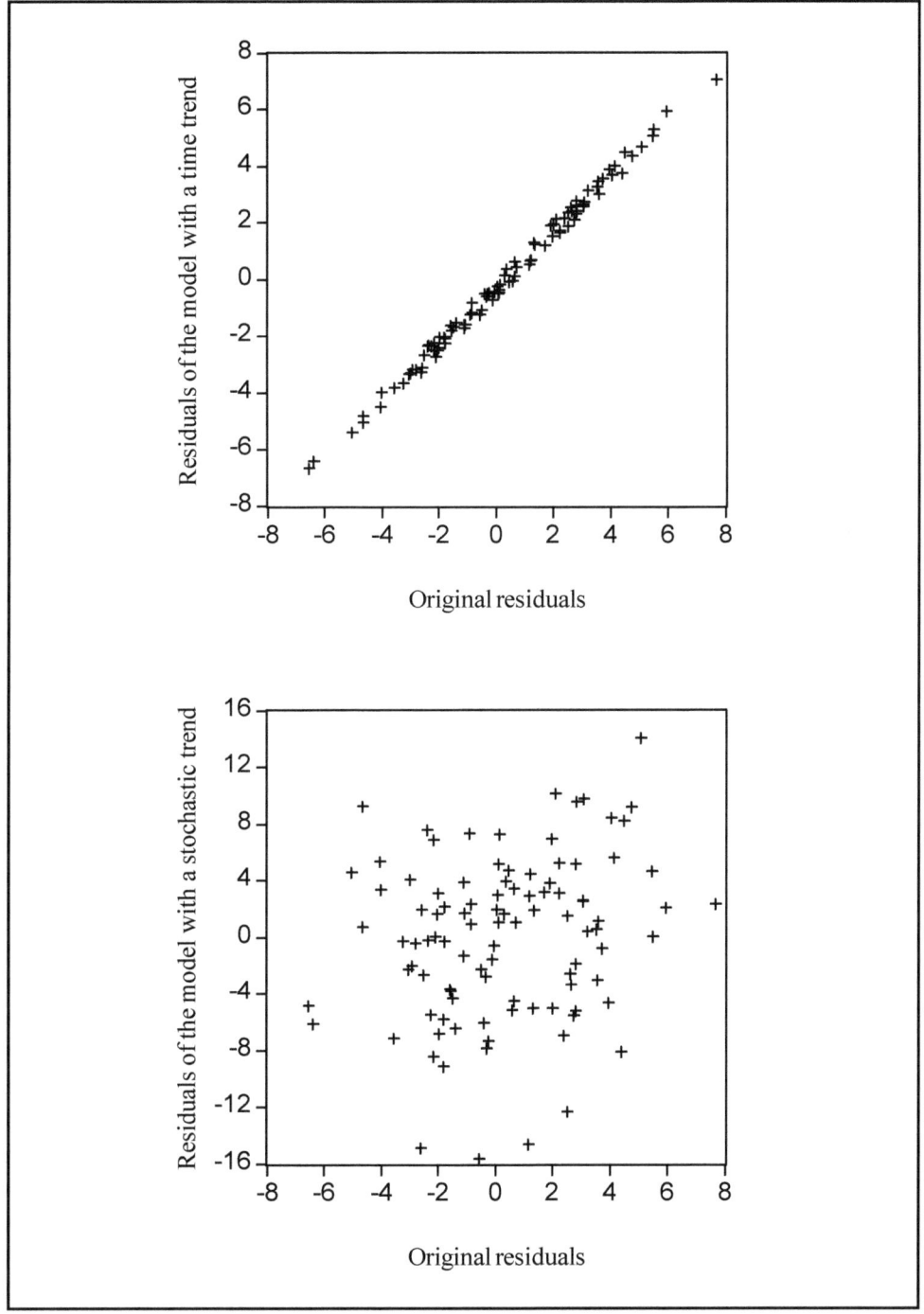

Figure 5.5: Scatter diagrams of the residuals of regressions on a time trend against the original residuals of nonstationary processes

164 5 Nonstationary Processes

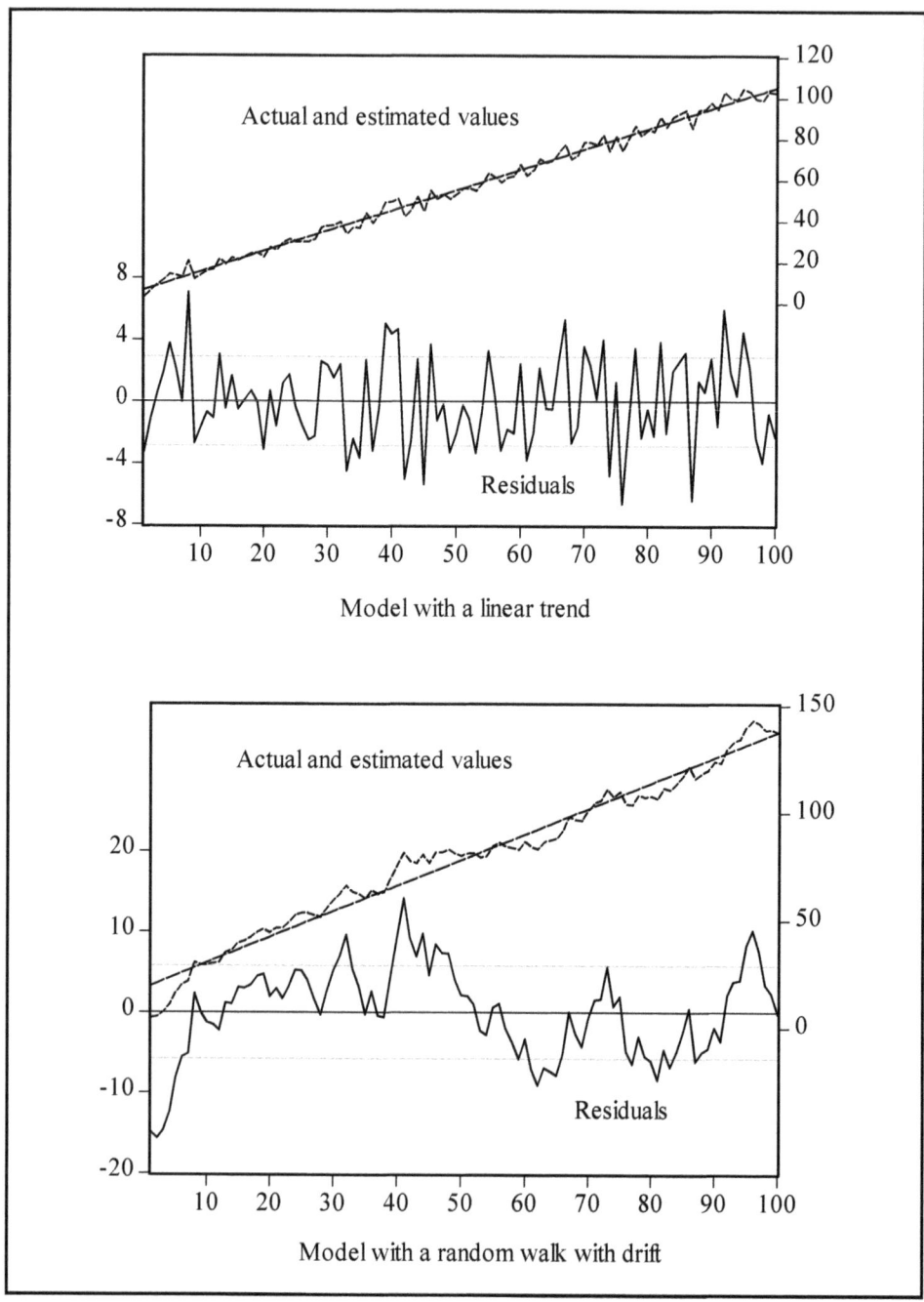

Figure 5.6   *Actual and estimated values and residuals of the models with linear deterministic and stochastic trends*

for the trend variable. Furthermore, we get – for the 'wrong' model – an acceptable value of the coefficient of determination, even if the Durbin-Watson statistic correctly indicates high first order autocorrelation. This is also true for the random walk with drift. However, the coefficient of determination and the t statistic of the regression coefficient of the trend variable are now much higher due to the fact that this process implicitly contains a linear trend.

Table 5.1: Results of Linear Trend Elimination (100 Observations)

	Model with a		
	linear trend	random walk	random walk with drift
Constant term	5.678 (9.79)	19.673 (16.89)	18.673 (16.03)
linear trend	0.993 (99.60)	0.191 (9.55)	1.191 (59.48)
$\bar{R}^2$	0.990	0.477	0.973
Durbin-Watson	2.085	0.247	0.247

Figure 5.6 shows the residuals, the actual and the estimated values of regressions of the model with linear trend and the random walk with drift on a linear trend. It is obvious that the residuals of the model of a random walk with drift still contain systematic variations which might be wrongly interpreted as genuine cycles.

These examples clearly indicate that the analysis of nonstationary time series requires a serious investigation of the trending behaviour, i.e. of the causes of the nonstationarity, as an inappropriate trend elimination procedure might generate artificial movements in the resulting time series. There is a risk that these statistical artefacts are interpreted in terms of economics.

5.3 Unit Root Tests

As we have seen, it is important to take the kind of nonstationarity into account, i.e. to ask whether the series contains a deterministic or a stochastic trend when it comes to transforming nonstationary into stationary time series. Otherwise, statistical artefacts might appear in the transformed series. Within the framework of the Box-Jenkins approach, nonstationary behaviour of time series is covered by ARIMA(p,d,q) models. Time series analysts have long tried to find the order of differencing, d, leading to a stationary ARMA process simply by considering the autocorrelation function. For these purposes, the estimated correlograms of the levels and the successive differences are investigated. If the autocorrelation coefficients decrease very slowly with increasing order, this is taken as evidence of nonstationarity. The following rule of thumb can be used for this procedure: Determine the order of differencing in such a way that the autocorrelation coefficients approach zero quite rapidly and that the variance of the resulting series is smallest compared to variances resulting from other orders of differencing. Generally, this guarantees that there is no overdifferencing: overdifferenced series often have a rather pronounced negative first order autocorrelation coefficient, and the estimated variance of the series is often increased by the transformation which actually leads to overdifferencing.

This descriptive procedure can be generalised if not only multiple unit roots are determined by successive differencing but when, quite generally, all roots with an absolute value of one are determined in the characteristic equation or in the lag polynomial of the autoregressive part.

This approach, which goes back to GEORGE C. TIAO and RUEY S. TSAY (1983), uses the following model as starting point:

(5.10) $$\eta(L)\alpha(L)y_t = \delta + \beta(L)u_t,$$

where all roots of $\eta(L) = 0$ are on the unit circle and all roots of $\alpha(L) = 0$ and $\beta(L) = 0$ are outside the unit circle. If, instead of the true model (5.10), autoregressive models with increasing order $k = 1, 2, ..., p^{max}$ are estimated with ordinary least squares,

(5.11) $$y_t = a_0 + a_1 y_{t-1} + ... + a_k y_{t-k} + v_t^{(k)},$$

it can be shown that all roots on the unit circle are consistently estimated. This is true despite the fact that the residuals of (5.11) will usually be autocorrelated because of the wrong AR order and/or the missing MA part. Due to the autocorrelation of the residuals, however, this consistency result does not hold for the roots of the stable part of the model. But even if the order of the estimated AR process exceeds the order of the nonstationary part $\eta(L)$, the number of the roots on the unit circle remains constant. This stability property can be used to determine all roots which cause nonstationarity. In order to do so, the roots of the characteristic equation (or the corresponding lag polynomial) of the AR(k) process in equation (5.11)

(5.12) $$\lambda^k - \hat{a}_1 \lambda^{k-1} - ... - \hat{a}_k = 0, \quad k = 1, 2, ..., p^{max},$$

5.3.1 The Dickey-Fuller Test

The procedures described so far neither provide a formal test nor do they allow to distinguish between trend stationary and difference stationary behaviour of a time series. Both demands can principally be satisfied by using unit root tests. Such tests have first been developed by WAYNE A. FULLER (1976, pp. 366 ff.) as well as by DAVID A. DICKEY and WAYNE A. FULLER (1979, 1981).

If we set m = 1 in relation (5.1) and if we suppose that we have a stationary AR(1) process in (5.2), we get

$$(5.13) \qquad y_t = \delta_0 + \delta_1 t + \frac{1}{1-\alpha_1 L} u_t$$

or

$$y_t = [(1-\alpha_1)\delta_0 + \alpha_1 \delta_1] + (1-\alpha_1)\delta_1 t + \alpha_1 y_{t-1} + u_t .$$

With $\alpha = (1-\alpha_1)\delta_0 + \alpha_1\delta_1$, $\beta = (1-\alpha_1)\delta_1$ and $\rho = \alpha_1$, this relation can be written as

$$(5.14) \qquad y_t = \alpha + \beta t + \rho y_{t-1} + u_t .$$

If the AR(1) process has a unit root, i.e. if $\alpha_1 = 1$,

$$(5.15) \qquad y_t = \delta_1 + y_{t-1} + u_t$$

leads to a random walk with drift, which can be used as the null hypothesis of a test, while the alternative hypothesis, $|\alpha_1| < 1$, leads to a trend stationary process.

If we want to distinguish between a stationary AR(1) process with a mean different from zero and a nonstationary AR(1) process with $\delta_0 \neq 0$ and $\delta_1 = 0$ and under the null hypothesis $\alpha_1 = 1$,

$$y_t = y_{t-1} + u_t$$

leads to a random walk without drift, while the alternative is a stationary AR(1) process with mean different from zero.

If we can assume a priori that the mean is zero, i.e. that $\delta_0 = 0$, the null hypothesis $\alpha_1 = 1$ again leads to a random walk without drift, whereas the alternative is

$$y_t = \rho y_{t-1} + u_t \text{ with } |\rho| < 1.$$

These distinctions with respect to the alternative hypotheses are necessary as in all three cases even the asymptotic distributions under the null hypothesis no longer correspond to standard distributions. They also depend on other parameters, especially on those of the trend and the mean. If we start from the general model (5.14), the null hypothesis is $\rho = 1$ in all three cases, i.e. the AR part has a unit root. It can be shown that, under the null hypothesis, the least squares estimator of ρ is downward biased and has a skewed left distribution. Thus, even if the null hypothesis $\rho = 1$ is true, we expect values smaller than one for $\hat{\rho}$. Correspondingly, the usual t statistic of $\hat{\rho} - 1$, which is normally used as test statistic, no longer follows a t distribution. Critical values for the t tests of all three cases have first been provided by WAYNE A. FULLER (1976, Table 8.5.2, p. 373). They were derived by using simulations. Today, slightly more precise critical values are usually employed which were derived through simulations by JAMES G. MACKINNON (1991, p. 275). Nowadays, these values are included in many computer programs. For a one-sided test against the alternative $\rho < 1$, a significance level of 5 percent and 100 observations, the critical values are -1.94 for a zero mean, -2.89 if the mean is different from zero and -3.46 if a linear trend is included in addition. As all these values are larger in absolute value than the critical value of the t statistic, which is -1.65, using this distribution would reject the null hypothesis far too often. The decision would mistakenly be in favour of a stationary or trend stationary process despite the fact that the series contains a random walk with or without drift. If the combined hypotheses $\alpha = \beta = 0$ and $\rho = 1$, or $\beta = 0$ and $\rho = 1$, respectively, are to be tested, the F tests proposed by DAVID A. DICKEY and WAYNE A. FULLER (1981) with the critical values tabulated by these authors (pp. 1062f.) can be used.

Example 5.1

To demonstrate the deviation of the distributions of the estimated parameters $\hat{\rho}$ and $\hat{t}$ from the standard distributions, we performed a Monte-Carlo simulation. We generated 100'000 realisations with T = 200 observations for the model

(E5.1) $$y_t = \rho y_{t-1} + u_t$$

with $\rho = 1.0$. Then, we estimated relation (5.14) with $\beta = 0$. The empirical distributions of $\hat{\rho}$ and $\hat{t}$ (which are smoothed with a kernel estimator) are given in *Figure 5.7*. First of all, we can see that $\hat{\rho}$ is not symmetrically distributed around its true value of one; the mean of the estimated coefficients is 0.973.

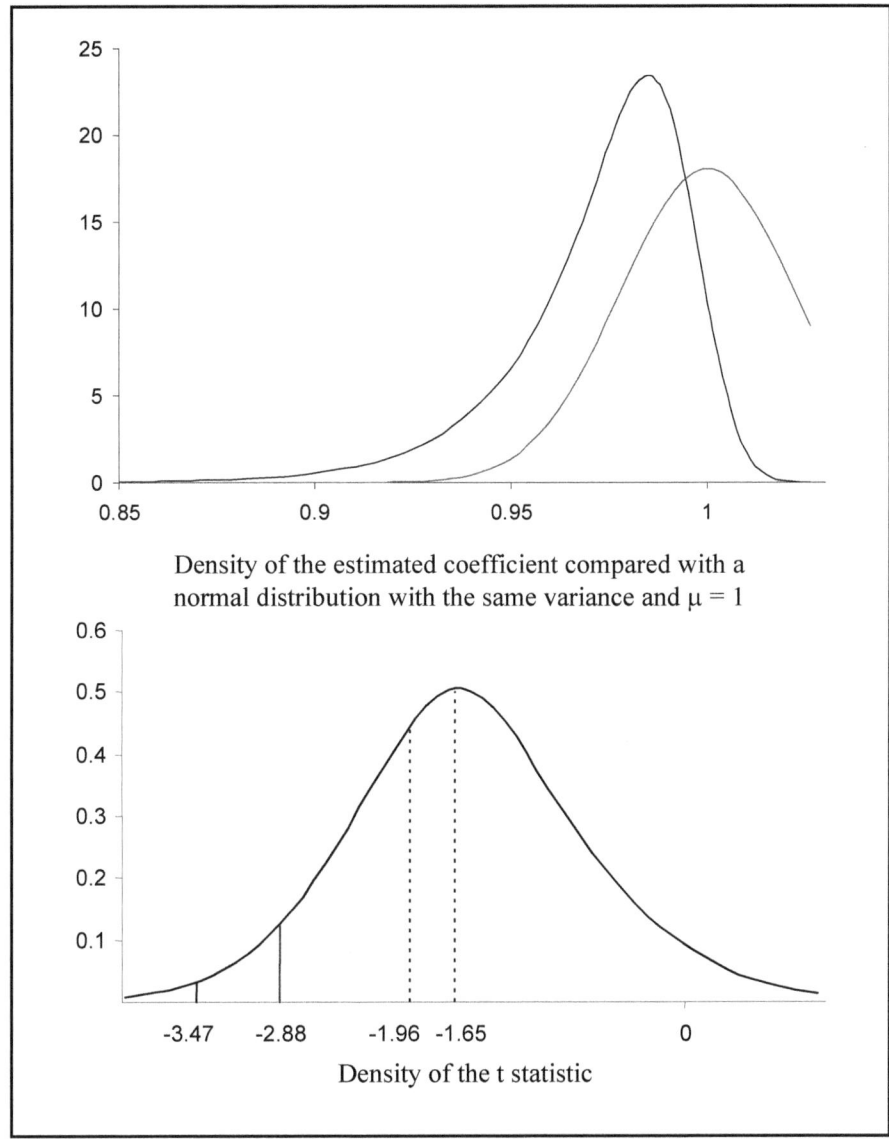

Figure 5.7: Density of the estimated autocorrelation coefficient and the t statistic under the null hypothesis of a random walk.

Thus, there is a systematic underestimation of the autoregressive parameter. Second, this leads to a strong deviation of the estimated t values under the null hypothesis H_0: $\rho = 1.0$ of the corresponding t distribution; the mean of the distribution of the estimated t statistic is -1.534 instead of the theoretical value of zero. The area under the density function left of -1.96, the critical value which is usually employed for this sample size, is not 2.5 percent but 30.18 percent. For a one-sided test, a significance level of 5 percent and the usual critical value of -1.65, the null hypothesis would be rejected in 35.58 percent of all cases. However, if we use the critical values of JAMES G. MACKINNON (1991), which, in this situation, are -2.876 at the 5 percent level and -3.465 at the 1 percent level, with rejection rates of 4.99 percent and 0.99 percent, the significance levels are almost exactly realised in our simulations.

In order to use the conventional t value directly, which implies a test of the estimated parameter of y_{t-1} against the null hypothesis of zero, relation (5.14) can be transformed by subtracting y_{t-1} on both sides:

(5.16) $$\Delta y_t = \alpha + \beta t + (\rho - 1) y_{t-1} + u_t.$$

5.3.2 The Augmented Dickey-Fuller Test

If the autoregressive process is of order higher than one, i.e. if we have an AR(p) process with $p > 1$, the tests can be generalised quite easily, because an AR(p) process

$$y_t = \alpha_1 y_{t-1} + \alpha_2 y_{t-2} + \ldots + \alpha_p y_{t-p} + u_t$$

can immediately be reparameterised as

$$y_t = \rho y_{t-1} + \theta_1 \Delta y_{t-1} + \theta_2 \Delta y_{t-2} + \ldots + \theta_{p-1} \Delta y_{t-p+1} + u_t$$

with

$$\rho = \theta_0 = \sum_{j=1}^{p} \alpha_j, \quad \theta_i = -\sum_{j=i+1}^{p} \alpha_j, \quad i = 1, 2, 3, \ldots, p-1.$$

If this AR(p) process has a unit root, it holds that $1 - \alpha_1 - \alpha_2 - \ldots - \alpha_p = 0$ or $\rho = 1$, respectively. All alternative hypotheses discussed so far can be applied to this more general situation. In addition, the same asymptotic distributions hold as in the AR(1) case. This allows us to use the same critical values. Thus, for the situation with deterministic trend the generalisation of the test equation (5.16) is

(5.17) $$\Delta y_t = \alpha + \beta t + (\rho - 1) y_{t-1} + \theta_1 \Delta y_{t-1} + \ldots + \theta_k \Delta y_{t-k} + u_t$$

for the *Augmented Dickey-Fuller* (ADF) test, where k is chosen to ensure that the residuals follow a pure random process.

If the data generating process is trend stationary but the unit root test is mistakenly performed without including a time trend, these tests have, as PIERRE PERRON (1988) showed, asymptotically disappearing power, i.e. the null hypothesis of a random walk is not rejected often enough, and is never rejected in the limiting case. Thus, the quality of a unit root test largely depends on whether the test is performed within the appropriate model. If the data suggest that a deterministic trend might exist, one should start with model (5.17) to perform the tests and use the simplified versions only if the null hypothesis $H_0: \beta = 0$ cannot be rejected and it is, therefore, not necessary to include a time trend into the test equation. The analogous argumentation holds for the constant term.

Correspondingly, PIERRE PERRON (1988) proposed the following strategy to perform unit root tests: We start with the general model (5.17)

$$\Delta y_t = \alpha + \beta (t - T/2) + (\rho - 1) y_{t-1} + \sum_{i=1}^{k} \theta_i \Delta y_{t-i} + u_t,$$

where the trend variable is centred, however, ensuring that it has no effect on the estimated constant term. (T denotes the sample size.) We can use the Dickey-Fuller t test with the null hypothesis $H_0: \rho = 1$ and the alternative hypothesis that y_t is trend stationary. We can also use an F test in order to test the combined hypothesis $H_0: (\alpha, \beta, \rho) = (\alpha, 0, 1)$. If this hypothesis is rejected, it might be assumed that a deterministic trend exists. In addition, we can test this with the null hypothesis $H_0: \beta = 0$. If both null hypotheses cannot be rejected, we can, in a second step, use the model

(5.17')
$$\Delta y_t = \alpha + (\rho - 1) y_{t-1} + \sum_{i=1}^{k} \theta_i \Delta y_{t-i} + u_t$$

and again perform a t test for the null hypothesis $H_0: \rho = 1$, i.e. we test for a unit root. In this situation, the alternative hypothesis is the existence of a stationary AR process with non-zero mean.

If, in addition, it has to be tested whether the constant term is zero, we can again perform an F test with $H_0: (\alpha, \rho) = (0, 1)$. If this null hypothesis cannot be rejected, we can use the model

(5.17'')
$$\Delta y_t = (\rho - 1) y_{t-1} + \sum_{i=1}^{k} \theta_i \Delta y_{t-i} + u_t,$$

in order to test $H_0: \rho = 1$.

Even if the residuals in model (5.14) are generated by a MA or ARMA process, test equation (5.17) can be used because invertible MA and ARMA processes can be approximated by higher order autoregressive processes. However, this might lead to a considerable reduction of the test power. Thus, with increasing k it is – ceteris paribus – increasingly difficult to reject the null hypothesis of nonstationarity.

If the true data generating process is an ARIMA(0,1,1) process, i.e. if

$$(1 - L) y_t = (1 - \beta L) u_t$$

with $0 < \beta < 1$, problems arise if β is close to (but still smaller than) one. Then, the unit root in the autoregressive part is nearly outweighed by the MA part. Using simulations, G. WILLIAM SCHWERT (1987, 1989) showed that in this case the true null hypothesis is rejected far too often. SAÏD E. SAÏD and DAVID A. DICKEY (1985) proposed a procedure that takes into account the MA component and thus reduces the bias of the test results considerably. In all cases, the critical values derived by JAMES G. MAC-KINNON (1991) for the t tests and by DAVID A. DICKEY and WAYNE A. FULLER (1981, p. 1063) for the F tests can be used.

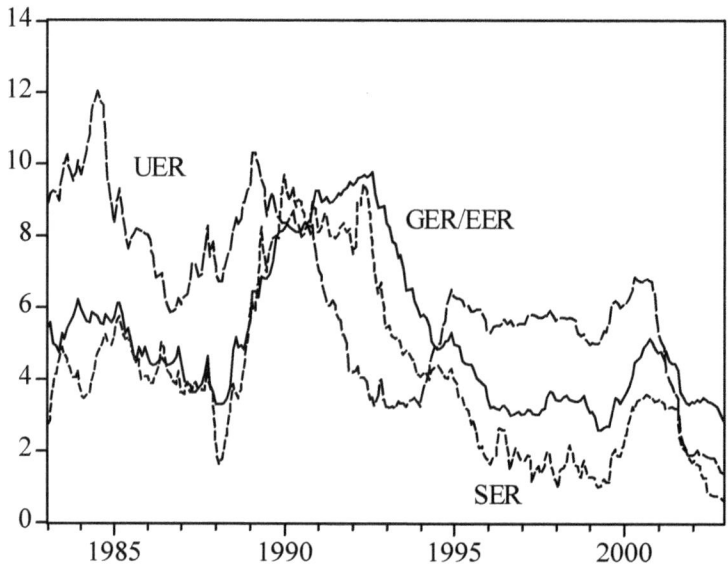

Figure 5.8:   Development of the Swiss, German/European and US Euromarket interest rates. Monthly data, January 1983 – December 2002

Example 5.2

Figure 5.8 shows the Euromarket three month interest rates of the United States (UER), Switzerland (SER) and 'Euroland' (GER/EER). As the Euro has been the common currency of the member countries of the European Monetary Union only since January 1, 1999, and as, in the period before, many of these countries pegged their currencies more or less to the strongest European currency, the German Mark, we use the German interest rate for the period up to December 1998. To test whether these series have a unit root, we performed ADF tests. As these data do not contain obvious trends - which, by the way, would be surprising in case of interest rates - we performed the tests with model (5.17'). To determine the lag length k, we used the Hannan-Quinn criterion.

The results are given in *Table 5.2*. It is obvious that the hypothesis of a unit root cannot be rejected for all three interest rates. In a second step, using model (5.17"), we applied the test on the first differences of these time series to determine the order of integration. Here, the null hypothesis of nonstationarity can clearly be rejected. Taking this into account, we assume that the interest rate series are integrated of order one (I(1)). It follows from this that ARIMA(p,1,q) processes are appropriate statistical models for such series. The interest rate series show high persistence and (at best) only very weakly pronounced mean reverting behaviour.

Table 5.2: Results of the Augmented Dickey-Fuller Tests 1/1983 – 12/2002, 240 Observations

Variable	Levels		1. Differences	
	k	Test Statistic	k	Test Statistic
SER	3	-1.194 (0.678)	2	-7.862 (0.000)
GER/EER	1	-0.957 (0.768)	0	-11.962 (0.000)
UER	1	-0.995 (0.755)	0	-11.220 (0.000)

The tests were performed for levels with as well as for first differences without a constant term. The numbers in parentheses are the p values. The number of lags, k, has been determined with the Hannan-Quinn criterion.

5.3.3 The Phillips-Perron Test

An alternative approach to consider autoregressive and/or heteroscedastic error terms in relation (5.14) goes back to PETER C.B. PHILLIPS and PIERRE PERRON (1988). Here, unlike in equation (5.17), these effects are not modelled by adding lagged differences in the systematic part of the equation. The test statistic for the hypothesis $\rho = 1$ is, however, rather adjusted by a non-parametric estimate of the long-run variance of the estimated parameter $\hat{\rho}$ that takes the autocorrelation of the residuals into account.

To estimate the long-run adjusted variance of the residuals the two authors propose

$$\text{(5.18)} \qquad s_{Tm}^2 = \frac{1}{T}\sum_{t=1}^{T}\hat{u}_t^2 + \frac{2}{T}\sum_{i=1}^{T-1}\left(w_{im}\sum_{t=i+1}^{T}\hat{u}_t\hat{u}_{t-i}\right),$$

where û are the least squares residuals of equation (5.14). The truncation parameter m denotes the maximal order up to which the autocovariances are included. With sample size T, m has to increase to infinity, but not as fast as T. The w_{im} are weights that do not only ensure the consistency of this long-run variance estimator but also its non-negativity. PIERRE PERRON (1988) proposed to use the following weights which go back to MAURICE STEVENSON BARTLETT (1948):

$$\text{(5.19)} \qquad w_{im} = \begin{cases} 1 - \dfrac{i}{m+1}, & i = 1, \ldots, m \\ 0, & i > m \end{cases}.$$

Using this adjusted variance, we get the following F Test with the null hypothesis H_0: $(\alpha, \beta, \rho) = (\alpha, 0, 1)$ for the model with time trend and constant term in equation (5.14):

$$\text{(5.20)} \qquad \tilde{F}_{Tr} = \frac{s}{s_{Tm}}\hat{F}_{Tr} - \frac{(s_{Tm}^2 - s^2)}{2 s_{Tm}^2}\left[T(\hat{\rho}-1) - \frac{T^6(s_{Tm}^2 - s^2)}{48\ |X'X|}\right],$$

where s is the estimated standard error of regression (5.14) and X the matrix of predetermined variables, i.e. the matrix X contains, besides the vector of ones, the two column vectors y_{t-1} and t:

$$X = [1\ y_{t-1}\ t].$$

$\hat{F}_{Tr}$ is the conventional F statistic for the null hypothesis given above. Instead of the usual t statistic to test the null hypothesis H_0: $\rho = 1$ in this model with trend, the following adjusted test statistic has been proposed:

$$\text{(5.21)} \qquad \tilde{t}_{Tr} = \frac{s}{s_{Tm}}\hat{t}_{Tr} - \frac{(s_{Tm}^2 - s^2)T^3}{4 s_{Tm}\sqrt{3}\ |X'X|}.$$

Here, $\hat{t}_{Tr}$ denotes the usual t statistic.

If the tests in (5.20) and (5.21) cannot reject the corresponding null hypotheses, it might be assumed that there is no deterministic trend. In this case, the stronger null hypothesis H_0: $(\alpha, \beta, \rho) = (0, 0, 1)$ can be tested with the following statistic:

$$(5.20') \quad \tilde{F}_{Tr} = \frac{s}{s_{Tm}}\hat{F}_{Tr} - \frac{(s_{Tm}^2 - s^2)}{3s_{Tm}^2}\left[T(\hat{\rho}-1) - \frac{T^6(s_{Tm}^2 - s^2)}{48\,|X'X|}\right].$$

Under the assumption that there is no deterministic trend in the data, the test statistic

$$(5.20'') \quad \tilde{F}_{\mu} = \frac{s}{s_{Tm}}\hat{F}_{\mu} - \frac{(s_{Tm}^2 - s^2)}{2s_{Tm}^2}\left[T(\hat{\rho}-1) - \frac{T^2(s_{Tm}^2 - s^2)}{4\sum_{t=1}^{T}(y_t - \bar{y})^2}\right]$$

tests the combined null hypothesis H_0: $(\alpha, \rho) = (0, 1)$. Here, $\hat{F}_{\mu}$ is the usual F statistic for this null hypothesis. If it cannot be rejected, we can check the null hypothesis H_0: $\rho = 1$ in the model without deterministic components with

$$(5.21') \quad \tilde{t}_{\rho} = \frac{s}{s_{Tm}}\hat{t}_{\rho} - \frac{0.5\,(s_{Tm}^2 - s^2)T}{s_{Tm}\sqrt{\sum_{t=2}^{T}y_{t-1}^2}}$$

i.e. we check whether the series contains a random walk without drift. If this hypothesis is rejected, with

$$(5.21'') \quad \tilde{t}_{\mu} = \frac{s}{s_{Tm}}\hat{t}_{\mu} - \frac{0.5\,(s_{Tm}^2 - s^2)T}{s_{Tm}\sqrt{\sum_{t=1}^{T}(y_t - \bar{y})^2}}$$

the hypothesis of a random walk with drift can be tested. $\hat{t}_{\mu}$ and $\hat{t}_{\rho}$ are again the usual t statistics. In all cases, the critical values derived by JAMES G. MACKINNON (1991) for the t tests and by DAVID A. DICKEY and WAYNE A. FULLER (1981, p. 1063) for the F tests can be used.

The augmented Dickey-Fuller test, which parametrically models the autocorrelation of the residuals, has the advantage that we can test whether the residuals of the estimated test equation are still autocorrelated. This is not possible with the Phillips-Perron test. On the other hand, the advantage of this nonparametric approach is that the results are less sensitive to small changes of the truncation parameter m. (However, as DONALD W. ANDREWS (1991) showed, the choice of m is not without problems when it comes to practical applications. Here, m is often chosen equal to approximately the fourth root of the sample size.) The power of the ADF test is reduced by too large a number of lagged differences. On the other hand, too small a number of lags has the effect that the test is no longer correctly applicable due to the autocorrelation of the estimated residuals. Firstly, for

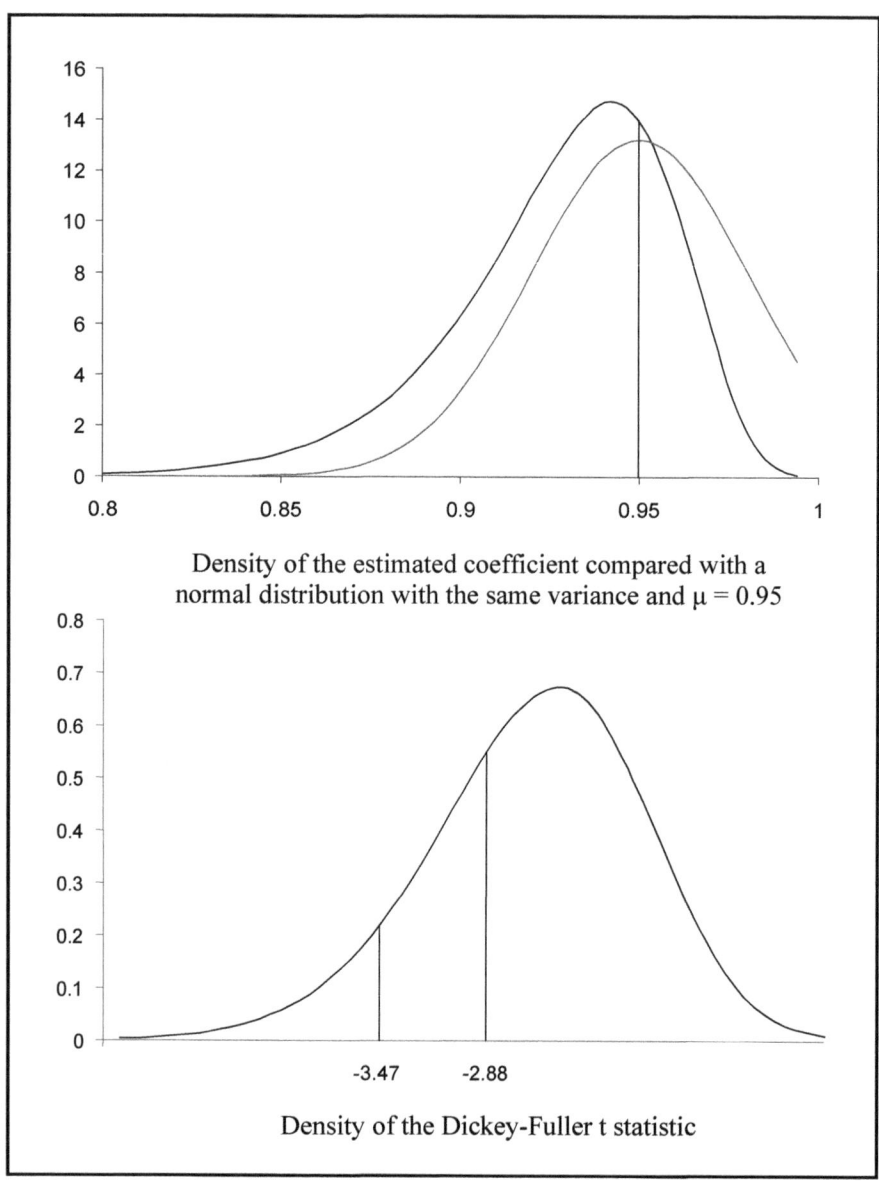

Figure 5.9a: Density of the estimated coefficient and of the t statistic for the null hypothesis of an AR(1) process with $\rho = 0.95$

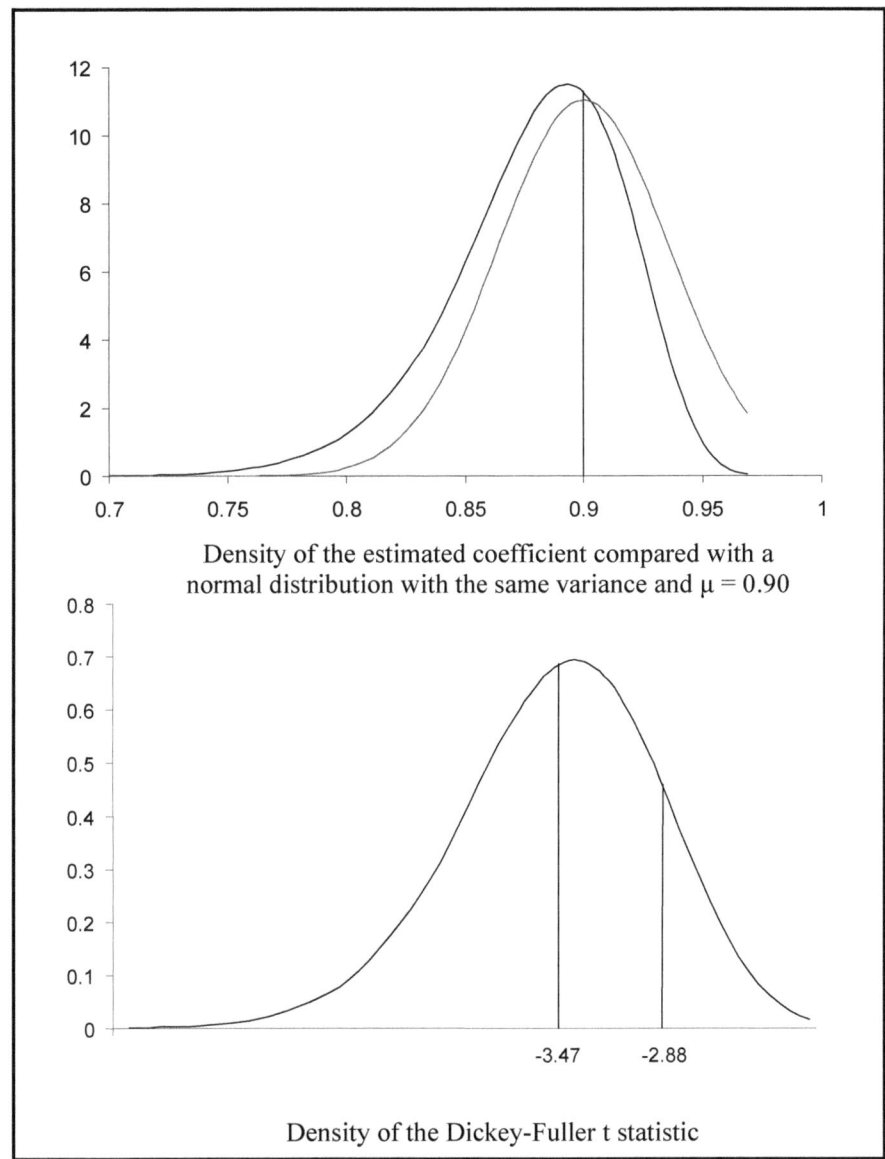

Figure 5.9b: Density of the estimated coefficient and of the t statistic for the null hypothesis of an AR(1) process with $\rho = 0.90$

the nonparametric tests, the number of lags has no impact on the estimated parameters, and, secondly, if the autocorrelation coefficients tend towards zero they have, at best, a small impact on the estimated variance. The increase of m does not reduce the sample size of the estimated equation.

Thus, one should assume that nonparametric tests are better suited to cope with the autocorrelation of the residuals. However, this holds only partly.

G. WILLIAM SCHWERT (1987, 1989) showed in a simulation study that, once the model contains an MA term with negative autocorrelation, the true null hypothesis is even more often rejected when using the Phillips-Perron test as compared to the augmented Dickey-Fuller test. Thus, the procedure proposed by SAÏD E. SAÏD and DAVID A. DICKEY (1985), which considers this problem, should definitely be applied in this case.

One problem with the ADF test as well as with the Phillips-Perron test is that their power is rather low if, under the alternative hypothesis, the first order autocorrelation coefficient is close to one, if, for example, $0.95 \leq \rho < 1$ holds for an AR(1) process. In such situations, i.e. if the mean reverting behaviour is only very weakly pronounced, very large sample sizes are necessary to reject the null hypothesis. With economic data, however, such a sample size is rare, at least as long as only monthly, quarterly or even annual data are available.

Example 5.3

To illustrate the problems with respect to the power of unit root tests, we once again performed Monte-Carlo simulations. In order to do so, we again generated 100'000 realisations with a sample size of 200 observations for model (E5.1). However, in this simulation we used the values $\rho = 0.95$ and $\rho = 0.90$ for the autoregressive parameter.

As *Figures 5.9a* and *5.9b* show, the estimated values are also shifted considerably to the left. The estimated means are 0.928 for $\rho = 0.95$ and 0.880 for $\rho = 0.90$. Thus, only 25.1 percent and 32.3 percent of the estimated values are on the right of the true value for $\rho = 0.95$ and for $\rho = 0.90$, respectively.

The density functions of the t statistics indicate the low test power for values of ρ close to 1.0. If the test is performed for the null hypothesis $\rho = 1.0$ and the true value is $\rho = 0.95$, even by applying the critical values of JAMES G. MACKINNON (1991), the (false) null hypothesis can only be rejected in 8.3 percent of all situations using the 1 percent significance level and in 30.5 percent of all situations using the 5 percent significance level. Thus, the type II error occurs in 91.7 or 69.5 percent of all situations. However, for $\rho = 0.90$ it occurs much less often: when testing at the 1 percent level we falsely accept the null hypothesis in 52.6 percent of all cases and at the 5 percent level in 14.7 percent of all cases.

5.3.4 Unit Root Tests and Structural Breaks

A further problem arises if (trend) stationary processes have a structural break. In such situations, the tests described so far are usually unable to reject the null hypothesis of a unit root even if the sample size increases: the power of the test tends asymptotically towards zero. If we know the date of the structural break and have enough observations in both periods, we can perform unit root tests separately for the time before and after the structural break. The problem is, however, that the power of these tests is reduced due to the smaller sample sizes.

An alternative to this procedure was proposed by PIERRE PERRON (1989, 1994). He assumes that the date of the structural break, t^*, is known. A typical example for such an assumption is the German Unification. He distinguishes two models: the first one is formulated in analogy to an additive outlier (AO model) and represents a sudden break in level or a change in the slope of the deterministic trend. The second model allows for an outlier in the innovations (OI model) and assumes a gradual adjustment to the new situation; the shocks on the trend function (the deterministic component of the model) have the same impact on the level of the series as regular shocks.

As most economic time series exhibit a trend, PIERRE PERRON uses AO models showing a coincidence of structural break with deterministic trend. Thus, in order to eliminate deterministic components, he first of all estimates the following relations with OLS:

(5.22) $\quad y_t = \alpha + \beta t + \delta_1 DV_t + x_t$,

(5.22') $\quad y_t = \alpha + \beta t + \delta_1 DV_t + \delta_2 DV_t(t-t^*) + x_t$,

(5.22'') $\quad y_t = \alpha + \beta t + \delta_2 DV_t(t-t^*) + x_t$,

where the dummy variable DV is zero up to the structural break which takes place in t^* and one afterwards. For the residuals of the equations (5.22) or (5.22'), $\hat{x}_t$, he performs the augmented Dickey-Fuller-Test based on the following regression:

(5.23) $\quad \Delta\hat{x}_t = (\rho-1)\hat{x}_{t-1} + \sum_{i=0}^{k} d_i \Delta DV_{t-i} + \sum_{i=1}^{k} \theta_i \Delta\hat{x}_{t-i} + u_t$.

JÜRGEN WOLTERS and UWE HASSLER (2006) demonstrate why it is necessary to include lagged ΔDV in (5.23).

For the residuals of equation (5.22''), PIERRE PERRON uses the regression

(5.23') $\quad \Delta\hat{x}_t = (\rho-1)\hat{x}_{t-1} + \sum_{i=1}^{k} \theta_i \Delta\hat{x}_{t-i} + u_t$.

For the OI model with a linear trend, however, we get the following test equation for a structural break in the level of the series

(5.24) $$\Delta y_t = \alpha + \beta t + \delta_1 DV_t + \delta_2 \Delta DV_t + (\rho - 1) y_{t-1} + \sum_{i=1}^{k} \theta_i \Delta y_{t-i} + u_t.$$

For the model with a structural break in the level of the series as well as in its deterministic trend we get

(5.24') $$\Delta y_t = \alpha + \beta t + \delta_1 DV_t + \delta_2 \Delta DV_t + \delta_3 DV_t (t - t^*) + (\rho - 1) y_{t-1} + \sum_{i=1}^{k} \theta_i \Delta y_{t-i} + u_t.$$

In the AO as well as in the OI model, the test statistic is the t value of $\hat{\rho} - 1$. Critical values which also depend on the date of the structural break are given in PIERRE PERRON (1989, pp. 1376ff.; 1994, pp. 137ff.).

5.3.5 A Test with the Null Hypothesis of Stationarity

An alternative procedure for testing the stationarity properties of time series was proposed by DENIS KWIATKOWSKI, PETER C.B. PHILLIPS, PETER SCHMIDT and YONGCHEOL SHIN (KPSS, 1992). They developed a test where the null hypothesis is not the existence of a unit root but – quite the contrary – stationarity. (This test is therefore often called a stationarity test contrary to the unit root tests discussed so far.)

Contrary to relation (5.14) where we assume high positive autocorrelation in the time series, the starting point of this *KPSS test* is the following model:

(5.25) $$y_t = \alpha_t + \beta t + u_t,$$

where now instead of the commonly used constant term, a random walk,

(5.25a) $$\alpha_t = \alpha_{t-1} + \varepsilon_t$$

is allowed.

The residuals of (5.25a), ε, are assumed to be independently and identically normally distributed. Under the null hypothesis that y is trend stationary, the variance of ε is zero, i.e. α_t is a constant. The problem is now to find a test procedure which can discriminate between a constant term and a random walk. Such a test is designed for situations in which a random walk might possibly be added to a (trend) stationary component. It is the purpose of the test to detect this random walk.

The KPSS test tries to discriminate as follows between a purely trend stationary process and a process with an additive random walk. In a first step, y is regressed on a constant term and possibly also on a deterministic trend, i.e. it is adjusted for deterministic components. In a second step, partial sums of the residuals û of these regressions are considered:

$$S_{t,j} = \sum_{i=1}^{t} \hat{u}_{i,j},$$

where $j = \mu$, Tr, indicates whether the original series is only adjusted for a constant term or also for a deterministic trend. If y is a stationary process, the sum of the residuals with zero mean, is integrated of order one. The sum of the squares of an I(1) process diverges with T^2. Therefore, the test statistic

$$(5.26) \qquad \hat{\eta}_j = \frac{1}{T^2} \frac{\sum_{t=1}^{T} (S_{t,j})^2}{s_u^2}, \quad j = \mu, Tr,$$

has a limiting distribution that does not depend on additional parameters. Critical values for this statistic, which are again derived with simulations, are given by DENIS KWIATKOWSKI, PETER C.B. PHILLIPS, PETER SCHMIDT and YONGCHEOL SHIN (1992, p. 166).

In this form, the test presupposes that the residuals of the original process (5.25) are white noise. As this is usually not the case, the possible autocorrelation must be taken into account. The authors suggest that instead of s_u^2, as with the Phillips-Perron test, the estimator for the long-run variance defined in (5.18), s_{Tm}^2 – adjusted for the impact of autocorrelation – should be employed. Asymptotically, the same critical values as in the model with white noise residuals are appropriate.

Example 5.4

UWE HASSLER and JÜRGEN WOLTERS (1995) asked whether the inflation rates of consumer prices (calculated with respect to the previous month) in the United States, the United Kingdom, France, Germany and Italy are weakly stationary. They used seasonally adjusted monthly data from January 1969 to September 1992. They employed the ADF test and the Phillips-Perron test, where the null hypothesis postulates a unit root, as well as the KPSS test, where we assume weak stationarity under the null hypothesis, and they performed the tests for different lag lengths k and different truncation parameters, m, respectively. All test equations contain a constant term but no trend variable.

The results are given in *Table 5.3*. Irrespective of the number of autocovariances included, the Phillips-Perron test always rejects the null hypothesis of a unit root at least at the 1 percent significance level. According to these results, the monthly inflation rates of all countries are stationary. On the other hand, the KPSS test nearly always rejects the null hypothesis of stationarity also at the 1 percent level. Thus, according to these results, the inflation rates exhibit nonstationary behaviour. The situation is different for the ADF test. The null hypothesis of a unit root is always rejected for k = 3, but only in three out of five cases for k = 6, and never for k = 12, not even at the 10 percent level. In this example, the results of the semi-parametric tests, the Phillips-Perron and the KPSS tests, are hardly influenced by the value of m, whereas the results of the ADF test are sensitive to changes of k. Moreover, the results of the two semi-parametric tests contradict each other.

Table 5.3: *Results of Unit Root and Stationarity Tests for Inflation 1/1969 – 9/1992, 285 Observations*

	m/k	United States	United Kingdom	France	Germany	Italy
Phillips-Perron	6	-8.95**	-9.30**	-5.82**	-10.32**	-6.40**
	12	-10.20**	-10.54**	-6.84**	-11.65**	-7.39**
KPSS	6	0.81**	1.02**	1.57**	1.26**	0.94**
	12	0.51*	0.65**	0.91**	0.80**	0.56**
ADF	3	-4.43**	-4.48**	-2.71(*)	-4.98**	-3.31*
	6	-3.06*	-2.97*	-1.71	-3.49**	-2.24
	12	-1.86	-2.27	-1.29	-1.75	-2.39

'(*)', '*' or '**' denote that the corresponding null hypothesis can be rejected at the 10, 5, or 1 percent significance level, respectively.

Source: U. HASSLER and J. WOLTERS (1995, *Tables 3* and *4*, p. 39).

As *Example 5.4* shows, problems arise whenever different test procedures produce different, contradictory results and when these results are to be interpreted. One reason for such contradictions might be the fact that the tests discussed so far can only differentiate between the integer orders of integration d = 0 and d = 1, which corresponds to the methodology of the ARIMA(p, d, q) models with d = 0, 1, 2, ..., . One possibility to handle the problem is to gain more flexibility by abandoning the restriction to integer orders of integration: d might be treated as a real number. How this is done within the framework of fractionally integrated ARMA models is discussed below in *Section 5.5*.

5.4 Decomposition of Time Series

If one takes into account that nonstationary time series might contain a stationary component along with the nonstationary one, the decomposition of the series into two components, a permanent and a transitory one, seems fairly obvious:

$$(5.27) \qquad y_t = y_t^p + y_t^t,$$

where y^p denotes the permanent (nonstationary) and y^t transitory (stationary) component. Such a decomposition makes it possible to find a measure of the persistence of the series, i.e. for the relative importance of changes in its permanent component compared to changes in the series itself.

Such a decomposition was proposed, for example, by STEPHEN BEVERIDGE and CHARLES R. NELSON (1981). They showed that every ARIMA model with d = 1 can be represented as the sum of a random walk, possibly with drift,

$$(5.28) \qquad y_t^p = \mu + y_{t-1}^p + v_t,$$

and a stationary component which is the difference between the process y itself and its nonstationary component y^p.

Starting point for the decomposition is the general ARIMA(p,1,q) model. To make things easier, we use the Wold decomposition of Δy, written in the following form:

$$y_t = \mu + \psi(L) u_t + y_{t-1}.$$

By backward substitution we get

$$\begin{aligned} y_t &= \mu + \psi(L) u_t + \mu + \psi(L) u_{t-1} + y_{t-2} \\ &= 2\mu + \psi(L)(u_t + u_{t-1}) + y_{t-2} \\ &\vdots \\ &= t\mu + \psi(L) \sum_{i=1}^{t} u_i + y_0. \end{aligned}$$

With the additional assumptions $y_0 = 0$ and $u_t = 0$ for $t \leq 0$ it follows that

$$y_t = t\mu + \sum_{j=0}^{\infty} \psi_j \left(\sum_{i=j}^{t-1} u_{t-i} \right).$$

This can be transformed to

$$y_t = t\mu + \sum_{j=0}^{\infty} \psi_j \left(\sum_{i=0}^{t-1} u_{t-i} \right) - \sum_{j=1}^{\infty} \psi_j \left(\sum_{i=0}^{j-1} u_{t-i} \right)$$

$$= t\mu + \psi(1) \left(\sum_{i=0}^{t-1} u_{t-i} \right) - \sum_{i=0}^{\infty} u_{t-i} \left(\sum_{j=i+1}^{\infty} \psi_j \right).$$

Defining

$$y_t^p = t\mu + \psi(1) \left(\sum_{i=0}^{t-1} u_{t-i} \right),$$

leads to the representation given in (5.28). Thus, we get

(5.29) $$y_t^p = \mu + y_{t-1}^p + \psi(1) u_t,$$

with $v_t = \psi(1) u_t$.

(5.30) $$y_t^t = \xi(L) u_t, \text{ with } \xi_i = -\sum_{j=i+1}^{\infty} \psi_j, \quad i = 0, 1, 2, \ldots$$

holds for the transitory component $y_t^t = y_t - y_t^p$.

The permanent component y^p can also be represented by the observed values of y. To show this, we start with the representation of an ARIMA(p,1,q) process,

$$\alpha(L) \Delta y_t = \delta + \beta(L) u_t \text{ with } \mu = \delta/\alpha(1),$$

where the roots of $\alpha(L) = 0$ and $\beta(L) = 0$ are all outside the unit circle and there are no identical roots in the two polynomials. Solving for u results in

$$u_t = \frac{\alpha(L)}{\beta(L)} \Delta y_t - \frac{\delta}{\beta(1)} = \frac{\alpha(L)}{\beta(L)} \Delta y_t - \frac{\alpha(1)}{\beta(1)} \mu.$$

Thus, (5.29) leads to

$$\Delta y_t^p = \mu + \psi(1) u_t$$

$$= \mu + \frac{\beta(1)}{\alpha(1)} \left[\frac{\alpha(L)}{\beta(L)} \Delta y_t - \frac{\alpha(1)}{\beta(1)} \mu \right],$$

or

(5.31) $$y_t^p = \frac{\beta(1)}{\alpha(1)} \cdot \frac{\alpha(L)}{\beta(L)} y_t,$$

respectively, i.e. the permanent component can be represented as a weighted average of the observed values.

As a measure of the persistence of the time series, P, we define

$$(5.32) \qquad P = \frac{\sigma_v^2}{\sigma_{\Delta y}^2} = \frac{(\psi(1))^2 \sigma_u^2}{\sigma_{\Delta y}^2}.$$

The problem with this decomposition, however, is that the residuals of the stationary and the nonstationary parts are perfectly negatively correlated, except for the degenerated case $\psi(1) = 0$, where the permanent component is the straight line μt. If we assume a different value for the correlation between these two parts, we get a different decomposition. (An obvious assumption would be that the innovations of the permanent and transitory parts are uncorrelated.) Thus, depending on the assumption about the correlation between the two innovation series, we can derive rather different decompositions leading to different values of the permanent component. However, as JOHN H. COCHRANE (1988) showed, the variance of the different estimates of the permanent component will always be the same, thus, leading to the same value of the persistence measure.

An alternative measure for the persistence of a time series was proposed by JOHN H. COCHRANE (1988). He considers the ratio of the variance of the changes that are accumulated over k periods to the variance of the one period change,

$$(5.33) \qquad V_k = \frac{1}{k+1} \frac{E(y_{t+k} - y_{t-1})^2}{E(y_t - y_{t-1})^2}, \quad k = 1, 2, \ldots.$$

As the changes (of an I(1) process) are stationary by definition, and because of

$$\rho(j) = \frac{E[(y_{t+j} - y_{t+j-1})(y_t - y_{t-1})]}{E[(y_t - y_{t-1})^2]},$$

we get

$$(5.34) \qquad V_k = 1 + 2 \sum_{j=1}^{k} \left(1 - \frac{j}{k+1}\right) \rho(j).$$

If k tends to infinity, we get

$$(5.35) \qquad \lim_{k \to \infty} V_k = 1 + 2 \sum_{j=1}^{\infty} \rho(j).$$

As ρ(k) tends towards zero with increasing k in stationary processes, JOHN H. COCHRANE (1988) proposed to increase k until V_k approaches its maximum and to use this k to estimate the persistence of a series.

Example 5.5

The special case of a random walk, $y_t = y_{t-1} + u_t$ results in:

$$E[(y_{t+k} - y_{t-1})^2] = E[(\Delta y_{t+k} + \Delta y_{t+k-1} + ... + \Delta y_t)^2]$$
$$= (k+1)\sigma_u^2.$$

According to (5.33), we thus get

$$V_k = 1, \quad k = 1, 2, ...,$$

i.e. this measure shows that the random walk does not contain any stationary (transitory) component besides the stochastic trend.

A different approach to decompose a time series into a permanent component y^p and a transitory (cyclical) component y^t goes back to ROBERT J. HODRICK and EDWARD C. PRESCOTT (1997). Contrary to the approach of STEPHEN BEVERIDGE and CHARLES R. NELSON (1981), which is based on an ARIMA(p,1,q) model, ROBERT J. HODRICK and EDWARD C. PRESCOTT (1997) do not presume an explicit model for the observed time series. The idea is rather to model the permanent component y^p sufficiently smooth. The sum of squares of the second differences of y^p is taken as a measure of the smoothness of the time path. On average, the cyclical component, $y^t = y - y^p$ should not deviate substantially from zero over the observation period. To approach these goals, the following objective function is minimised with respect to y^p

$$(5.36) \quad Z(y_t^p; \lambda) = \sum_{t=1}^{T}(y_t - y_t^p)^2 + \lambda \sum_{t=2}^{T-1}\left((y_{t+1}^p - y_t^p) - (y_t^p - y_{t-1}^p)\right)^2.$$

The smoothness of y^p can be controlled for with the penalty parameter λ. The larger λ is chosen, the smoother is the time path of y^p. For $\lambda \to \infty$, y^p follows a linear trend. The values of λ depend on the frequency of the data. In practical applications, the following values are often chosen: $\lambda = 100$ for annual data, $\lambda = 1'600$ for quarterly data, and $\lambda = 14'400$ for monthly data. The result of this minimisation is the so-called Hodrick-Prescott (HP) filter which provides the permanent or trend component, respectively.

In empirical macroeconomics, the HP filter is today the standard approach to estimate the permanent component of a time series. Although quite different values for the penalty parameter λ are suggested depending on the frequency of the data, the result of the cyclical components is rather robust with respect to the choice of λ. In contrast to this, the estimates for the last values of the permanent component of the observation period are quite sensitive to the choice of the sample endpoint.

Example 5.6

The permanent component of the annual German inflation rate is to be determined by using the Beveridge-Nelson approach and the HP filter. We investigate the period from the first quarter of 1975 to the last quarter of 1998, as this corresponds to the period when the German Bundesbank used the quantity of money as its target. To measure inflation, we use the implicit deflator of the gross national product (PGNP), i.e. $IR_t = 100 \cdot (\ln(PGNP_t) - \ln(PGNP_{t-4}))$. Estimating an ARIMA model leads to the following result:

(E5.2) $\quad \Delta IR_t = - \underset{(-3.29)}{0.308} \Delta IR_{t-4} + \hat{u}_t + \underset{(2.68)}{0.275} \hat{u}_{t-2},$

$$\bar{R}^2 = 0.145, \quad SE = 0.571, \quad Q(6) = 4.233 \ (p = 0.645),$$

where the t values are again indicated in parentheses. Both estimated coefficients differ significantly from zero at the 1 percent level, and the Box-Ljung Q statistic, calculated with 8 correlation coefficients (6 degrees of freedom), does not indicate any remaining autocorrelation of the residuals. For the ARIMA(4, 1, 2) model in (E5.2) we get:

(E5.3a) $\quad\quad\quad\quad \alpha(L) = 1 + 0.308 \ L^4$, and

(E5.3b) $\quad\quad\quad\quad \beta(L) = 1 + 0.275 \ L^2.$

The Wold representation ψ(L) is derived by a series expansion of β(L)/α(L). This results in

$$\psi(L) = \frac{\beta(L)}{\alpha(L)},$$

$$\psi(1) = \frac{\beta(1)}{\alpha(1)} = \frac{1.275}{1.308} = 0.975.$$

Because of the parameters estimated in (E5.2), we get $\sigma^2_{\Delta IR} = 1.188 \ \sigma^2_u$ for the variance of ΔIR. (See for this *Section 2.3.2*.) According to (5.31), the permanent component $IR^{p,BN}$ is

$$IR_t^{p,BN} = 0.975 \ \frac{1 + 0.308 \ L^4}{1 + 0.275 \ L^2} \ IR_t, \quad \text{or}$$

$$IR_t^{p,BN} = -0.275 \ IR_{t-2}^{p,BN} + 0.975 \ IR_t + 0.300 \ IR_{t-4}.$$

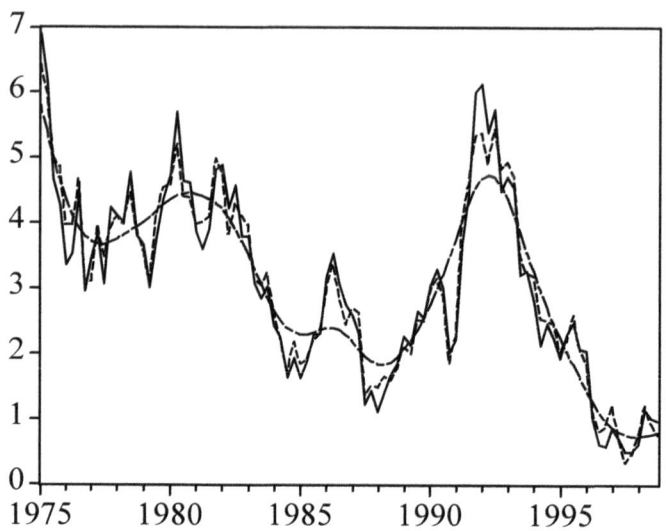

Figure 5.10a: *German Inflation Rate: Actual values (——), permanent component according to S. BEVERIDGE and CH.R. NELSON (-------), permanent component according to R.J. HODRICK and E.C. PRESCOTT (– – – – –)*

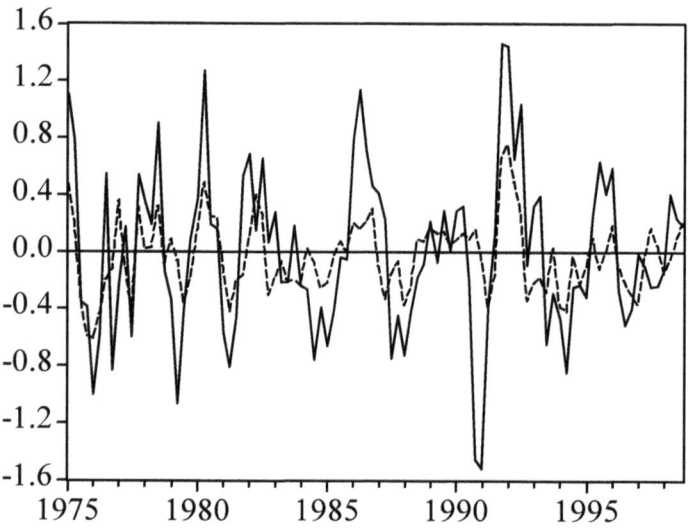

Figure 5.10b: *German Inflation Rate: cyclical component according to S. BEVERIDGE and CH.R. NELSON (--------), cyclical component according to R.J. HODRICK and E.C. PRESCOTT (——)*

Figure 5.10a shows the observed inflation rate IR, together with the permanent component $IR^{p,BN}$ which was calculated according to the Beveridge-Nelson approach. The development of the permanent component is quite similar to the actual inflation rate. The only difference is that it does not exhibit the extreme amplitudes of the original series. Contrary to this, when using the HP filter, the permanent component of the series, $IR^{p,HP}$, which is also shown in *Figure 5.10a*, is much smoother. It must be taken into account that it was not calculated with $\lambda = 1'600$, which is normally used for quarterly data, but with $\lambda = 100$, because otherwise the development would have been too smooth.

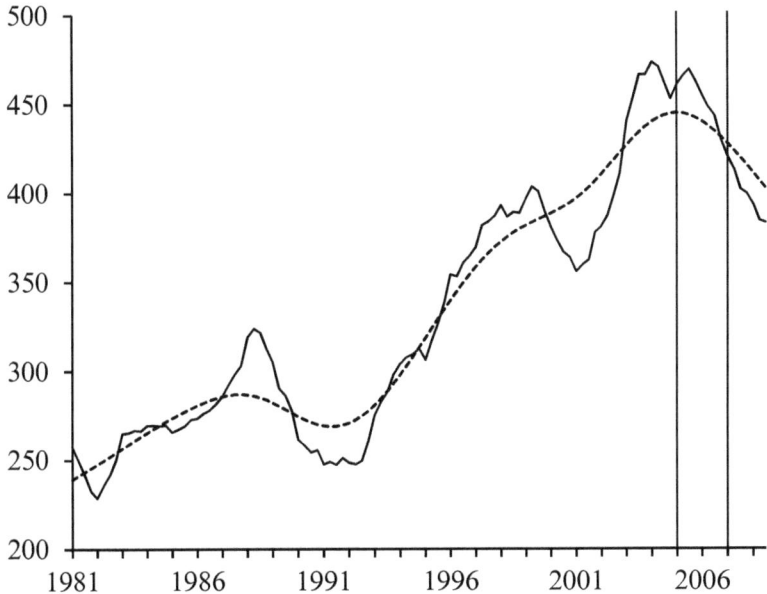

Figure 5.11a: Swiss real money balances M2, 1981 – 2008: Actual values (———) and permanent component (--------) due to the Hodrick-Prescott filter

Example 5.7

To demonstrate how the choice of the sample endpoint influences the slope of the permanent component estimated by the HP filter near the end of the observation period, we use quarterly seasonally adjusted data for Swiss real money balances in the definition of M2 for the period starting in 1981. We use $\lambda = 1600$ which is the suggested value for quarterly data. To analyse the behaviour of the long-run component in the pre-crisis years 2005 and 2006, we used two different endpoints: the third quarter of 2008 and the fourth quarter of 2006.

Figure 5.11a shows the result if we use the data up to the third quarter of 2008. The long-run component is clearly decreasing during the years 2005 and 2006. In contrast to this result, when the observation period only extends to 2006, the estimated long-run trend is increasing during this period, as shown in *Figure 5.11b*, whereas, the estimates of the permanent component for the period before 2005 are hardly influenced by the choice of the endpoint. This indicates that it is highly problematic to rely on the final estimates of the trend for economic policy decisions.

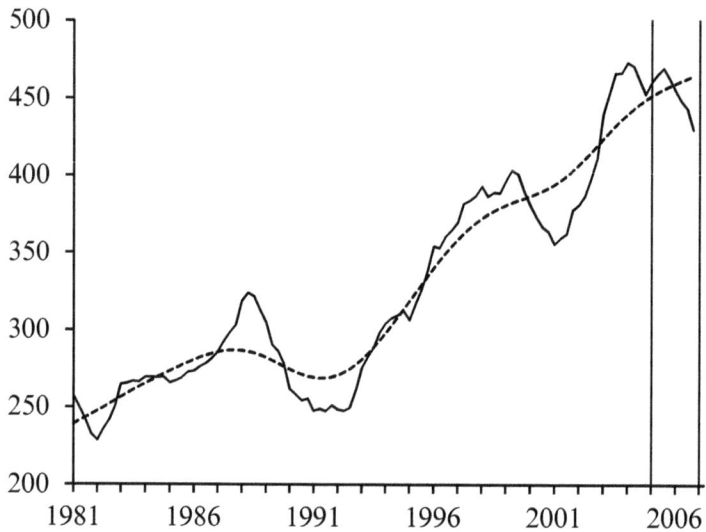

Figure 5.11b: Swiss real money balances M2, 1981 – 2006: Actual values (———) and permanent component (-------) due to the Hodrick-Prescott filter

5.5 Further Developments

As shown above in *Example 5.4*, problems arise if tests lead to systematically contradictory results which cannot be interpreted as being statistical artefacts. This indicates that the approach used so far to handle nonstationarities is not flexible enough. The fractionally integrated models discussed below are one possibility to cope with this problem and to get more flexible solutions.

A further, not yet discussed problem arises whenever fourth differences (for quarterly data) or twelfth differences (for monthly data) are performed in order to transform a nonstationary into a stationary time series. This procedure is often used when annual growth rates are calculated (with quarterly or monthly data). The problem of *seasonal integration* which is presupposed by this procedure shall also be discussed in the following.

5.5.1 Fractional Integration

As mentioned above, the concept of integrated time series should be extended to that effect that the order of integration, d, is no longer restricted to be an integer number. It might be any real number. By forming first differences, we can always reduce the value of d by one. In the following, we therefore only consider the interval $0 \leq d \leq 1$. In analogy to the definition of integrated variables given in *Section 5.1* the following definition holds:

- A stochastic process y is *fractionally integrated of order d*, $0 < d < 1$, if it can be transformed into a weakly stationary invertible process using the filter $(1 - L)^d$, i.e.

$$(1 - L)^d y_t = \delta + x_t,$$

where x is an ARMA(p,q) process. The transformation $(1 - L)^d$ results from the binomial series development

(5.37) $\quad (1 - L)^d = 1 - dL - \dfrac{d(1-d)}{2!}L^2 - \dfrac{d(1-d)(2-d)}{3!}L^3 - ...$

$$= \sum_{j=0}^{\infty} d_j L^j \text{ with } d_j = \frac{j-1-d}{j} d_{j-1}, \ d_0 = 1.$$

The original process y is then denoted as an ARFIMA(p,d,q) process (Autoregressive Fractional Integrated Moving Average Process).

The coefficients d_j are quadratically summable for $|d| < 0.5$. For this reason, the process is stationary for $0 < d < 0.5$ and displays *long memory* in the sense that the autocorrelations are not summable. For $d \geq 0.5$ the process is nonstationary. Thus, there is a whole range of values of d ($0.5 \leq d \leq 1$) that generate persistent processes and not only the single value $d = 1$ (or integer multiples of it), like with the ARIMA(p,d,q) models.

The inverse filter $(1 - L)^{-d}$ is given by the substitution of d by -d in (5.37),

(5.37') $\quad (1 - L)^{-d} = 1 + dL + \dfrac{d(1+d)}{2!}L^2 + \dfrac{d(1+d)(2+d)}{3!}L^3 + ... \ .$

Thus, if we apply the filter $(1 - L)^{-d}$ on the stationary and invertible ARMA(p,q) process with the representation $\alpha(L) x_t = \beta(L) u_t$, we get an ARFIMA process with

$$y_t = (1-L)^{-d} x_t.$$

If $x_t = u_t$, i.e. a pure random process, we get the model of a pure, fractionally integrated noise:

(5.38) $\qquad (1-L)^d y_t = u_t \quad \text{or} \quad y_t = (1-L)^{-d} u_t.$

The series expansion in (5.37) or (5.37'), respectively, indicates that this process might be represented as a special $AR(\infty)$ or $MA(\infty)$ process. Relation (5.38) gives the most parsimonious parameterisation of it, employing only one single parameter.

The unit root tests discussed in *Sections 5.3.1, 5.3.2* and *5.3.3* test the null hypothesis $d = 1$ against the alternative hypothesis $d = 0$, while the KPSS test, described in *Section 5.3.5*, tests the null hypothesis $d = 0$ against the alternative hypothesis $d = 1$. If the 'true' d is between zero and one, both null hypotheses might be rejected, as was the case in *Example 5.4*. The reason for this apparent contradiction between the results of the two tests is that the modelling approach only allowed for zero and one to be possible orders of integration and was thus too restrictive.

Example 5.8

Due to the contradicting results with respect to the stationarity properties presented in *Example 5.4*, UWE HASSLER and JÜRGEN WOLTERS (1995) estimated ARFIMA models for the inflation rates of these countries. They showed that according to (5.38), the monthly inflation rates of all these countries can be modelled as purely fractionally integrated white noise. The values of d vary from $d = 0.40$ for Germany, $d = 0.41$ for the United States, $d = 0.51$ for the United Kingdom, $d = 0.54$ for France up to $d = 0.57$ for Italy. The null hypothesis that the order of integration equals 0.5 can in no case be rejected. As fractional processes with $d \geq 0.5$ are nonstationary, at least the inflation rates of the United Kingdom, France and Italy show persistent behaviour, even if they are not I(1).

A so-called fractional integration test, or long memory test, in the tradition of the Augmented Dickey-Fuller test has been proposed by MATEI DEMETRESCU, VLADIMIR KUZIN and UWE HASSLER (2008) who refined a suggestion by JÖRG BREITUNG and UWE HASSLER (2002). Notwithstanding the analogy to the ADF test, the limiting distribution is standard, however. The null hypothesis is that the order of integration equals some value δ with $0 \leq \delta \leq 1$:

$$H_0: d = \delta.$$

Then one computes the corresponding fractional differences under the null hypothesis: $(1-L)^\delta y_t$. In practice, the infinite expansion of the differences has to be truncated at the beginning of the sample. With the expansion from (5.37), we write

$$(1-L)^\delta = \sum_{j=0}^{\infty} \delta_j L^j,$$

and define the approximate differences

$$z_t = \sum_{j=0}^{t-1} \delta_j y_{t-j} \approx (1-L)^\delta y_t, \quad t = 1, \ldots, T.$$

The lag-augmented regression equation becomes in analogy to the ADF test

$$z_t = \varphi z^*_{t-1} + \sum_{i=1}^{k} \theta_i z_{t-i} + u_t, \quad t = k+1, \ldots, T,$$

where the variable of interest is the weighted sum:

$$z^*_{t-1} = \sum_{j=1}^{t-1} \frac{z_{t-j}}{j}.$$

Note for $\delta = 1$ that $z_t = \Delta y_t$, such that the new regression corresponds to (5.17"), only the nonstationary regressor $y_{t-1} = \sum_{j=1}^{t-1} \Delta y_{t-j}$, where we assume a starting value of zero, is replaced by the regressor $z^*_{t-1} = \sum_{j=1}^{t-1} j^{-1} \Delta^\delta y_{t-j}$.

Under the null hypothesis, z^*_{t-1} is *asymptotically stationary*, such that the least-squares regression results in a limiting normal distribution of $\hat{\varphi}$. It was shown that the null hypothesis of interest, $d = \delta$, translates into $\varphi = 0$ in the regression equation. Hence, the test relies on the usual t statistic, t_φ, compared with the conventional critical values from the standard normal distribution. It is possible to perform a one-sided test, where H_1: $d > \delta$ corresponds to H_1: $\varphi > 0$, and analogously for H_1: $d < \delta$. Null hypotheses of particular interest are $\delta = 0$ (short memory) against $d > 0$ (long memory), where the test is a powerful competitor to the KPSS test against fractional alternatives, and similarly for the traditional unit root case $\delta = 1$ versus $d < 1$. Moreover, one may test against (non)stationarity for H_0: $\delta = 0.5$ against $d < 0.5$ (stationarity) or $d > 0.5$ (nonstationarity).

5.5.2 Seasonal Integration

The integrated processes discussed so far exhibit nonstationary behaviour because there is a unit root in the lag polynomial of the autoregressive part. This can be eliminated by forming first differences. One might ask whether there are additional roots on the unit circle which imply nonstationarity and can be economically interpreted. As shown in *Sections 1.2* and *1.3*, the application of the filter $1-L^4$ generated developments of quarterly data which no longer exhibit seasonal variations. The factorisation

$$(1-z^4) = (1-z^2)(1+z^2) = (1-z)(1+z) \cdot (1-iz)(1+iz)$$

where $i^2 = -1$, immediately shows that $1-z^4$ has four roots on the unit circle, i.e.

$$z_1 = 1, \quad z_2 = -1, \quad z_{3,4} = \pm i.$$

Using the filter $1 - L$, the following process can be generated with u_t as white noise

$$(1-L)\, y_t = u_t,$$

or

$$y_t = y_{t-1} + u_t.$$

This corresponds to a random walk which can be used to model stochastic trend behaviour. Applying the filter $1 + L$, the process

(5.39) $$y_t = -y_{t-1} + u_t$$

can be similarly generated. For large values of t, the correlation between two adjacent elements of this process approaches -1, i.e. the process exhibits regular two-period fluctuations which correspond to fluctuations within a period of half a year for quarterly data. This also becomes clear if y_{t-1} is substituted in (5.39), which leads to

$$y_t = y_{t-2} + u_t - u_{t-1}.$$

If we only considered every second observation, we would again get a random walk.

The roots $\pm\, i$ correspond to the filter $1 + L^2$, which can generate the process

(5.40) $$y_t = -y_{t-2} + u_t.$$

Here, all adjacent elements are uncorrelated, while the correlation between the values of y which are two periods apart from each other converges to -1 for large values of t. Thus, the process exhibits fluctuations with a length of four periods, corresponding to the annual cycle in the context of quarterly data. This also becomes clear if y_{t-2} in (5.40) is substituted. This leads to

$$y_t = y_{t-4} + u_t - u_{t-2}.$$

5.5 Further Developments

If we only considered every fourth period, we would again get a random walk.

Thus, the processes with roots -1 and ± i capture the nonstationary seasonal fluctuations of quarterly data. To eliminate such fluctuations, the filter

$$(1 + L)(1 + L^2) = 1 + L + L^2 + L^3,$$

must be used, i.e. a third order moving average eliminates nonstationary seasonal fluctuations of quarterly data. Because of

$$(1 - L^4) = (1 - L)(1 + L + L^2 + L^3),$$

forming annual differences also eliminates any stochastic trend, as *Figures 1.4* and *1.5* in *Chapter 1* already showed.

In analogy to the ADF test, SVEND HYLLEBERG, ROBERT F. ENGLE, CLIVE W.J. GRANGER and BYUNG SAM YOO (1990) (HEGY) developed a procedure which not only tests for the stochastic trend but also for the different seasonal roots. In order to perform this test, the quarterly series y has to be transformed in the following way:

$$y_{1,t} = (1 + L + L^2 + L^3) y_t,$$
$$y_{2,t} = -(1 - L + L^2 - L^3) y_t,$$
$$y_{3,t} = -(1 - L^2) y_t,$$
$$y_{4,t} = (1 - L^4) y_t.$$

y_1 is a series which no longer contains any seasonal unit root. y_2 is a series which does not contain a stochastic trend, nor any annual fluctuations, whereas the stochastic trend as well as the half annual cycle have been eliminated from y_3. Finally, y_4 does not have any root on the unit circle. Disregarding all deterministic terms like the constant term, a time trend or seasonal dummies, the following equation is estimated by OLS in order to perform the HEGY test:

$$\theta^*(L) y_{4,t} = \pi_1 y_{1,t-1} + \pi_2 y_{2,t-1} + \pi_3 y_{3,t-1} + \pi_4 y_{3,t-2} + u_t,$$

where the order of the lag polynomial $\theta^*(L)$ is chosen in a way that the estimated residuals û are white noise.

The null hypothesis that there is a stochastic trend is stated as

$$H_0: \pi_1 = 0,$$

the null hypothesis that there is a nonstationary semi-annual component as

$$H_0: \pi_2 = 0$$

and the null hypothesis that there is a nonstationary annual component as

$$H_0: \pi_3 = \pi_4 = 0.$$

The test statistics are the corresponding t or F values, respectively. As with the 'usual' unit root test, the classical t and F distributions do not hold for this test. Depending on which deterministic terms are included, different critical values are appropriate. The corresponding values for the HEGY test, derived again with simulations, are provided in SVEND HYLLEBERG et al. (1990, *Tables 1a* and *1b*, pp. 226f).

5.6 Deterministic versus Stochastic Trends in Economic Time Series

It has hardly ever been disputed that economic time series are trending, even though procedures for stationary variables have mostly been applied. As mentioned in *Chapter 1*, even the *classical* time series analysis distinguished between trend, (business) cycle, seasonal variation and irregular movements. However, the 'nature' of the trend has hardly ever been considered. Depending on the kind of procedure, either high order moving averages were calculated or linear or polynomial (deterministic) trends estimated and subtracted from the original series. Series transformed in this way were used for further investigations.

Whether such a trend is deterministic or stochastic, however, is not only important for the application of the appropriate statistical procedures but also has an impact on the economic interpretation. If, for example, the logarithm of the gross national product follows a linear deterministic trend, the model not only implies a constant long-run growth rate but also the fact that all deviations from the long-run equilibrium path are only temporary; all deviations are counter-balanced in the long-run. Contrary to this, when the series follows a stochastic trend, singular changes have permanent consequences: the series has a (long) *memory*. Even if the long-run growth rate is fixed, a variable deviating from the growth path it has followed so far will hardly ever return to the path: from this new initial point, the development continues with the same (average) growth rate but along a new path (with a different level). Thus, these kinds of shocks are called permanent contrary to the transitory shocks in the model with a deterministic trend.

Permanent and transitory shocks have a different economic meaning. Permanent shocks are usually attributed to the supply side, transitory shocks rather to the demand side of the economy. Correspondingly, unexpected changes of the quantity of money are typically interpreted as transitory shocks: They might have real effects in the short run, but they have no

5.6 Deterministic versus Stochastic Trends in Economic Time Series

long-run impact, at least as long as the classical dichotomy is accepted. Therefore, monetary policy might be stabilising in the short run, but has hardly any long-run effect on economic growth, at least as long as inflation is 'moderate'. One indication for this is that empirical studies on the relation between (moderate) inflation and economic growth do not exhibit conclusive results. Contrary to this, a technology shock is usually seen as permanent: The development of a new technology which has not been available so far has a permanent effect on the production possibilities in an economy and might, therefore, shift the economy to a new growth path with a higher initial position. Against this background it is understandable that it has been extensively discussed in the United States whether GNP has a unit root or not, a question which at first glance seems to be a purely statistical one.

The distinction between permanent and transitory shocks has, above all, an impact on business cycle theory. Traditional Keynesian as well as Monetarist approaches assume that cyclical fluctuations are caused by transitory shocks. As shown in *Chapter 2*, given a specific structure of the economic system (or the time series representing this system), uncorrelated random shocks can generate cycles with certain frequencies. If the necessary information is available, (anti-cyclical) stabilisation policy can counteract and thus smooth the economic development. Correspondingly, in their discussion on the possibility of the government to perform an active stabilisation policy, Monetarists and Keynesians focused on two questions: (i) Which one is the better instrument, monetary or fiscal policy? (ii) Does the government (or the central bank, respectively) have the information necessary to perform a successful stabilisation policy or does it even make things worse because interventions often take place at the wrong point of time? The general possibility of stabilisation policy was not called into question.

A quite different stance has been taken by the *Real Business Cycle Theory* which belongs to the *New Classical Macroeconomics*. It attempts to interpret business cycles as results of technology shocks. In such a model, any economic policy that tries to stabilise business cycles is useless in the first place. Theoretical models with such properties have been developed. However, the empirical evidence is not very convincing. Even if the existence of permanent shocks is taken into account, it is sensible to assume that there are both temporary and permanent shocks on the supply side as well as on the demand side. The question no longer is whether such impacts exist at all but rather how strong the different impacts (shocks) are in relation to each other. Recent empirical research goes in this direction.

This implies, however, that the same model has to allow for transitory as well as permanent shocks. While the traditional models of a deterministic trend do not have this possibility, as they only know transitory deviations of the fixed long-run equilibrium path, models with a stochastic trend usually also contain a transitory component. It is the purpose of the procedures discussed in *Section 5.4* to differentiate between these two components.

Finally, it must be mentioned that, given the existence of permanent shocks, the distinction between trend and cycle is dubious. From an economic perspective, this implies that a distinct separation between economic growth on the one hand and the development of the business cycle on the other hand is no longer possible; if the economic system has the appropriate structure, economic growth occurs in cycles. This is a new way to take up an old idea, which was already developed by JOSEPH A. SCHUMPETER in his *"Theory of Economic Development"*. In this theory, business cycles are also generated by supply shocks and not by demand shocks.

For all this, however, we should take into account that we always use samples for empirical analysis and that the 'true' data generating processes are different from the ones assumed in our models. For example, we often assume that the investigated variables are normally distributed. This implies that the occurring values can be both very high and very low, even if we know that this would be impossible in a concrete situation. Body heights are a classical example of this. The same is true for the differentiation between stationary and nonstationary variables. If a variable is really stationary, the estimator for the mean of the coming year might be better provided by the mean of some past observations with long distances between each other than by the mean of the last three months. On the other hand, the assumption of nonstationarity implies that, with increasing time horizon, the variable will almost certainly exceed any limit. Both assumptions are, for example, invalid for interest rates. When we investigate samples, perform tests and finally decide to (preliminarily) regard the variable as stationary or nonstationary, we assume that the chosen model is the best available approximation on the unknown data generating process of the model classes we considered. This might be different in case of a different time period or a different frequency of data.

References

A useful informal **introduction** to the econometrics of nonstationary time series is given by

JAMES H. STOCK and MARK W. WATSON, Variable Trends in Economic Time Series, *Journal of Economic Perspectives* 2, issue 3/1988, pp. 147 – 174.

The procedure to identify **the number of unit roots** when calculating the solutions of the characteristic equation of an AR model was proposed by

GEORGE C. TIAO and RUEY S. TSAY, Consistency of Least Squares Estimates of Autoregressive Parameters in ARIMA Models, *Annals of Statistics* 11 (1983), pp. 856 – 871.

The **Dickey-Fuller test** goes back to

WAYNE A. FULLER, *Introduction to Statistical Time Series*, Wiley, New York 1976, and

DAVID A. DICKEY and WAYNE A. FULLER, Distribution of the Estimators for Autoregressive Time Series with a Unit Root, *Journal of the American Statistical Association* 74 (1979), pp. 427 – 431,

In WAYNE A. FULLER (1976), **critical values of the t statistics** of unit root tests were indicated for the first time. Today, the more precise values presented in

JAMES G. MACKINNON, Critical Values for Cointegration Tests, in: R.F. ENGLE and C.W.J. GRANGER (eds.), *Long-Run Economic Relationships: Reading in Cointegration*, Oxford University Press, Oxford et al. 1991, pp. 267 – 276,

or, the further improved ones in

JAMES G. MACKINNON, ALFRED HAUG and LEO MICHELIS, Numerical Distribution Functions of Likelihood Ratio Tests for Cointegration, *Journal of Applied Econometrics* 14 (1999), pp. 563 – 577,

are usually employed. Critical values of the **F statistics** are given in

DAVID A. DICKEY and WAYNE A. FULLER, Likelihood Ratio Statistics for Autoregressive Time Series with a Unit Root, *Econometrica* 49 (1981), pp. 1057 – 1072.

The **Phillips-Perron test** is presented in

PETER C.B. PHILLIPS and PIERRE PERRON, Testing for a Unit Root in Time Series Regression, *Biometrika* 75 (1988), pp. 335 – 346, as well as in

PIERRE PERRON, Trends and Random Walks in Macroeconomic Time Series: Further Evidence from a New Approach, *Journal of Economic Dynamics and Control* 12 (1988), pp. 297 – 332.

The latter paper provides a good survey of the different test statistics and the sources of the corresponding critical values. The window used by the Phillips-Perron test was proposed by

MAURICE STEVENSON BARTLETT, Smoothing Periodograms from Time Series with Continuous Spectra, *Nature* 161 (1948), pp. 686 – 687,

The question of how to determine the **optimal number** of correlation coefficients, m, used for this estimator, is discussed in

DONALD W. ANDREWS, Heteroskedasticity and Autocorrelation Consistent Covariance Matrix Estimation, *Econometrica* 59 (1991), pp. 817 – 858.

The testing procedure for unit roots in the presence of **structural breaks** was first proposed by

PIERRE PERRON, The Great Crash, The Oil Price Shock, and the Unit Root Hypothesis, *Econometrica* 57 (1989), pp. 1361 – 1401.

Some extensions are given in

PIERRE PERRON, Trend, Unit Root and Structural Change in Macroeconomic Time Series, in: B.B. RAO (ed.), *Cointegration for the Applied Economist,* St. Martin Press, New York 1994, pp. 113 – 146.

A survey about more recent developments for situations when the date of a structural break is unknown is given by

BRUCE E. HANSEN, The New Econometrics of Structural Change: Dating Breaks in U.S. Labor Productivity, *Journal of Economic Perspectives* 15, Issue 4/2001, pp. 117 – 128.

The fact that the null hypothesis of a unit root is rejected too often with the Dickey-Fuller as well as with the Phillips-Perron test if the process contains a MA part with negative first order autocorrelation was first mentioned in

G. WILLIAM SCHWERT, Effects of Model Specification on Tests for Unit Roots in Macroeconomic Data, *Journal of Monetary Economics* 20 (1987), pp. 73 – 103, as well as

G. WILLIAM SCHWERT, Tests for Unit Roots: A Monte Carlo Investigation, *Journal of Business and Economic Statistics* 7 (1989), pp. 147 – 159.

A testing procedure which is more appropriate in such a situation is given by

SAÏD E. SAÏD and David A. DICKEY, Hypothesis Testing in ARIMA(p,1,q) Models, *Journal of the American Statistical Association* 80 (1985), pp. 369 – 374.

The test that applies the **stationarity of a time series as null hypothesis** was developed by

DENIS KWIATKOWSKI, PETER C.B. PHILLIPS, PETER SCHMIDT and YONGCHEOL SHIN, Testing the Null Hypothesis of Stationarity Against the Alternative of a Unit Root, *Journal of Econometrics* 54 (1992), pp. 159 – 178.

The different philosophies behind the tests with nonstationarity or stationarity as the null hypothesis are discussed by

ULRICH K. MÜLLER, Size and Power of Tests for Stationarity in Highly Autocorrelated Time Series, *Journal of Econometrics* 128 (2005), pp. 195 – 213.

The different results of unit root and stationarity tests applied to inflation rates of different countries are given in

UWE HASSLER and JÜRGEN WOLTERS, Long Memory in Inflation Rates: International Evidence, *Journal of Business and Economic Statistics* 13 (1995), pp. 37 – 45.

A **survey** of the different test procedures is given in

JAMES H. STOCK, Unit Roots, Structural Breaks and Trends, in: R.F. ENGLE and D.L. MCFADDEN (eds.), *Handbook of Econometrics*, Volume IV, Elsevier, Amsterdam et al. 1994, pp. 2739 – 2841.

PETER C. B. PHILLIPS and ZHIJIE XIAO, A Primer on Unit Root Testing, *Journal of Economic Surveys* 12 (1998), pp. 423 – 470, and in

JÜRGEN WOLTERS and UWE HASSLER, Unit Root Testing, *Allgemeines Statistisches Archiv* 90 (2006), pp. 43 – 58; reprinted in: O. HÜBLER and J. FROHN (eds.), *Modern Econometric Analysis*, Springer, Berlin 2006, pp. 41 – 56.

An important paper introducing the application of unit root tests to many **economic time series of the United States** for the first time is

CHARLES R. NELSON and CHARLES I. PLOSSER, Trends and Random Walks in Macroeconomic Time Series: Some Evidence and Implications, *Journal of Monetary Economics* 10 (1982), pp. 139 – 162.

An application to German **real interest rates** is presented in

GEBHARD KIRCHGÄSSNER and JÜRGEN WOLTERS, Are Real Interest Rates Stable?, An International Comparison, in: H. SCHNEEWEISS and K.F. ZIMMERMANN (eds.), *Studies in Applied Econometrics*, Physica, Heidelberg 1993, pp. 214 – 238.

Theoretical considerations about what happens when **trend eliminations** are 'wrongly' performed are to be found in

K. HUNG CHAN, JACK C. HAYYA and J.-KEITH ORD, A Note on Trend Removal Methods: The Case of Polynomial Regression versus Variate Differencing, *Econometrica* 45 (1977), pp. 737 – 744,

CHARLES R. NELSON and HEEJOON KANG, Spurious Periodicity in Inappropriately Detrended Time Series, *Econometrica* 49 (1981), pp. 741 – 751, as well as in

CHARLES R. NELSON and HEEJOON KANG, Pitfalls in the Use of Time as an Explanatory Variable in Regression, *Journal of Business and Economic Statistics* 2 (1984), pp. 73 – 82.

The first procedure for a **decomposition of a time series** into its nonstationary and its stationary component and a measure for the **persistence of a time series** was proposed by

STEPHEN BEVERIDGE and CHARLES R. NELSON, A New Approach to the Decomposition of Economic Time Series into Permanent and Transitory Components with Particular Attention to Measurement of the Business Cycle, *Journal of Monetary Economics* 7 (1981), pp. 151 – 174,

Alternative procedures for the decomposition of time series were proposed by

ANDREW C. HARVEY, *Forecasting, Structural Time Series Models, and the Kalman Filter*, Cambridge University Press, Cambridge (England) et al. 1989, and

ROBERT J. HODRICK and EDWARD C. PRESCOTT, Post-War U.S. Business Cycles: A descriptive Empirical Investigation, *Journal of Money, Credit, and Banking* 29 (1997), pp. 1 – 16.

An alternative measure for the persistence comes from

JOHN H. COCHRANE, How Big is the Random Walk in GNP?, *Journal of Political Economy* 96 (1988), pp. 893 – 920.

An introduction into the theory and estimation of **fractionally integrated models** is to be found in the paper by UWE HASSLER and JÜRGEN WOLTERS (1995) mentioned above but also in

CLIVE W.J. GRANGER and ROSELYNE JOYEUX, An Introduction to Long-Memory Time Series Models and Fractional Differencing, *Journal of Time Series Analysis* 1 (1980), pp. 15 – 29; or

RICHARD T. BAILLIE, Long Memory Processes and Fractional Integration in Economics, *Journal of Econometrics* 73 (1996), pp. 5 – 59.

Tests for fractionally integrated series which are analogous to the Dickey-Fuller test for unit roots, based on a suggestion of

JÖRG BREITUNG and UWE HASSLER, Inference on the Cointegration Rank in Fractionally Integrated Processes, *Journal of Econometrics* 110 (2002), 167 – 185,

are proposed by

MATEI DEMETRESCU, VLADIMIR KUZIN and UWE HASSLER, Long Memory Testing in the Time Domain, *Econometric Theory* 24 (2008), pp. 176 – 215.

The concept of **seasonal integration** of time series has been developed by

SVEND HYLLEBERG, ROBERT F. ENGLE, CLIVE W.J. GRANGER and BYUNG SAM YOO, Seasonal Integration and Cointegration, *Journal of Econometrics* 44 (1990), pp. 215 – 238; reprinted in: S. HYLLEBERG (ed.), *Modelling Seasonality*, Oxford University Press, Oxford et al. 1992, pp. 425 – 466.

In this volume edited by S. HYLLEBERG, there are further papers about the econometric handling of time series with seasonal variations.

For the discussion of stochastic versus deterministic trends and their implication for **macroeconomic theorizing** as well as for econometric work, see the above-mentioned paper by JAMES H. STOCK and MARK W. WATSON (1988). It was of special interest whether the **gross national product** has a unit root, i.e. whether permanent shocks have an impact on its development. See for this

JAMES H. STOCK and MARK W. WATSON, Does GNP Have a Unit Root?, *Economics Letters* 22 (1986), pp. 147 – 151,

PIERRE PERRON and PETER C.B. PHILLIPS, Does GNP Have a Unit Root?, A Re-Evaluation, *Economics Letters* 23 (1987), pp. 139 – 145, or

GLENN D. RUDEBUSCH, Trends and Random Walks in Macroeconomic Time Series: A Re-Examination, *International Economic Review* 33 (1992), pp. 661 – 680.

The **theory of real business cycles**, which goes back to

FINN E. KYDLAND and EDWARD PRESCOTT, Time to Build and Aggregate Fluctuations, *Econometrica* 50 (1982), pp. 1345 – 1370,

is surveyed in

GEORGE W. STADLER, Real Business Cycles, *Journal of Economic Literature* 32 (1994), pp. 1750 – 1783.

An empirical test of Real Business Cycle Theory but, however, without conclusive results, is given in

MARK W. WATSON, Measures of Fit for Calibrated Models, *Journal of Political Economy* 101 (1993), pp. 1011 – 1041.

A more recent paper in this area that investigates the business cycles of five European countries and the United States is

PETER R. HARTLEY and JOSEPH A. WHITT, Macroeconomic Fluctuations: Demand or Supply, Permanent or Temporary, *European Economic Review* 47 (2003), pp. 61 – 94.

A supply side theory of the business cycle is already included in

JOSEPH A. SCHUMPETER, *Theorie der wirtschaftlichen Entwicklung: eine Untersuchung über Unternehmergewinn, Kapital, Kredit, Zins und den Konjunkturzyklus*, Duncker und Humblot, Berlin 1912; English translation: *The Theory of Economic Development*, Harvard University Press, Cambridge (Mass.) 1934.

A survey of papers on the relation between **inflation** and **economic growth** is given in

JONATHAN TEMPLE, Inflation and Growth: Stories Short and Tall, *Journal of Economic Surveys* 14 (2000), pp. 395 – 426.

6 Cointegration

In the preceding chapter, we used stochastic trends to model nonstationary behaviour of time series, i.e. the variance of the data generating process increases over time, the series exhibits persistent behaviour and its first difference is stationary. For many economic time series, such a data generating process is a sufficient approximation, so that, in the following, we only consider processes which are integrated of order one (I(1)).

For a long time, econometricians have not taken into account that economic time series might be integrated. They applied traditional statistical procedures developed for the investigation of stationary stochastic series. CLIVE W.J. GRANGER and PAUL NEWBOLD (1974) showed that this might lead to severe problems. In a simulation study, they regressed two independently generated random walks on each other. They observed that the least-squares regression parameters do not converge towards zero but towards random variables with a non-degenerated distribution. Testing these parameters by employing the critical values of the usual t distribution, the null hypothesis of a zero coefficient is (wrongly) rejected much too often. Furthermore, the coefficient of determination does not converge towards the theoretically correct value of zero but towards a non-degenerated distribution. The estimated residuals show I(1) behaviour as expected for theoretical reasons. This implies that the Durbin-Watson statistic of the residuals converges towards zero.

Example 6.1

We performed Monte Carlo simulations to illustrate the problem of spurious regressions. First, we generated 100'000 replications with a sample size of T = 200 observations for two independent random walks x and y. Then we estimated the following equation:

$$y_t = a + b x_t + v_t$$

using ordinary least squares. As both series are independently generated, the slope coefficient as well as the R^2 should be zero. In this case, v follows a random walk, i.e. the first order autocorrelation coefficient is one and the value of the Durbin-Watson statistic zero. *Figure 6.1* shows the density functions of the t statistic of $\hat{b}$, R^2 and the Durbin-Watson statistic (smoothed by a kernel estimator).

206   6 Cointegration

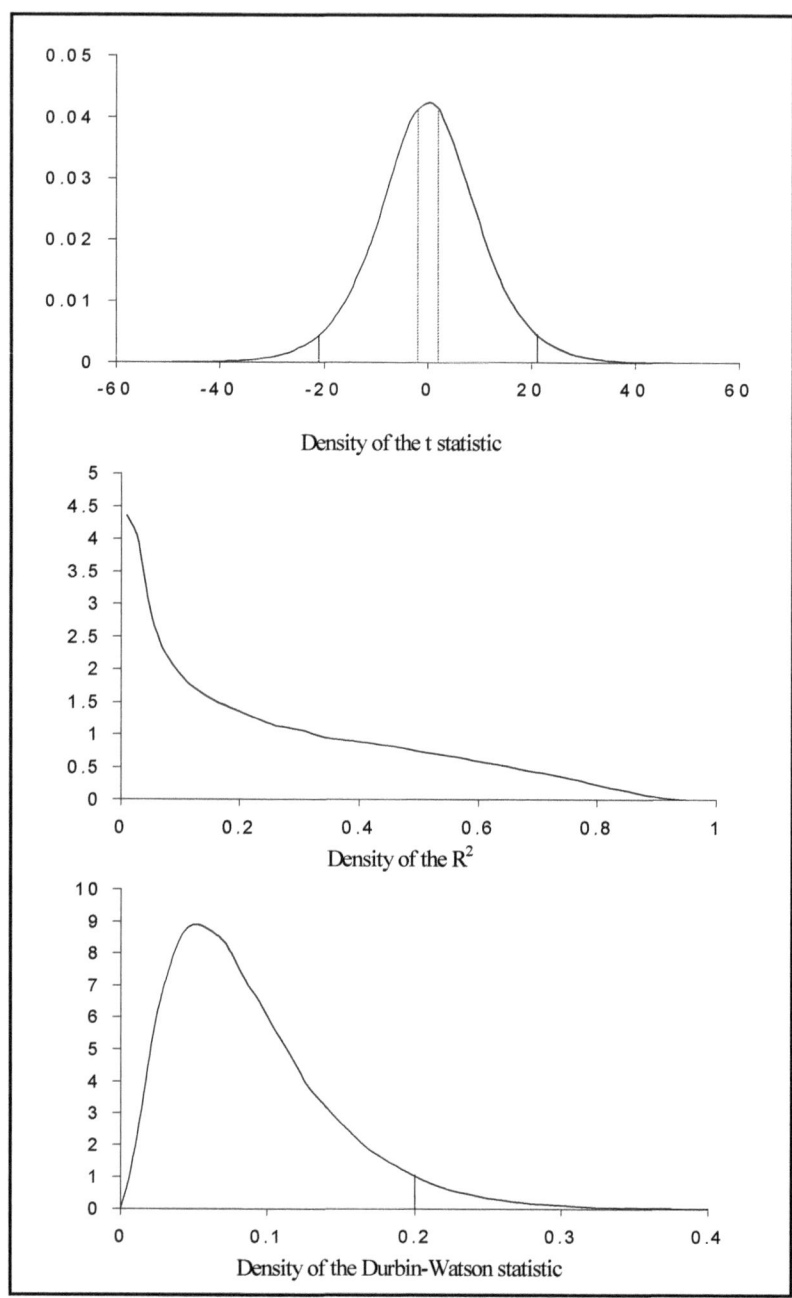

Figure 6.1: Densities of the estimated t value, R^2, and the Durbin-Watson statistic

The test statistic $\hat{t}$ has a symmetric density function, which, however, has a much larger variance than the standard normal distribution. The vertical dashed lines show the critical values of the normal distribution for the 2.5 and 97.5 percentiles, ± 1.96. If the classical distribution theory would be used (wrongly), a significant result would not only arise in 5 percent but in 83.32 percent of all cases. The correct values for the 2.5 and 97.5 percentiles are ± 21.06, indicated by solid lines. The density function of R^2 shows that values greater than 0.2 (0.5) have a probability of 46.13 (16.13) percent despite the fact that the true R^2 should be zero. The classical F distribution for the null hypothesis $H_0: R^2 = 0$, applied with 200 observations, leads to a critical value of 0.019 at the 5 percent level. Thus, when using this wrong distribution, almost all estimates would be accepted as being significant.

In these simulations, the estimated values of the Durbin-Watson statistic are between zero and 0.4. The probability that a value greater than 0.2 occurs is 4.62 percent. This almost corresponds to the figures given by ROBERT F. ENGLE and BYUNG SAM YOO (1987, *Table 4*); they report a critical value of 0.20 at the five percent significance level for a sample size of 200 observations. Thus, contrary to the t and F tests, the Durbin-Watson test provides the expected results.

To avoid such spurious relations, time series analysts advised against the use of the original series but recommended that they should be transformed in such a way that they can be considered as realisations of weakly stationary processes. GEORGE E.P. BOX and GWILYM M. JENKINS (1970, pp. 378f.), for example, recommended that, in order to estimate the dynamic relations between time series, one had to difference the series until their correlograms no longer indicated nonstationarity, and that after these transformations the cross-correlation functions should be used to identify the relation. This is one possible reason for the spurious independence results of Granger causality tests mentioned in *Chapter 3*.

Example 6.2

The following example illustrates how differencing leads to an underestimation of the true relation between I(1) variables. Given the following relations:

(E6.1) $\qquad y_t = w_t + u_{y,t}$,

(E6.2) $\qquad x_t = w_t + u_{x,t}$,

(E6.3) $\qquad w_t = w_{t-1} + u_{w,t}$,

where u_i, $i = \{x, y, w\}$ are three pure random processes and u_w is independently generated from u_x and u_y. Thus, y and x are I(1) processes; they contain a common stochastic trend.

To eliminate this trend, first differences are performed. The following regression is estimated to capture the relation between the two variables:

$$\Delta y_t = b\,\Delta x_t + v_t, \quad t = 1, \ldots, T.$$

The least squares estimator gives the following result:

$$\hat{b} = \frac{\sum_{t=1}^{T} \Delta x_t \, \Delta y_t}{\sum_{t=1}^{T} (\Delta x_t)^2} = \frac{\sum_{t=1}^{T} (u_{w,t} + \Delta u_{x,t})(u_{w,t} + \Delta u_{y,t})}{\sum_{t=1}^{T} (u_{w,t} + \Delta u_{x,t})^2}$$

$$= \frac{\sum_{t=1}^{T} u_{w,t}^2 + \sum_{t=1}^{T} u_{w,t}\Delta u_{x,t} + \sum_{t=1}^{T} u_{w,t}\Delta u_{y,t} + \sum_{t=1}^{T} \Delta u_{x,t}\Delta u_{y,t}}{\sum_{t=1}^{T} u_{w,t}^2 + 2\sum_{t=1}^{T} u_{w,t}\Delta u_{x,t} + \sum_{t=1}^{T} \Delta u_{x,t}^2}.$$

Thus, the probability limit of $\hat{b}$ is

$$\text{plim}\ \hat{b} = \frac{\sigma_{u_w}^2 + 2\sigma_{u_x u_y}}{\sigma_{u_w}^2 + 2\sigma_{u_x}^2}.$$

Contrary to the true one to one relation between the levels of x and y, the estimation in differences leads to a slope parameter which is smaller than one if u_x and u_y are uncorrelated. The larger the variance $\sigma_{u_x}^2$ is compared to the variance $\sigma_{u_w}^2$ the smaller is this estimate. This holds even more if u_x and u_y are negatively correlated. If their correlation is positive, both, under- or overestimations might occur.

This example reveals two problems. Firstly, estimated regression coefficients may not be significantly different from zero, although the respective relation exists. Secondly, estimated regression coefficients might be biased downwards because of errors-in-variables, even if they are statistically significant. To evade the Skylla of spurious independence as well as the Charybdis of spurious regressions, i.e. to render the type I and type II errors as unlikely as possible, CLIVE W.J. GRANGER and PAUL NEWBOLD (1974, p. 118) recommended to estimate the relations in the levels as well as in first differences, in order to be better able to (economically) interpret the results.

To solve this problem, it is necessary to develop statistical procedures which are suited for capturing relations between nonstationary variables correctly. This solution is provided by the theory of *cointegrated* relations developed in the 1980s. The idea goes back to CLIVE W.J. GRANGER (1981, 1986) and was popularised in papers by ROBERT F. ENGLE and CLIVE W.J. GRANGER (1987), JAMES H. STOCK (1987) as well as SØREN JOHANSEN (1988). Today, these procedures have become standard instruments for every time series econometrician. There are two main reasons for the rapid dissemination of this approach: First, the estimated cointegrating relations are closely connected to economic equilibrium relations. Second, in many applications it is sufficient to use ordinary least squares to get consistent estimates. Thus, traditional programme packages can be used further on.

A quite simple approach to avoid the spurious regression problem with I(1) variables is to include lagged values of the dependent and independent variables into the regression since, in this case, parameter values exist for which the residuals are I(0). Applying OLS results in consistent estimates of all parameters. (See JAMES D. HAMILTON (1994, pp. 561ff.).)

In the following, we define cointegrated processes and present their properties (*Section 6.1*). *Section 6.2* shows how single equation models with integrated variables can be estimated and how cointegration tests can be performed. The handling of systems of such equations using vector autoregressions as discussed in *Chapter 4* is described in *Section 6.3*. *Section 6.4.* discusses the importance of these procedures for the analysis of long-run economic (equilibrium) relations.

6.1 Definition and Properties of Cointegrated Processes

Quite generally, cointegration might be characterised by two or more I(1) variables indicating a common long-run development, i.e. they do not drift away from each other except for transitory fluctuations. This defines a statistical equilibrium which, in empirical applications, can often be interpreted as a long-run economic relation.

ROBERT F. ENGLE and CLIVE W.J. GRANGER (1987) defined *cointegration* as follows:

- The elements of a k-dimensional vector Y are cointegrated of order (d, c), Y ~ CI(d, c), if all elements of Y are integrated of order d, I(d), and if there exists at least one non-trivial linear combination z of these variables, which is I(d-c), where $d \geq c > 0$ holds, i.e. iff

$$\beta_i' Y_t = z_{i,t} \sim I(d-c), \quad i = 1, ..., r.$$

The vectors β_i are denoted as cointegration vectors. The *cointegration rank* r is equal to the number of linearly independent cointegration vectors. The cointegration vectors are the columns of the cointegration matrix B, with

$$B' Y_t = Z_t.$$

If all variables of Y are I(1), it holds that $0 \leq r < k$. For $r = 0$, the elements of the vector Y are not cointegrated. Correspondingly, the appropriate model is a system of first differences.

Important properties of cointegrated relations were summarised in the *Granger Representation Theorem*, presented by ROBERT F. ENGLE and CLIVE W.J. GRANGER (1987, pp. 255f.). The most important part of this theorem is:

- If the k-dimensional vector Y is cointegrated of order CI(1, 1) with cointegration rank r, besides the AR representation

$$A(L) Y_t = U_t,$$

with U_t being white noise, there also exists an error correction representation (as discussed in *Section 4.1*)

$$A^*(L) (1 - L) Y_t = -\Gamma Z_{t-1} + U_t,$$

with

$$A(1) = \Gamma B',$$

Γ and B being k×r matrices of rank r, $0 < r < k$, and

$$Z_t = B' Y_t$$

being an r-dimensional vector of I(0) variables. The reverse is also true: The existence of an error correction representation of I(1) variables implies cointegration.

In addition to this theorem, the following two lemmata hold:

Lemma 1:   If x_t and y_t are I(1) and cointegrated, x_t and $y_{t+\tau}$ are also cointegrated for any τ.

Lemma 2:   If x and y are I(1) and cointegrated, x is Granger causal to y and/or y is Granger causal to x.

Lemma 1 holds because

$$y_{t+\tau} = y_t + \Delta y_{t+1} + \ldots + \Delta y_{t+\tau},$$

implying that $y_{t+\tau}$ differs from y_t only by a stationary term, which does not change the cointegration relation. *Lemma 2* holds because an error correction representation exists for at least one of any two cointegrated variables, and error correction representations always imply Granger causal relations.

However, the reverse – that Granger causality between integrated variables implies cointegration – does not hold.

6.2 Cointegration in Single Equation Models: Representation, Estimation and Testing

In the following, we start with the simple case of a bivariate model, i.e. a regression relation between two I(1) variables. Then, we extend the analysis to a multivariate (single equation) regression model.

6.2.1 Bivariate Cointegration

Let x and y be two I(1) processes. In general, any linear combination of these two variables will again be an I(1) process. However, if there exists a parameter b so that the linear combination

(6.1) $$y_t - b\, x_t = z_t + a$$

is stationary, then x and y are cointegrated. The I(0) process z has an expectation of zero. The parameter a defines the level of the corresponding equilibrium relation which is given by

(6.2) $$y = a + b\, x.$$

The vector $\beta' = [1\ -b]$ is the cointegration vector. It is unique only because of its normalisation, as $\alpha\, \beta'$ with $\alpha \neq 0$ also leads to a stationary linear combination of y and x. The stationary process z describes the deviations from the equilibrium, the *equilibrium error*. Because of the finite variance of z, the deviations from the equilibrium are bounded; the system is always returning to its equilibrium path. Thus, relation (6.2) is an *attractor*.

Cointegration of x and y implies that both variables follow a common stochastic trend which can be modelled as a random walk,

(6.3a) $$w_t = w_{t-1} + u_t,$$

where u is again a white noise process. Thus, the two cointegrated I(1) processes can, for example, be represented as

(6.3b) $$y_t = b\, w_t + \tilde{y}_t \ \text{ with } \ \tilde{y}_t \sim I(0)$$

and

(6.3c) $$x_t = w_t + \tilde{x}_t \ \text{ with } \ \tilde{x}_t \sim I(0).$$

The linear combination

(6.3d) $$y_t - b x_t = \tilde{y}_t - b \tilde{x}_t = z_t$$

is stationary, as a linear combination of stationary processes is again stationary. Thus, (6.3d) is a cointegrating relation.

According to the Granger representation theorem, there exists an error correction representation for any cointegrating relation. In the bivariate case its reduced form can be written as:

(6.4a) $$\Delta y_t = -\gamma_y(y_{t-1} - a - b x_{t-1}) + \sum_{j=1}^{n_x} a_{xj}\Delta x_{t-j} + \sum_{j=1}^{n_y} a_{yj}\Delta y_{t-j} + u_{y,t},$$

(6.4b) $$\Delta x_t = +\gamma_x(y_{t-1} - a - b x_{t-1}) + \sum_{j=1}^{k_x} b_{xj}\Delta x_{t-j} + \sum_{j=1}^{k_y} b_{yj}\Delta y_{t-j} + u_{x,t},$$

with u_x and u_y as pure random processes. If x and y are cointegrated, at least one γ_i, i = x, y, has to be different from zero. It is obvious that, in this case, a relation exists between the levels of the variables. A model estimated only in first differences would be misspecified because the term $y_{t-1} - b\, x_{t-1}$ is missing. The representation (6.4) has the advantage that it only contains stationary variables although the underlying relation is between non-stationary (I(1)) variables. Thus, if the variables are cointegrated and the cointegration vector in (6.4) is known, the traditional statistical procedures can be applied for estimating and testing. The parameterisation in system (6.4) provides a separation of the short-run adjustment processes modelled by the lagged differences of the variables from the adjustment to the long-run equilibrium because the system also reacts to the deviations from the equilibrium relation which are lagged by one period.

System (6.4) is stable whenever $0 \leq \gamma_y < 2$ and also $0 \leq \gamma_x < 2$ hold, and if at least one of the two parameters is different from zero. This implies that – ceteris paribus – a positive deviation from the long-run equilibrium leads to a reduction of y and an increase of x and, therefore, to a reduction of the initial equilibrium error: the system tends towards its attractor (6.2). If the initial equilibrium error is negative, a corresponding adjustment process is initiated. If one of the two adjustment coefficients is zero, i.e. if $\gamma_x = 0$, the adjustment is only possible via changes in y. The development of the I(1) variable x is independent of the equilibrium error, it is – so to speak – the stochastic trend driving the system. In this situation, x is called *weakly exogenous*. If $\gamma_x > 0$ and γ_y is negative, or if $\gamma_y > 0$ and γ_x is negative, the system might also be stable. According to SØREN JOHANSEN (1995, p. 54), however, this depends on the relative moduli of the adjustment coefficients.

Thus, in a bivariate system with two I(1) variables, only the following two situations can occur:

(i) The two variables are not cointegrated, i.e. $\gamma_x = \gamma_y = 0$. Then, the system contains two (different) stochastic trends.

(ii) The two variables are cointegrated, i.e. at least one γ_i, $i = x, y$, is positive. Then the system contains one cointegrating relation and one common stochastic trend. It follows from *Lemma 2* that at least one simple Granger-causal relation exists between x and y.

Example 6.3

Let the ARIMA(1,1,0) process

(E6.4) $\qquad (1 - \alpha L)\Delta x_t = u_t$ with $|\alpha| < 1$,

be given, and the relation

(E6.5a) $\qquad y_t = b x_t + z_t, \quad b \neq 0$,

with

(E6.5b) $\qquad z_t = \rho z_{t-1} + v_t$,

where u_t and v_t are uncorrelated white noise processes. Because of (E6.4) x_t is I(1) and this also holds for y_t. According to the definition of cointegration, it is obvious that x and y are cointegrated for $|\rho| < 1$. However, if $\rho = 1$, there is no cointegration. In this case, the development of y is determined by the two stochastic trends x and z.

To derive the error correction model corresponding to (E6.4) and (E6.5a,b), we first insert (E6.5b) in (E6.5a). This leads to

$$y_t = \rho y_{t-1} + b x_t - \rho b x_{t-1} + v_t.$$

Subtracting y_{t-1} on both sides of this equation and adding as well as subtracting the term $b x_{t-1}$ on the right hand side, we get the (conditional) structural form of the error correction representation,

$$\Delta y_t = -(1 - \rho) y_{t-1} + b(1 - \rho) x_{t-1} + b \Delta x_t + v_t.$$

This holds because x is weakly exogenous. By taking (E6.4) into account, the reduced form of the error correction model is given by

(E6.6a) $\qquad \Delta x_t = \alpha \Delta x_{t-1} + u_{x,t}$,

(E6.6b) $\qquad \Delta y_t = -(1 - \rho)(y_{t-1} - b x_{t-1}) + b \alpha \Delta x_{t-1} + u_{y,t}$,

where $u_{x,t} = u_t$ and $u_{y,t} = v_t + b u_t$.

The error correction equation of x, (E6.6a), does not contain the equilibrium error $y - bx$. Thus, the weakly exogenous x drives the whole system. If there is cointegration, i.e. for $-1 < \rho < 1$, it holds that $0 < \gamma_y < 2$ for the adjustment parameter $\gamma_y = (1 - \rho)$. Thus, the system is stable; y is adjusting to the long-run equilibrium. For $\rho = 1$, i.e. if there is no cointegration, (E6.6b) no longer contains the error-correction term. The system contains two stochastic trends. In any case, the error correction model only contains stationary variables, the differences of I(1) variables and the stationary equilibrium error.

6.2.2 Cointegration with More Than Two Variables

If there are only two I(1) variables after normalisation, there are either only one (unique) cointegrating relation and one common stochastic trend or two stochastic trends without cointegration. The situation is much more complicated if there are more than two I(1) variables which are cointegrated.

Let us consider the situation of three I(1) variables, y_i, $i = 1, 2, 3$. Then two independent cointegrating relations could exist, as, for example, by assuming zero expectations for all variables:

$$y_{1,t} = b_2 y_{2,t} + z_{1,t}, \quad b_2 \neq 0,$$

$$y_{2,t} = b_3 y_{3,t} + z_{2,t}, \quad b_3 \neq 0.$$

In this case, $\beta_1' = [1\ -b_2\ 0]$ and $\beta_2' = [0\ 1\ -b_3]$ are linearly independent. However, linear combinations of β_1 and β_2 provide cointegration vectors which include all three I(1) variables, $Y' = [y_1\ y_2\ y_3]$,

$$\beta_\gamma = \gamma \beta_1 + (1 - \gamma) \beta_2 = \begin{bmatrix} \gamma \\ 1 - \gamma(1 + b_2) \\ -(1 - \gamma) b_3 \end{bmatrix}, \quad 0 \leq \gamma \leq 1.$$

β_γ are again cointegrating vectors. This follows from

$$\beta_\gamma' Y_t = \gamma y_{1,t} + (1 - \gamma(1 + b_2)) y_{2,t} - (1 - \gamma) b_3 y_{3,t}$$

$$= \gamma (y_{1,t} - b_2 y_{2,t}) + (1 - \gamma)(y_{2,t} - b_3 y_{3,t})$$

$$= \gamma z_{1,t} + (1 - \gamma) z_{2,t} = z_{\gamma,t},$$

where z_γ as a linear combination of the two I(0) processes z_1 and z_2 is also stationary. For $\gamma = 1$, we get the cointegration vector β_1, and for $\gamma = 0$ the cointegration vector β_2. These two vectors form the basis of the cointegration space with dimension two, $r = 2$, because there are only two linearly independent cointegration vectors. However, as there exists an infinite number of bases for this space, the representation of the equilibrium relations is not unique. Thus, we again face the well-known identification problem of traditional econometrics; only additional a priori restrictions (which are not contained in the data) can lead to a unique representation.

With k = 3 I(1) variables and r = k − 1 = 2 cointegrating relations, the system contains just one stochastic trend; otherwise the supposed pairwise cointegration between y_1 and y_2, y_2 and y_3, as well as y_1 and y_3 would be impossible.

On the other hand, if a system of three I(1) variables contains two stochastic trends, there can only be one cointegrating relation, and the corresponding cointegration vector is again unique after normalisation, e.g. for $\beta' = [1 \; -\tilde{b}_2 \; -\tilde{b}_3]$. Then the long-run equilibrium relation is

$$y_{1,t} = \tilde{b}_2 y_{2,t} + \tilde{b}_3 y_{3,t}.$$

According to the definition in *Section 6.1*, a vector with k integrated variables of order one, I(1), is cointegrated of rank r, 0 < r < k, if there exist exactly r linearly independent cointegration vectors $\beta_i \neq 0$, i = 1, 2, ..., r. Combining the cointegration vectors as columns of the cointegration matrix B,

$$B = [\beta_1 \; \beta_2 \; ... \; \beta_r]$$

indicates the deviations of the r statistical equilibria $Z' = [z_1 \; z_2 \; ... \; z_r]$ as

(6.5) $$B' Y_t = Z_t.$$

In case of I(1) variables, the system contains k − r common stochastic trends. The cointegration rank r must always be smaller than the number of I(1) variables k, because otherwise the cointegration matrix B would be invertible and $Y_t = B'^{-1} Z_t$ would be a linear combination of stationary processes. This contradicts the assumption that all k variables are I(1). If r = k − 1, we get the special case of only one common stochastic trend in the system. Therefore, pairwise cointegrating relations exist between all components of Y.

6.2.3 Testing Cointegration in Static Models

In order to handle cointegrating relations in single equation models correctly, it has to be presupposed that there exists at most one cointegrating relation between k I(1) variables which comprehends all variables. In this case, unit root tests can be used to test for cointegration by applying them to the residuals of an estimated (static) equilibrium relation. If y_1 is taken to be the dependent variable and if there exists no cointegration relation between $y_2, \ldots, y_k$, the following equation is estimated by OLS:

$$(6.6) \qquad y_{1,t} = a_0 + a_1 t + \sum_{j=2}^{k} b_j y_{j,t} + z_t$$

for the k I(1) variables, where (in the case of cointegration) z is again the equilibrium error. In most applications no time trend is included, i.e. $a_1 = 0$. The parameters $b_2, b_3, \ldots, b_k$ can be estimated consistently with the least squares approach. This method minimises the residual variance. If the estimated parameters differ from the true cointegration parameters, the residual process is nonstationary, i.e. its variance is increasing with increasing sample size T. On the other hand, the residual process is stationary for the cointegrating parameters and, therefore, has a finite variance. Apparently, this is the minimum.

Table 6.1: *Critical Values of the Dickey-Fuller Test on Cointegration in the Static Model*

α	k			
	1	2	3	4
	Model with constant term			
0.10	-2.57	-3.05	-3.45	-3.81
0.05	-2.86	-3.34	-3.74	-4.10
0.01	-3.43	-3.90	-4.30	-4.65
	Model with constant term and time trend			
0.10	-3.13	-3.50	-3.83	-4.15
0.05	-3.41	-3.78	-4.12	-4.43
0.01	-3.96	-4.33	-4.67	-4.97

The values for k = 1 are the critical values of the Dickey-Fuller unit root test.

Source: J.G. MACKINNON (1991, *Table 1*, p. 275).

Following this logic, ROBERT F. ENGLE and CLIVE W.J. GRANGER (1987) proposed a testing procedure for the null hypothesis that there is no cointegrating relation and, therefore, the residual process is nonstationary, H_0: z_t

~ I(1), against the alternative of cointegration, i.e. that this process is stationary, H_1: z_t ~ I(0). It requires two steps to perform this test. Firstly, relation (6.6) is estimated with OLS. Secondly, the augmented Dickey-Fuller test, as presented in *Section 5.3.1*, is applied to the estimated residuals. As OLS residuals have a zero mean by construction, the version without deterministic terms, (5.17"), is used. However, the critical values are different because the test is applied to a 'generated' and not to an observed time series. They depend on the number of I(1) variables k but also on the deterministic components of the equilibrium relation, i.e. on whether a constant term and/or a deterministic time trend is included in model (6.6).

Table 6.1 shows some asymptotic critical values derived through simulations by JAMES G. MACKINNON (1991). The null hypothesis of no cointegration is rejected for too small values of the test statistic. The values for k = 1 are those of the augmented Dickey-Fuller unit root test. Following the considerations in UWE HASSLER (2004), the critical values for the model with a constant term are valid if and only if the regressors in (6.6) contain a unit root but no linear trend. If, on the other hand, the data generating process of at least one (single) regressor in (6.6) also contains a linear trend, the correct critical values are those in the lower part of *Table 6.1* for the case k-1. However, these values are hardly different from those of the model without a trend.

The test is correct if and only if the explanatory variables, y_2, y_3, ..., y_k, themselves are not cointegrated and the unique cointegration relation includes y_1. In practical applications, it is recommended to start with small models in relation (6.6) and to add additional variables only as long as the null hypothesis of no cointegration cannot be rejected. Due to the invariance property of cointegration, i.e. that two or more variables do not change their cointegration property if further I(1) variables are added, the *specific-to-general* approach is appropriate in this framework.

In the case of cointegration, the parameter estimates $\hat{b}_2, \hat{b}_3, ..., \hat{b}_k$, in equation (6.6) are *super consistent*, i.e. they converge with a rate of T towards their true values, and therefore their convergence is faster than the one of parameters estimated in regressions with stationary variables, which converge with a rate of $\sqrt{T}$. Contrary to the stationary case, simultaneity of the variables or errors in variables do not inhibit this consistency result. However, the estimates are even asymptotically not normally distributed and biased for finite samples. ANINDYA Banerjee, JUAN J. DOLADO, DAVID F. HENDRY and GREGOR W. SMITH (1986) showed that the bias is proportional to $1 - R^2$. The reason for this is that in the case of cointegration R^2 tends towards one with increasing sample size, because the vari-

6 Cointegration

ances of the nonstationary regressors, which increase with the sample size, dominate the finite variance of the stationary error term.

Example 6.4

The situation of a simple regression can be used to demonstrate the finite sample bias. Let y and x be cointegrated I(1) variables, i.e. the relation

(E6.7a) $$y_t = a + b x_t + z_t,$$

holds and z_t is stationary. As explained above, this relation can be estimated super-consistently with OLS. The same holds for the reverse regression

(E6.7b) $$x_t = \tilde{a} + \tilde{b} y_t + v_t.$$

The product of the two regression coefficients estimated with OLS leads to:

$$\hat{b}\hat{\tilde{b}} = \frac{(\widehat{Cov}[y,x])^2}{\hat{V}[y] \cdot \hat{V}[x]} = R^2 \leq 1.$$

If the variables are cointegrated, R^2 tends towards one, i.e. $\hat{\tilde{b}}$ tends towards $\hat{b}^{-1}$. To the extent that R^2 is smaller than one for finite samples, the product of the two estimated coefficients is systematically underestimated.

Moreover, standard inference procedures are not possible as, in general, the t statistics do not have asymptotically normal distributions. However, following PENTTI SAIKKONEN (1991) as well as JAMES H. STOCK and MARK W. WATSON (1993), a simple correction can be applied to equation (6.6) ensuring that the estimation is still super consistent and that the estimated t statistics are, nevertheless, asymptotically normally distributed: Additional lagged and future differences of the regressors are included to ensure that the I(1) regressors are uncorrelated with the residuals:

(6.7) $$y_{1,t} = a_0 + a_1 t + \sum_{j=2}^{k} b_j y_{j,t} + \sum_{j=-k_2}^{k_1} \pi_{2,j} \Delta y_{2,t-j} + \ldots$$
$$+ \sum_{j=-k_2}^{k_1} \pi_{k,j} \Delta y_{k,t-j} + \tilde{z}_t.$$

Information criteria might be used to determine the maximal lag and lead k_1 and k_2. The t statistics of $\hat{b}_2, \hat{b}_3, \ldots, \hat{b}_k$ converge towards a normal distribution with the corresponding true parameters as expectations and the variance $\omega^2/V[\tilde{z}_t]$, with

$$\omega^2 = V[\tilde{z}_t] + 2\sum_{\tau=1}^{\infty}\text{Cov}[\tilde{z}_t, \tilde{z}_{t+\tau}].$$

This long-run variance can be estimated according to (5.18). In case of no autocorrelation of the residuals $\tilde{z}$ the t statistics are asymptotically standard normal since $\omega^2 = V[\tilde{z}_t]$.

Example 6.5

Figure 6.2 shows the logarithm of the real quantity of money M1 in per capita terms, m, the logarithm of the real per capita Gross National Product (GNP), y, and the long-run interest rate, r, for the Federal Republic of Germany. We use quarterly data from the first quarter of 1961 to the last quarter of 1989, i.e. for the period before the German Unification. Unit root tests clearly indicate that all three time series are I(1). The Engle-Granger approach is used to investigate whether cointegration relations exist between these variables. However, this approach can only be applied if there exists just one cointegrating relation. Thus, we start by checking whether the time series are pairwise cointegrated. The null hypothesis of no cointegration can never be rejected in all three possible cases.

In the next step we regress the quantity of money, m, on GNP, y, and the interest rate, r. We chose m as the dependent variable as we are interested in a long-run money demand function. When estimating this relation with OLS, we include seasonal dummies along with the constant term because m as well as y exhibit strong seasonal variations. To ensure that the constant term really captures the level effect, we use *centred seasonal dummies* s_i, i = 1, 2, 3, which take on the value 0.75 for the i-th quarter and -0.25 elsewhere. Thus, we have an annual mean of zero. The estimated relation (with the standard errors in parentheses) is:

(E6.8) $\quad m_t = \underset{(0.142)}{-1.370} + \underset{(0.016)}{1.133\, y_t} - \underset{(0.260)}{3.059\, r_t} + \underset{(0.010)}{0.036\, s_{1,t}} + \underset{(0.010)}{0.036\, s_{2,t}}$

$\quad\quad\quad\quad - \underset{(0.010)}{0.018\, s_{3,t}} + \hat{z}_t,$

$\bar{R}^2 = 0.977$, SE = 0.038, T = 116.

The Dickey-Fuller unit root test for the estimated residuals $\hat{z}$ provides the following test equation (with t values in parentheses):

$$\Delta\hat{z}_t = \underset{(-3.85)}{-0.231\hat{z}_{t-1}} + \underset{(4.62)}{0.376\,\Delta\hat{z}_{t-4}} + \hat{u}_t.$$

The estimated test statistic is -3.85. m and y contain a linear trend as we can see from *Figure 6.2*. Due to economic reasons, (E6.8) does not include a trend. Therefore we have to take the critical values for k = 2 from the lower part of *Table 6.1*.

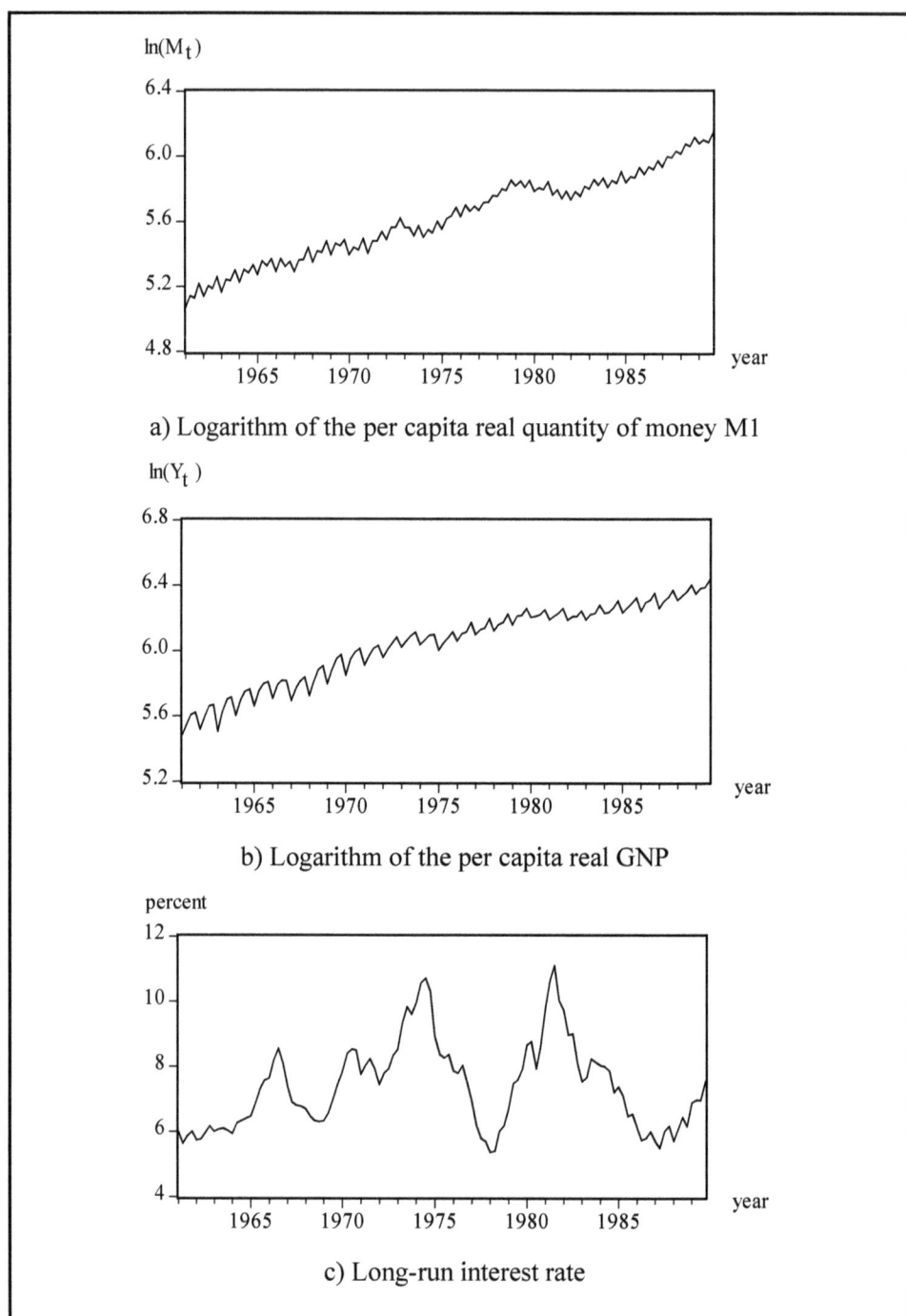

Figure 6.2: Data for the Federal Republic of Germany, 1961 – 1989

The critical value is -3.78 at the 5 percent significance level. Thus, the null hypothesis of no cointegration can be rejected at the 5 per cent level. Economically, the estimated parameters are meaningful and can be interpreted in the sense of a long-run money demand function. The estimated income elasticity of the money demand function is close to one and the interest rate elasticity is negative; at an interest rate level of 5 percent, for example, it has the value of -0.15 (= - 3.059 · 0.05).

6.2.4 Testing Cointegration in Dynamic Models

Despite the super consistency of the estimates, the static approach has the disadvantage that with a finite number of observations the estimated cointegration parameters might be seriously biased. This bias is only slightly reduced with an increasing number of observations. One possible reason for the bias are highly autocorrelated residuals due to the fact that the dynamic is neglected in relation (6.6). It is explicitly captured in the error correction equations. Because of the Granger representation theorem mentioned above, assuming weak exogeneity of $y_2, \ldots, y_k$ ($\gamma_2 = \ldots = \gamma_k = 0$) a cointegration test can also be performed in the unconditional error correction equation of y_1,

$$(6.8) \quad \Delta y_{1,t} = a_0 - \gamma_1 y_{1,t-1} + \sum_{j=2}^{k} \theta_j y_{j,t-1} + \sum_{j=1}^{k_1} a_{1j} \Delta y_{1,t-j} + \ldots + \sum_{j=1}^{k_k} a_{kj} \Delta y_{k,t-j} + u_{1,t},$$

or

$$(6.8') \quad \Delta y_{1,t} = -\gamma_1 \left(y_{1,t-1} - \frac{a_0}{\gamma_1} - \sum_{j=2}^{k} \frac{\theta_j}{\gamma_1} y_{j,t-1} \right) + \sum_{j=1}^{k_1} a_{1j} \Delta y_{1,t-j} + \ldots + \sum_{j=1}^{k_k} a_{kj} \Delta y_{k,t-j} + u_{1,t},$$

respectively. With

$$(6.9) \quad a = \frac{a_0}{\gamma_1} \quad \text{and} \quad b_j = \frac{\theta_j}{\gamma_1}, \quad j = 2, \ldots, k,$$

the expression in parentheses in (6.8') can be written as

$$(6.10) \quad y_{1,t-1} - a - b_2 y_{2,t-1} - \ldots - b_k y_{k,t-1} = z_{t-1}.$$

If all y_i, $i = 1, \ldots, k$, are $I(1)$, the first differences of these variables are stationary. Thus, equations (6.8) or (6.8') are only balanced, i.e. the stationary variable Δy_1 is explained by stationary variables, if (6.10) is a stationary linear combination which reflects deviations from the long-run equilibrium or, if this is not the case, it does not contribute to the explanation of Δy_1, i.e. if $\gamma_1 = 0$. Thus, for the cointegration test in the error correction framework we get the null hypothesis

$$H_0: (y_1, y_2, \ldots, y_k) \text{ are not cointegrated, i.e. } \gamma_1 = 0,$$

against the alternative

$$H_1: \text{the variables are cointegrated, i.e. } \gamma_1 > 0.$$

If there is cointegration, the adjustment parameter has to be positive, $\gamma_1 > 0$, as the model would otherwise not be stable; there would be no adjustment towards the equilibrium. The test is performed in such a way that equation (6.8) is estimated by using ordinary least squares and the lag lengths $k_1, \ldots k_k$ are chosen so that the estimated residuals $\hat{u}$ do not exhibit significant autocorrelation.

The test statistic is the t value of $-\hat{\gamma}_1$. The null hypothesis that there is no cointegration is rejected if these values are too small. The corresponding critical values are given in ANINDYA BANERJEE, JUAN J. DOLADO and RICARDO MESTRE (1998, *Table 1*, pp. 276f.). Again, these values depend on whether relation (6.8) is estimated with or without a constant term or a trend and, of course, on the number of I(1) variables included in the test equation. Selected asymptotically valid critical values are given in *Table 6.2*.

UWE HASSLER (2000) showed that in the case that relation (6.8) contains only a constant term, the critical values are only correct if the I(1) regressors do not contain a deterministic trend. If at least one of the k I(1) variables contains a deterministic trend, we get the correct critical values from the lower part of *Table 6.2* (for the model with constant term and trend), now choosing the critical values for the case $k - 1$. If (6.8) contains only two I(1) variables, the appropriate critical values are those of unit root tests when the test equation includes a deterministic trend, i.e. the critical values for the model with constant term and trend for $k = 1$ are given in *Table 6.1*.

When these tests are applied in empirical research, it is not clear from the outset which equations of the multivariate error correction model contain the error correction term. Thus, the described test procedure must also be applied with the dependent variables $y_2, y_3, \ldots, y_k$.

Table 6.2: Critical Values of the Cointegration Test in the Error Correction Model

α	k		
	2	3	4
Model with constant term			
0.10	-2.89	-3.19	-3.42
0.05	-3.19	-3.48	-3.74
0.01	-3.78	-4.06	-4.46
Model with constant term and time trend			
0.10	-3.39	-3.62	-3.82
0.05	-3.69	-3.91	-4.12
0.01	-4.27	-4.51	-4.72

Source: A. BANERJEE, J.J. DOLADO and R. MESTRE (1998, *Table 1*, pp. 276f.)

In relation (6.8), the instantaneous changes of $y_2, y_3, ..., y_k$ might also be included if the adjustment parameters in the corresponding equations are zero, i.e. that $\gamma_2 = \gamma_3 = ... = \gamma_k = 0$. This means that $y_2, y_3, ..., y_k$ are weakly exogenous for the estimation of the parameters in the long-run relation. In a Monte Carlo study, UWE HASSLER and JÜRGEN WOLTERS (2006) showed that using the conditional error correction equation, i.e. including the instantaneous changes of $\Delta y_2, \Delta y_3, ..., \Delta y_k$ in equation (6.8), results in a more powerful cointegration test than without these variables. The general finding is that in any case, the conditional error correction regression outperforms the unconditional one.

If there is cointegration, equation (6.10) provides an estimation of the long run relation if the theoretical values in (6.9) are substituted by their least squares estimates. This is the non-linear cointegration estimator going back to JAMES H. STOCK (1987) which is also super consistent. The representation (6.8') gives the corresponding error correction equation.

Example 6.6

Now we use the data of *Example 6.5* to test for cointegration in the error correction model (6.8). This approach avoids the possible bias in the Engle-Granger procedure since the short-run dynamic is not neglected. It serves as a starting point for the estimation of a complete money demand function. To capture the strong seasonal movements in m and y, the maximal lag for the changes in the explanatory variables is four. Centred seasonal dummies are also included. The transition to floating the DMark with respect to the Dollar in March 1973 was followed by a rather restrictive monetary policy. We take into account this episode by introducing the impulse dummy D_{7302} which takes on the value of one in the second quarter of 1973 and zero elsewhere. Eliminating the variable with the lowest t value successively leads to the following parsimonious model (with t values in parentheses):

$$(E6.9) \quad \Delta m_t = \underset{(-2.09)}{-0.143} - \underset{(-4.10)}{0.160\ m_{t-1}} + \underset{(3.99)}{0.177\ y_{t-1}} - \underset{(-4.29)}{0.740\ r_{t-1}}$$

$$\underset{(-2.47)}{-0.184\ \Delta m_{t-1}} + \underset{(2.41)}{0.173\ \Delta m_{t-2}} + \underset{(4.29)}{0.304\ \Delta m_{t-4}} - \underset{(-2.25)}{0.200\ \Delta y_{t-1}}$$

$$\underset{(-5.98)}{-0.475\ \Delta y_{t-2}} - \underset{(-3.42)}{0.271\ \Delta y_{t-3}} - \underset{(-2.05)}{0.170\ \Delta y_{t-4}} - \underset{(-4.57)}{1.314\ \Delta r_{t-1}}$$

$$\underset{(-4.30)}{-0.055\ D_{7302}} - \underset{(-2.83)}{0.044\ s_{1,t}} + \underset{(0.52)}{0.004\ s_{2,t}} - \underset{(-2.47)}{0.037\ s_{3,t}} + \hat{u}_t,$$

$\bar{R}^2 = 0.946$, SE $= 0.012$, T $= 115$, JB $= 1.906\ (p = 0.386)$,
LM(1) $= 1.050\ (p = 0.308)$, LM(2) $= 2.092\ (p = 0.129)$,
LM(4) $= 1.116\ (p = 0.354)$, LM(8) $= 1.135\ (p = 0.348)$.

The Jarque-Bera test (JB) does not reject the null hypothesis of normality of the residuals at any conventional significance level. The residuals do not show deviations from white noise according to the Lagrange Multiplier tests (LM(n)) that test autocorrelation up to order n. This means that the specification in (E6.9) captures the short- and long-run dynamics of the variables in a reasonable way.

There exists a cointegrating relation between m, y, and r if the estimated coefficient of m_{t-1} is significantly negative. In this case, where m and y contain deterministic trends, as can be seen from *Figure 6.2*, and no trend term is included in (E6.9), the correct critical value is found in the lower part of *Table 6.2* for the case k = 2. Thus, the critical value with a 5 percent significance level is -3.69. Since the estimated t value is -4.10, the null hypothesis of no cointegration can be rejected at the 5 percent level. Equation (E6.9) is balanced. According to (6.8') and

(6.9), this leads to the following long-run money demand equation (with the standard errors of the parameters in parentheses):

$$\text{(E6.10)} \quad m = -0.889 + 1.102\,y - 4.610\,r.$$
$$\,(0.31)\quad(0.03)\quad\,(0.72)$$

Comparing this result with the static long-run money demand function in (E6.8), we see that the income elasticity is about the same but that we get a stronger interest rate effect. Assuming an interest rate of 5 percent, the long-run interest rate elasticity is -0.23, contrary to -0.15 in the static approach.

6.3 Cointegration in Vector Autoregressive Models

Assuming that the k variables, $y_1, y_2, \ldots, y_k$, collected in the vector Y, are integrated of order one, the following cases are possible: Either there is no cointegration at all or there exist one or two up to $k-1$ linear independent cointegration vectors. In this case we cannot use single equation procedures which allow at most for one cointegration relation. We no longer get unique relations as seen in *Section 6.2.2*. If we have more than two I(1) variables we must at first estimate the cointegration rank r, i.e. the number of linear independent cointegration vectors. This can be done with a procedure developed by SØREN JOHANSEN (1988).

6.3.1 The Vector Error Correction Representation

Starting point of this approach is an adequate statistical description of the linear relations between the k nonstationary variables. The usual way is the modelling as a vector autoregressive process of finite order p. We can use the techniques for stationary processes presented in *Chapter 4*. Therefore, we have

$$\text{(6.11)} \quad Y_t = \sum_{j=1}^{p} A_j Y_{t-j} + D_t + U_t,$$

where U denotes a normally distributed k-dimensional white noise process, D represents the deterministic terms, and A_j, $j = 1, 2, \ldots, p$, are k×k-dimensional parameter matrices. The reparametrisation as a vector error correction model as described in *Sections 4.1* and *6.1* leads to

$$\text{(6.12)} \quad \Delta Y_t = -\Pi Y_{t-1} + \sum_{j=1}^{p-1} A_j^* \Delta Y_{t-j} + D_t + U_t,$$

with

$$\Pi = A(1) = I - \sum_{j=1}^{p} A_j \text{ and } A_j^* = -\sum_{i=j+1}^{p} A_i, \ j = 1, 2, ..., p-1.$$

The matrix Π represents the long-run relations between the variables.

Since all components of Y_t are I(1) variables, each component of $\Delta Y_{t,...}$, ΔY_{t-p+1} is stationary and each component of Y_{t-1} is also integrated of order one. This makes relation (6.12) unbalanced as long as Π has a full rank of k. In this case the inverse matrix Π^{-1} exists and we could solve equation (6.12) for Y_{t-1} as a linear combination of stationary variables. However, this would be a contradiction. Therefore, Π must have a reduced rank of r < k. Then, the following decomposition exists:

(6.13)
$$\underset{(k \times k)}{\Pi} = \underset{(k \times r)}{\Gamma} \underset{(r \times k)}{B'},$$

where all matrices have rank r. $B'Y_{t-1}$ are r stationary linear combinations which guarantee that the equations of system (6.12) are balanced. The columns of B contain the r linearly independent cointegration vectors and the matrix Γ contains the so-called loading coefficients which measure the contributions of the r long-run relations in the different equations of the system. The adjustment processes to the equilibria can be derived from these coefficients.

If there is no cointegration, i.e. if $r = 0$, Π is the zero matrix and (6.12) is a VAR of order p-1 in ΔY. This system possesses k unit roots, i.e. k different stochastic trends. If $r = k - 1$, the system contains exactly one common stochastic trend and all the variables of the system are pairwise cointegrated. As a general rule, the system (6.12) contains $k - r$ common stochastic trends and r linearly independent cointegration vectors for a cointegration rank r with $0 < r < k$.

Example 6.7

Let the following three-dimensional VAR(3) without deterministic terms be given:

$$Y_t = \begin{bmatrix} 1.3 & 0 & 0.8 \\ 0.2 & 0.4 & 0 \\ 0 & -0.3 & 1.2 \end{bmatrix} Y_{t-1} + \begin{bmatrix} -0.7 & 0 & -0.2 \\ -0.1 & 0.3 & 0 \\ 0 & 0.6 & -0.2 \end{bmatrix} Y_{t-2}$$

$$+ \begin{bmatrix} -0.5 & 0 & -0.3 \\ -0.1 & 0.3 & 0 \\ 0 & 0 & -0.2 \end{bmatrix} Y_{t-3} + U_t,$$

6.3 Cointegration in Vector Autoregressive Models

with

$$E[u_{i,t}\, u_{j,t-k}] = 0 \quad \text{for } i \neq j \text{ and } k \neq 0,$$

$$E[u_{i,t}\, u_{i,t-k}] = \begin{cases} 0 & \text{for } k \neq 0 \\ \sigma_i^2 & \text{for } k = 0 \end{cases}, \quad i = 1, 2, 3.$$

Using (6.12) we find the error correction representation:

$$\Delta Y_t = -\begin{bmatrix} 0.9 & 0 & -0.3 \\ 0 & 0 & 0 \\ 0 & -0.3 & 0.2 \end{bmatrix} Y_{t-1} + \begin{bmatrix} 1.2 & 0 & 0.5 \\ 0.2 & -0.6 & 0 \\ 0 & -0.6 & 0.4 \end{bmatrix} \Delta Y_{t-1}$$

$$+ \begin{bmatrix} 0.5 & 0 & 0.3 \\ 0.1 & -0.3 & 0 \\ 0 & 0 & 0.2 \end{bmatrix} \Delta Y_{t-2} + U_t.$$

The matrix Π contains the long-run equilibrium relations

$$\Pi = \begin{bmatrix} 0.9 & 0 & -0.3 \\ 0 & 0 & 0 \\ 0 & -0.3 & 0.2 \end{bmatrix}.$$

Since the rank of Π is two, we have two cointegrating relations and one common stochastic trend. Thus, any two variables are pairwise cointegrated. Normalising the first cointegration vector on y_1 and the second one on y_3, we find the following decomposition of the 3x3 matrix Π in the 3x2 loading matrix Γ and the 2x3 cointegration matrix B':

$$\begin{bmatrix} 0.9 & 0 \\ 0 & 0 \\ 0 & 0.2 \end{bmatrix} \begin{bmatrix} 1 & 0 & -\frac{1}{3} \\ 0 & -\frac{3}{2} & 1 \end{bmatrix} = \begin{bmatrix} 0.9 & 0 & -0.3 \\ 0 & 0 & 0 \\ 0 & -0.3 & 0.2 \end{bmatrix}.$$

Thus, the two long-run relations are

(E6.11a) $$y_{1,t} - \frac{1}{3} y_{3,t} = z_{1,t},$$

(E6.11b) $$y_{3,t} - \frac{3}{2} y_{2,t} = z_{2,t}.$$

Substituting (E6.11b) into (E6.11a) transforms the first equilibrium relation into

$$y_{1,t} - \frac{1}{2} y_{2,t} = z_{1,t} + \frac{1}{3} z_{2,t} = \tilde{z}_{1,t}.$$

This leads to the following decomposition

$$\begin{bmatrix} 0.9 & -0.3 \\ 0 & 0 \\ 0 & 0.2 \end{bmatrix} \begin{bmatrix} 1 & -\frac{1}{2} & 0 \\ 0 & -\frac{3}{2} & 1 \end{bmatrix} = \begin{bmatrix} 0.9 & 0 & -0.3 \\ 0 & 0 & 0 \\ 0 & -0.3 & 0.2 \end{bmatrix}.$$

This example shows that the decomposition in (6.13) is not unique, as we get

(6.14) $\quad\quad\quad \Pi = \Gamma B' = \Gamma H^{-1} H B' = \tilde{\Gamma} \tilde{B}'$

for any regular r×r matrix H. We are confronted with the usual identification problem for structural econometric systems. The cointegration vectors describing the economic long-run equilibria can only be estimated if meaningful economic restrictions are imposed.

6.3.2 The Johansen Approach

The approach proposed by SØREN JOHANSEN (1988) is a maximum likelihood estimation of (6.12) that considers restriction (6.13). Assuming first of all that the system (6.11) does not contain deterministic terms, we can write

(6.15) $\quad \Delta Y_t + \Gamma B' Y_{t-1} = A_1^* \Delta Y_{t-1} + ... + A_{p-1}^* \Delta Y_{t-p+1} + U_t$.

We get the maximum likelihood estimation of A_j^*, $j = 1, ..., p-1$, by applying ordinary least squares on (6.15) if Γ and B are given. Eliminating the influence of the short-run dynamics on ΔY_t and Y_{t-1} by regressing ΔY_t (Y_{t-1}) on the lagged differences, we get the residuals R_{0t} (R_{1t}) for which

(6.16) $\quad\quad\quad R_{0t} = -\Gamma B' R_{1,t} + \hat{U}_t$

holds. Here, R_0 is a vector of stationary and R_1 a vector of nonstationary processes. The idea of the Johansen approach is to find those linear combinations $B' R_1$ which show the highest correlations with R_0. The optimal values of Γ and the variance-covariance matrix Σ of U can be derived for known B by ordinary least squares estimation of (6.16). We get

(6.17) $\quad\quad\quad \hat{\Gamma}(B) = -S_{01} B (B' S_{11} B)^{-1}$

and

(6.18) $\quad\quad\quad \hat{\Sigma}(B) = S_{00} - S_{01} B (B' S_{11} B)^{-1} B' S_{10}$

6.3 Cointegration in Vector Autoregressive Models

with

(6.19) $$S_{ij} = T^{-1} \sum_{t=1}^{T} R_{i,t} R'_{j,t} \quad \text{for } i, j = 0, 1.$$

It can be shown that the likelihood function concentrated with (6.17) and (6.18) is proportional to $|\hat{\Sigma}(B)|^{-T/2}$. Therefore, the optimal values of B result from minimising the determinant

$$\left| S_{00} - S_{01} B (B' S_{11} B)^{-1} B' S_{10} \right|$$

with respect to B. SØREN JOHANSEN (1995, pp. 91f.) showed that this is equivalent to the solution of the following eigenvalue problem

(6.20) $$\left| \lambda S_{11} - S_{10} S_{00}^{-1} S_{01} \right| = 0$$

with the eigenvalues λ_i and the corresponding k-dimensional eigenvectors v_i, $i = 1, 2, ..., k$, for which

$$\lambda_i S_{11} v_i = S_{10} S_{00}^{-1} S_{01} v_i.$$

Using the arbitrary normalisation

$$\begin{bmatrix} v'_1 \\ \vdots \\ v'_k \end{bmatrix} S_{11} [v_1 \ ... \ v_k] = I_k,$$

with I_k being the k-dimensional identity matrix, leads to a unique solution. $1 \geq \hat{\lambda}_1 \geq ... \geq \hat{\lambda}_k \geq 0$ holds for the ordered estimated eigenvalues. The λ_i, $i = 1, ..., k$, are measures of the correlation between ΔY_t and the linear combinations $B' Y_{t-1}$. Since ΔY_t is stationary this measure only gives positive values if $B' Y_{t-1}$ is also stationary, implying that we have a cointegrating relation. In case of a nonstationary $B' Y_{t-1}$ the corresponding λ_i are zero. Therefore, if we have k I(1) variables and exactly r eigenvalues are positive while the remaining k-r eigenvalues are zero, these k nonstationary variables are said to have cointegration rank r.

The cointegrating vectors are estimated by the corresponding eigenvectors and combined in the k×r matrix

$$\hat{B} = [\hat{v}_1 \ ... \ \hat{v}_r].$$

The number of significantly positive eigenvalues determines the rank r of the cointegration space. This leads to two different likelihood ratio test procedures:

(i) The so-called trace test has the null hypothesis

H_0: There are at most r positive eigenvalues

against the alternative hypothesis that there are more than r positive eigenvalues. The test statistic is given by

$$(6.21) \qquad Tr(r) = -T \sum_{i=r+1}^{k} \ln(1-\hat{\lambda}_i).$$

(ii) The so-called λ_{max} test analyses whether there are r or r + 1 cointegrating vectors. The null hypothesis is

H_0: There are r positive eigenvalues

against the alternative hypothesis that there are r + 1 positive eigenvalues. The corresponding test statistic is given by

$$(6.22) \qquad \lambda_{max}(r, r+1) = -T \ln(1-\hat{\lambda}_{r+1}).$$

The series of tests starts with r = 0 and is performed until the first time the null hypothesis cannot be rejected. The cointegration rank is given by the corresponding value of r. The null hypothesis is rejected for too large values of the test statistic. Since the test statistics do not follow standard asymptotic distributions, the critical values are generated by simulations. The critical values depend on the included deterministic terms in the VAR(p) of relation (6.11) and the specification of the deterministics in the long-run relations of the corresponding error-correction model. To present the possible situations, we substitute (6.13) into (6.12) and generalise the resulting vector error correction model to

$$(6.23) \qquad \Delta Y_t = -\Gamma B^{*\prime} Y^*_{t-1} + c + dt + \sum_{j=1}^{p-1} A^*_j \Delta Y_{t-j} + U_t,$$

with

$$B^{*\prime} = \begin{bmatrix} \beta_{11} & \cdots & \beta_{1k} & \tilde{c}_1 & \tilde{d}_1 \\ \vdots & \ddots & \vdots & \vdots & \vdots \\ \beta_{r1} & \cdots & \beta_{rk} & \tilde{c}_r & \tilde{d}_r \end{bmatrix} = [B' \; \tilde{c} \; \tilde{d}]$$

and

$$Y^*_{t-1} = \begin{bmatrix} y_{1,t-1} \\ \vdots \\ y_{k,t-1} \\ 1 \\ t-1 \end{bmatrix} = \begin{bmatrix} Y_{t-1} \\ 1 \\ t-1 \end{bmatrix}.$$

If we use seasonally unadjusted data, centred seasonal dummies should also be included as regressors in (6.23).

The following five parameterisations of the deterministic terms in (6.23) are possible:

(i) The levels Y do not contain deterministic trends and the cointegrating relations do not contain constant terms:

$$\Gamma B^{*\prime} Y^*_{t-1} - c - dt = \Gamma B' Y_{t-1}.$$

(ii) The levels Y do not contain deterministic trends but the cointegrating relations contain constant terms:

$$\Gamma B^{*\prime} Y^*_{t-1} - c - dt = \Gamma (B' Y_{t-1} + \tilde{c}).$$

(iii) The levels Y contain linear deterministic trends and the cointegrating relations contain constant terms:

$$\Gamma B^{*\prime} Y^*_{t-1} - c - dt = \Gamma (B' Y_{t-1} + \tilde{c}) + \Gamma_{\perp}\mu.$$

In this case (and the following cases), the decomposition of the constants is arbitrary. SØREN JOHANSEN (1995) chooses the orthogonal complement matrix $\Gamma_{\perp}$ of Γ with $\Gamma'\Gamma_{\perp} = 0$ and $[\Gamma \vdots \Gamma_{\perp}]$ invertible for the decomposition.

(iv) The levels Y and the cointegrating relations contain linear deterministic trends:

$$\Gamma B^{*\prime} Y^*_{t-1} - c - dt = \Gamma (B' Y_{t-1} + \tilde{c} + \tilde{d}(t-1)) + \Gamma_{\perp}\mu.$$

In this case, the deterministic trends of the levels Y are not cancelled by the linear combination $B' Y_{t-1}$ as in (iii). Therefore, additional linear trends are included in the long-run relations.

(v) The levels Y contain quadratic deterministic trends and the cointegrating relations contain linear deterministic trends:

$$\Gamma B^{*\prime} Y^*_{t-1} - c - dt = \Gamma (B' Y_{t-1} + \tilde{c} + \tilde{d}(t-1)) + \Gamma_{\perp}(\mu + \delta t).$$

By using simulations, critical values for these five situations were derived by MICHAEL OSTERWALD-LENUM (1992) and SØREN JOHANSEN (1995, *Tables 15.1* to *15.5*, pp. 214ff.).

Because of (6.14), the cointegration vectors are not identified. They are simply stationary linear combinations which do not necessarily have meaningful economic interpretations. They might, however, represent linear combinations of economic equilibrium conditions. Thus, the question is how to test linear restrictions in the r cointegrating vectors. SØREN JOHANSEN (1988) developed a method to test restrictions on B which have the following form

(6.24) $$H_0: \; B = G \Phi,$$

where G is a given k×s matrix with full rank s, s < k, and Φ is an s×r matrix of free parameters. Estimating the vector error correction model under the restriction (6.24) with the Johansen approach results in r positive eigenvalues $\lambda_1^* > \lambda_2^* > ... > \lambda_r^*$. A likelihood ratio test compares the unrestricted with the restricted model, both with cointegration rank r. The corresponding likelihood ratio statistic is given by

(6.25) $$LR = T \sum_{i=1}^{r} \ln \frac{(1-\lambda_i^*)}{(1-\hat{\lambda}_i)} \; .$$

It is asymptotically χ^2 distributed with $r \cdot (k - s)$ degrees of freedom. Restrictions can also be formulated with respect to the adjustment parameters. The property of *weak exogeneity* is of special interest:

- A variable is weakly exogenous with respect to the cointegration parameters if and only if no cointegrating relation is included in the equation of this variable, i.e. if the corresponding row of the matrix Γ contains only zeros.

Example 6.8

From January 1986 to December 1998, the German Bundesbank published monthly money market rates with time to maturity of one month, z_1, three month, z_3, and six month, z_6. *Figure 6.3* shows the three month money market rate. (The development of the two other interest rates is quite similar.) Theoretically, the relation between these interest rates can be described by the expectations hypothesis of the term structure. Its linearized version is:

(E6.12) $$z_{m,t} = \frac{1}{m} \sum_{i=0}^{m-1} E_t[z_{1,t+i}] + \varphi_m \; .$$

6.3 Cointegration in Vector Autoregressive Models

z_m, $m = 1, 3, 6$, denote nominal interest rates with time until expiration of m months, φ_m a risk premium, and $E_t[\cdot]$ the conditional expectation, given all information up to time t. Because of

$$z_{1,t+i} = z_{1,t} + \Delta z_{1,t+1} + \Delta z_{1,t+2} + \ldots + \Delta z_{1,t+i},$$

(E6.12) can be written as

(E6.12') $$z_{m,t} = z_{1,t} + \frac{1}{m}\sum_{i=1}^{m-1}\frac{m-i}{m} E_t[\Delta z_{1,t+i}] + \varphi_m.$$

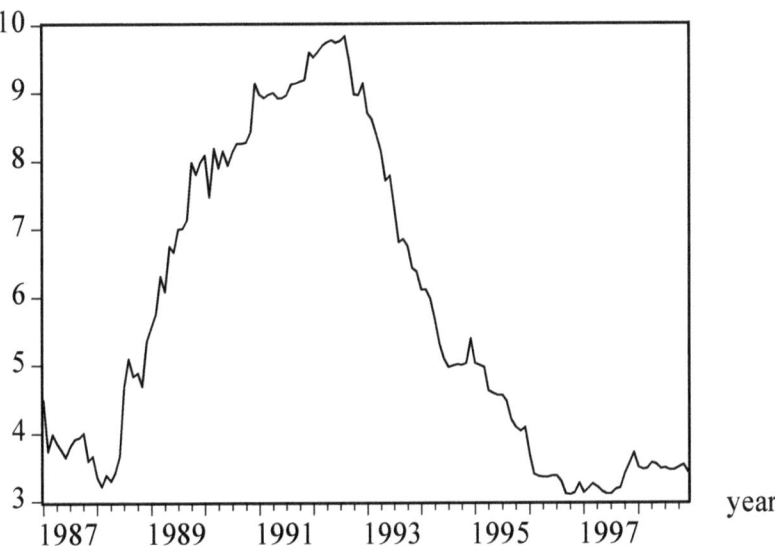

Figure 6.3: *German three month money market rate in Frankfurt*

Performing unit root tests for the interest rates z_1, z_3 and z_6, the null hypothesis of nonstationarity cannot be rejected for the levels of these variables, but it can be rejected for their first differences. Thus, the interest rates should be treated as I(1) variables. Because of (E6.12') it is obvious that

$$z_{m,t} - z_{1,t} \sim I(0), \quad m = 3, 6,$$

i.e. we have stationary interest rate spreads as implied by the expectations hypothesis. Therefore, any other difference between the interest rates is also stationary. Consequently, the three interest rates should contain one stochastic trend and generate two cointegrating relations. Possible linearly independent cointegration vectors are

$$\beta_1' = [1\ 0\ -1], \quad \beta_2' = [0\ 1\ -1].$$

Other representations are also possible, like, for example,

$$\tilde{\beta}_1' = \beta_2' - \beta_1' = [-1\ 1\ 0], \quad \tilde{\beta}_2' = -\tilde{\beta}_1' - \beta_1' = -\beta_2' = [0\ -1\ 1].$$

We use monthly data from January 1987 until December 1998 for the empirical analysis. We start with two bivariate models. The first model includes z_3 and z_1 and the second one z_6 and z_1. First we estimate VARs in the levels of the variables using the information criteria given from (4.10a) to (4.10d). The Hannan-Quinn criterion as well as the Schwarz criterion suggest a lag of two months.

For the parameterisation of the corresponding first order vector error correction models (VECM(1)), we assume that the variables do not contain a linear deterministic trend. Thus, the constant terms are elements of the cointegrating relations. The results of the trace and the λ_{max} tests are given in *Table 6.3*.

Table 6.3: Results of the Johansen Cointegration Test

Model	Hypotheses	Eigenvalues	Trace Test	λ_{max} Test
z_1, z_3	$r = 0$	0.257	43.559 (0.00)	42.715 (0.00)
	$r \leq 1$	0.006	0.843 (0.97)	0.843 (0.97)
z_1, z_6	$r = 0$	0.205	34.276 (0.00)	33.010 (0.00)
	$r \leq 1$	0.009	1.267 (0.91)	1.267 (0.91)

The numbers in parentheses are the p values of the corresponding statistics.

As theoretically expected, we find one cointegrating vector in each model; both are significant at least at the 0.1 per cent level. For the two long run relations we get:

$$z_{1,t} = \underset{(0.059)}{-0.080} + \underset{(0.009)}{0.998}\, z_{3,t} + \hat{\varsigma}_{1,t},$$

$$z_{1,t} = \underset{(0.132)}{-0.247} + \underset{(0.021)}{1.018}\, z_{6,t} + \hat{\varsigma}_{2,t}.$$

(The standard errors are given in parentheses.) The estimated coefficients of z_3 and z_6 are very close to one. Therefore, we test the theoretical long-run restriction leading to the cointegrating vector $[1, -1]$ using the approach described in equations (6.24) and (6.25). For the (z_1, z_3)-system we cannot reject this restriction with a p value of 0.86, and for the (z_1, z_6)-system with a p value of 0.39. From these results it follows that all possible spreads of the three interest rates are stationary.

Combining both systems to a three-dimensional VAR we expect one common stochastic trend and two cointegrating vectors. The Schwarz criterion suggests a lag length of one, whereas the Hannan-Quinn criterion leads to a lag length of two. The constant terms are again included in the long-run relations. The results of the Johansen approach are presented in *Table 6.4*.

Table 6.4: Results of the Johansen Cointegration Test

Model	Hypotheses	Eigenvalues	Trace Test	λ_{max} Test
z_1, z_3, z_6 VECM(0)	$r = 0$	0.448	116.587 (0.00)	85.500 (0.00)
	$r \leq 1$	0.187	31.087 (0.00)	29.883 (0.00)
	$r \leq 2$	0.008	1.204 (0.92)	1.204 (0.92)
z_1, z_3, z_6 VECM(1)	$r = 0$	0.384	98.883 (0.00)	69.711 (0.00)
	$r \leq 1$	0.177	29.172 (0.00)	27.999 (0.00)
	$r \leq 2$	0.008	1.173 (0.93)	1.173 (0.93)

The numbers in parentheses are the p values of the corresponding statistics.

We get very stable results regardless of the lag order. As expected, there are two cointegrating relations. Both are significant at the 0.1 per cent level. Thus, one common stochastic trend drives the system of the three interest rates. The estimated cointegrating vectors are again in line with the theoretical ones $[1, -1, 0]$ and $[1, 0, -1]$.

The mapping from the I(1) space of the three interest rates into a VECM with stationary variables needs to take the first differences of the interest rates and, for example, the two spreads

$$SP31 = r_3 - r_1, \quad SP63 = r_6 - r_3.$$

In the following, we estimate a parsimoniously parameterised vector error correction model using the Zellner seemingly unrelated regressions approach. Starting point is a reduced form VECM(1). First, we tested which of the two spreads has a significant impact in the different equations. Performing a Wald test, with a p value of 0.814 we cannot reject the combined hypothesis that SP31 does not have an influence on Δz_3 and Δz_6 while SP63 does not have an influence on Δz_1. We then successively eliminated the least significantly variables and finally got the following system of equations, with the estimated t statistics given in parentheses:

(E6.13a) $\quad \Delta z_{1,t} = \underset{(11.79)}{0.823} (SP31_{t-1} - \underset{(-4.48)}{0.090}) - \underset{(-2.74)}{0.403} \Delta z_{3,t-1}$

$\quad\quad\quad\quad + \underset{(2.42)}{0.376} \Delta z_{6,t-1} + \hat{u}_{1,t},$

$\bar{R}^2 = 0.241, \quad SE = 0.227, \quad LM(2) = 0.28 \ (p = 0.76),$
$LM(4) = 0.39 \ (p = 0.82), \quad LM(8) = 1.42 \ (p = 0.19).$

(E6.13b) $\quad \Delta z_{3,t} = \underset{(7.42)}{0.573} (SP63_{t-1} - \underset{(-1.87)}{0.047}) + \underset{(2.06)}{0.158} \Delta z_{6,t-1} + \hat{u}_{3,t},$

$\bar{R}^2 = 0.164, \quad SE = 0.223, \quad LM(2) = 0.31 \ (p = 0.73),$
$LM(4) = 0.25 \ (p = 0.91), \quad LM(8) = 0.89 \ (p = 0.52).$

(E6.13c) $\quad \Delta z_{6,t} = \underset{(5.07)}{0.445} (SP63_{t-1} - \underset{(-1.87)}{0.047}) + \underset{(2.98)}{0.238} \Delta z_{6,t-1} + \hat{u}_{6,t},$

$\bar{R}^2 = 0.150, \quad SE = 0.230, \quad LM(2) = 0.80 \ (p = 0.45),$
$LM(4) = 0.71 \ (p = 0.58), \quad LM(8) = 1.31 \ (p = 0.24).$

The estimated residuals of this system do not exhibit significant autocorrelation. The negative constant terms indicate that the term structure is on average (or in equilibrium) 'normal', i.e. the long-run rates are higher than the short-run ones. The estimated t values in the parentheses show that these constants are significant. Moreover, no interest rate is weakly exogenous. All adjustment parameters are highly significant. We find a unidirectional adjustment of the one month rate to the three month rate. Whereas the adjustment process of the three month rate and the six month rate show feedback relations, the two longer term rates are not influenced by the one month rate.

6.3.3 Analysis of Vector Error Correction Models

In the following, we discuss several concepts which are important for the interpretation of error correction models, like, for example, the concept of weak exogeneity or the implementation of Granger causality tests. In any case, a vector error correction model can be transformed into the corresponding vector autoregressive model. This allows to calculate the impulse response functions and to decompose the variances.

Stochastic Trend Representation

Taking the cointegration restriction (6.13) into account and neglecting the deterministic terms, the reduced form of an error correction model (6.12) can be written as

$$(6.26a) \quad \Delta Y_t = -\Gamma B' Y_{t-1} + \sum_{j=1}^{p-1} A_j^* \Delta Y_{t-j} + U_t, \quad U_t \sim N(0, \Sigma).$$

The necessary and sufficient condition for Y not to be integrated of order 2 is that

$$\tilde{C} = \Gamma'_\perp \left(I_k - \sum_{j=1}^{p-1} A_j^* \right) B_\perp$$

has full rank with $\Gamma_\perp$ and $B_\perp$ being the orthogonal complements of Γ and B. In this case, we can solve (6.26a) by deriving its moving average representation

$$(6.26b) \quad Y_t = C \sum_{i=1}^{t} U_i + C^*(L) U_t + y_0^*$$

where $C = B_\perp \tilde{C}^{-1} \Gamma'_\perp$ and y_0^* denote the initial values. $C^*(L)$ is an infinite-order polynomial in the lag operator with coefficient matrices C_j^* that go to zero with j going to infinity. C has the rank $k - r$ if (6.26a) has cointegration rank r. Therefore, equation (6.26b) indicates the stochastic trend representation of Y with $k - r$ common trends given as $\Gamma'_\perp \sum_{j=1}^{t} U_i$.

Conditional Error Correction Representation

In the following, we will derive the conditional error correction representation by partitioning the vector Y in (6.26a) into two sub-vectors X and Z, i.e. Y' = [X', Z']. This leads to

$$(6.27) \quad \begin{bmatrix} \Delta X_t \\ \Delta Z_t \end{bmatrix} = - \begin{bmatrix} \Gamma_x \\ \Gamma_z \end{bmatrix} B' Y_{t-1} + \sum_{j=1}^{p-1} \begin{bmatrix} A^*_{x_j} \\ A^*_{z_j} \end{bmatrix} \Delta Y_{t-j} + \begin{bmatrix} U_{x,t} \\ U_{z,t} \end{bmatrix},$$

with vectors and matrices having the appropriate dimensions and the variance-covariance matrix

$$\Sigma = \begin{bmatrix} \Sigma_{xx} & \Sigma_{xz} \\ \Sigma_{zx} & \Sigma_{zz} \end{bmatrix}, \quad \Sigma_{zx} = \Sigma'_{xz}.$$

If Z is interpreted as a vector of conditioning variables, even the current changes of Z, i.e. ΔZ_t, can be applied as explanatory variables for ΔX. Following SØREN JOHANSEN (1992) or H. PETER BOSWIJK (1995), the equivalent transformation of (6.27) leads to

$$(6.28a) \quad \Delta X_t = A^*_0 \Delta Z_t - \Gamma_{x|z} B' Y_{t-1} + \sum_{j=1}^{p-1} A^*_{x|z_j} \Delta Y_{t-j} + U_{x|z,t},$$

$$(6.28b) \quad \Delta Z_t = - \Gamma_z B' Y_{t-1} + \sum_{j=1}^{p-1} A^*_{z_j} \Delta Y_{t-j} + U_{z,t}.$$

Here, it holds that

$$A^*_0 = \Sigma_{xz} \Sigma_{zz}^{-1}, \quad \Gamma_{x|z} = \Gamma_x - A^*_0 \Gamma_z, \quad A^*_{x|z_j} = A^*_{x_j} - A^*_0 A^*_{z_j},$$

$$j = 1, 2, ..., p-1, \quad U_{x|z,t} = U_{x,t} - A^*_0 U_{z,t}.$$

In its systematic part, representation (6.28a) contains the contemporaneous correlation between ΔX and ΔZ. If $\Sigma_{xz} = 0$, then X and Z are block recursive and (6.28a, b) is identical with (6.27).

If either (6.27) or (6.28a,b) is the true data generating process, the cointegrating matrix B can be estimated efficiently by using the Johansen approach or performing a simultaneous estimation of (6.28a,b). However, the question of whether the cointegration vectors estimated in this way have an economic interpretation as long-run equilibrium relations remains open because of (6.14).

SØREN JOHANSEN (1992), H. PETER BOSWIJK (1995) and NEIL R. ERICSSON (1995) showed that it is possible to estimate B efficiently from

(6.28a) without using (6.28b), (i) if Z is weakly exogenous, i.e. $\Gamma_z = 0$, (ii) if none of the cointegrating relations of (6.28b) is also part of (6.28a), or (iii) if the system is block recursive, i.e. if $\Sigma_{xz} = 0$ holds.

If one of these conditions is fulfilled and if the sub-vector X contains only one single variable, the conditional error correction equation (6.28a) is a structural equation and the long-run relation has a structural interpretation. However, if the sub-vector X contains more than one single variable, the conditional error correction equations (6.28a) – in general – no longer have a structural interpretation because possible instantaneous relations between the endogenous variables are not covered. Thus, the cointegration vectors may no longer represent structural relations.

If, on the other hand, Z is weakly exogenous, (6.28a) can be used to derive a *structural error correction model* by multiplying it with a regular and correspondingly normalised matrix Γ_0, which, in addition, contains the identifying restrictions:

$$(6.29) \quad \Gamma_0 \Delta X_t = \tilde{A}_0^* \Delta Z_t - \tilde{\Gamma}_{x|z} B' Y_{t-1} + \sum_{j=1}^{p-1} \tilde{A}_{x|z_j}^* \Delta Y_{t-j} + \tilde{U}_{x|z,t},$$

with

$$\tilde{A}_0^* = \Gamma_0 A_0^*, \quad \tilde{\Gamma}_{x|z} = \Gamma_0 \Gamma_{x|z}, \quad \tilde{A}_{x|z_j}^* = \Gamma_0 A_{x|z_j}^*, \quad j = 1, 2, \ldots, p-1,$$

$$\tilde{U}_{x|z,t} = \Gamma_0 U_{x|z,t}.$$

The efficient estimation of B in (6.29) generates structural long-run relations. Only the estimation of structural error correction models leads to long-run relations with a structural interpretation, as these relations are exactly determined by the identifying restrictions. Every other situation leads to cointegrating vectors for which we cannot normally expect a direct economic interpretation. Usually, however, linear combinations of the cointegrating vectors can be interpreted as economic long-run equilibrium relations.

If there is only one endogenous variable in (6.28a) and if all explanatory variables are weakly exogenous, the parameters of the long-run relation can be estimated efficiently by using OLS, and the usual test statistics can be applied. If, on the other hand, the explanatory variables are not weakly exogenous and if we have identified cointegrating relations, OLS can still be applied to get super consistent estimates. However, the asymptotic efficiency is lost and the usual test statistics are no longer applicable.

Example 6.9

In *Example 6.5* and *Example 6.6* we assumed that only one cointegrating relation between real money m, real income y, and the bond yield r may exist and that it then can be interpreted as a long-run money demand relation. Applying the Johansen approach to this three dimensional system we have the possibility to check whether these assumptions are correct. The FPE, the AIC, and the HQ criteria, compare (4.10a,b,c) lead to a VAR(5) for the levels of m, y, r, including a constant term, centred seasonal dummies as well as the dummy D7302. Allowing for linear deterministic trends for the levels but only a constant term in the cointegrating relations we find with a p value of 0.082 (trace test) only one long-run relation (with standard errors in parentheses):

(E6.14) $\qquad$ m = − 1.003 + 1.106 y − 4.822 r.
$\qquad\qquad\qquad\qquad\qquad$ (0.04) $\quad$ (0.71)

The adjustment coefficients are -0.151(.037) for Δm, -0.011(.042) for Δy, and -0.013(.014) for Δr. According to the standard errors we can conclude that the equations for Δy and Δr do not contain the long-run relation (E6.14). A formal test of the weak exogeneity hypothesis of y and r shows that this cannot be rejected; we get a p value of 0.698. Testing additionally a unit income elasticity reduces the p value to 0.103. Since the hypothesis of weak exogeneity of y and r cannot be rejected in this system, we can conclude that the error correction equation (E6.9) estimated in *Example 6.6* is a structural equation and the derived long-run relation which is very similar to (E6.14) has a structural interpretation.

Granger Causality

The concept of Granger causality in the VAR framework has been discussed in *Chapter 4*. If vector error correction models are transformed into VAR models, the considerations in *Section 4.2* hold. On the other hand, tests for Granger causality can also be performed using error correction models. CLIVE W.J. GRANGER and JIN-LUNG LIN (1995) showed that the advantage of this procedure is that it allows to differentiate between long-run and short-run causal relations.

Example 6.10

Let the following error correction model with two cointegrated I(1) variables be given,

$$\Delta y_{1,t} = -\gamma_1 (y_{1,t-1} - \beta y_{2,t-1}) + a_{11} \Delta y_{1,t-1} + a_{12} \Delta y_{2,t-1} + u_{1,t},$$
$$\Delta y_{2,t} = \gamma_2 (y_{1,t-1} - \beta y_{2,t-1}) + a_{21} \Delta y_{1,t-1} + a_{22} \Delta y_{2,t-1} + u_{2,t}.$$

Here,

$$z_t = y_{1,t} - \beta y_{2,t}$$

represents the long-run relation. The variable y_2 is not Granger causal to y_1 if its lagged values are not included in the equation for y_1. Thus, there is no causal relation from y_2 to y_1 if $\gamma_1 = 0$ and $a_{12} = 0$ holds. There exists only 'short-run' causality if $\gamma_1 = 0$ but $a_{12} \neq 0$, and only 'long-run' causality if $\gamma_1 \neq 0$ but $a_{12} = 0$. Similar considerations hold for the question of whether y_1 is Granger causal to y_2.

Cointegration always implies the existence of a Granger causal relation. Thus, if cointegration exists, at least one γ_i, $i = 1,2$, is different from zero. Apparently, the opposite relation does not hold.

When testing for Granger causality, problems can arise when it is open whether the nonstationary variables are cointegrated or not. For this situation, HIRO Y. TODA and TAKU YAMAMOTO (1995) (and in a similar way also JUAN J. DOLADO and HELMUT LÜTKEPOHL (1996)) propose the following procedure: Starting point is a VAR in levels. Using the usual criteria described in *Chapter 4*, its optimal lag length p is determined. Then, a VAR of order p+d is estimated, where d is the (assumed) maximum degree of integration of the variables. Using this VAR, Wald tests for simple Granger causality are performed, and only the first p coefficients are employed to perform the test. The disadvantage of this procedure is that, compared with the error correction representation, the estimates of the VAR are less efficient due to the additionally included lagged variables. It avoids, however, misspecifications that might invalidate the test results.

Forecasting

At a first glance, everything said about forecasts with vector autoregressive processes in *Section 4.1* holds for the use of cointegrated systems for forecasting, as every error correction model can be transformed into a VAR in levels. Here, it also holds that

$$\hat{Y}_t(h) = E_t[Y_{t+h}], \quad h = 1, 2, \ldots.$$

Moreover, it is also possible to calculate impulse response functions and decompose variances in cointegrated systems. Because of the unit roots, these statistics converge – if at all – considerably more slowly than in stationary models. The error correction representation which is possible for systems of stationary or cointegrated variables interprets the possible parameters in a more informative way but does not change anything with respect to the relations between the variables. Thus, their explicit consideration does neither lead to different forecasts nor to different impulse-response functions or different variance decompositions compared to those of the VAR in levels.

This is different if there are restrictions in the deterministic part of the model. Then, the use of error correction models should lead to better forecasts. This was already presented by ROBERT F. ENGLE and BYUNG SAM YOO (1987). However, this is not necessarily the case, as, for example, PETER F. CHRISTOFFERSEN and FRANCIS X. DIEBOLD (1998) or MICHAEL P. CLEMENTS and DAVID F. HENDRY (2001) showed. The reason for this is that, in the long-run, even very small deviations in the constant term of the cointegrating relation might produce large deviations of the predicted from the realised values. A possible alternative to forecasts with error correction models are, therefore, forecasts with a VAR in first differences. As the first differences eliminate the long-run relations, the implied long-run forecasts for the levels are more or less the status quo.

Thus, the question arises what is to be predicted. The (unconditional) long-run development of variables with stochastic trend (without strong drift) cannot be predicted. This still holds when employing error correction models. On the other hand, the knowledge of the long-run equilibrium relations given by the error correction representation is necessary for conditional long-run forecasts. Short- to medium-term forecasts can be performed with models in first differences as well as with error correction models. Using the development of German money market interest rates, UWE HASSLER and JÜRGEN WOLTERS (2001) showed that (in this case) forecasts with an error correction model, with a constant term only in the cointegration relation, were superior to forecasts based on a VAR in first differences. It is, however, impossible to say how far this result can be generalised. Quite generally, models without restrictions on the constant term seem to produce inferior forecasts for variables without trend than alternative approaches restricting constant terms to zero.

6.4 Cointegration and Economic Theory

Macroeconomic theory is mainly based on long-run equilibrium relations, like the quantity equation, purchasing power parity, or uncovered interest rate parity. Economic theory rarely tells us anything about short-run dynamics. Although these relations hardly ever hold exactly in reality, some of them are part of nearly all usual models. They play a role as, for example, purchasing power parity and uncovered interest rate parity in monetary international economics. It is usually argued that we only observe *short-run* deviations from the equilibrium, which is compatible with the *long-run* validity of these relations.

The error correction models introduced in *Chapter 4* allow for a representation which differentiates between long-run equilibrium relations and short-run adjustment processes. Nevertheless, if the variables are stationary, the short-run dynamic has to be correctly specified in order to estimate the long-run relations consistently. Given that economic theory does mostly not consider short-run dynamics, these adjustment processes are usually modelled ad hoc, using statistical criteria.

If variables are nonstationary but cointegrated, it is possible that the parameters of long-run relations are estimated (super) consistently without considering the short-run dynamics. Taking the short-run dynamics into account improves the efficiency of the estimates (and the power of the corresponding tests) but does not change the consistency properties. Thus, a misspecification of the short-run dynamics (or the omission of stationary variables) does not lead to inconsistent estimates of the equilibrium relations between the nonstationary variables. The same holds for simultaneity problems and for errors in the (explanatory) variables. Contrary to estimates with stationary variables, these problems do not lead to inconsistent estimates.

All these aspects facilitate the empirical examination of economic theories. In order to estimate long-run equilibrium relations consistently, we no longer need the complete and fully specified model. It is sufficient to know which (nonstationary) variables are elements of these relations. It is even possible to estimate a model with OLS. Thus, the propagation of cointegration analysis also leads to a kind of renaissance of OLS estimations.

A further advantage is that these cointegrating properties are invariant to extensions of the information set. As KATARINA JUSELIUS (2006, p. 349) writes: "If cointegration is found between a set of variables this result will remain valid even if more variables are added to the analysis."
However, if tests are to be performed for the estimated relations, the price for these more 'simple' estimation procedures becomes easily obvious: Most test statistics do not follow their usual distributions, there are even massive deviations in some cases. This also holds asymptotically. Moreover, in most cases the exact distributions for finite samples are unknown. Thus, we have to resort to simulated critical values, as presented in many papers, or generate them by bootstrapping.

This does not mitigate the fact that the development of cointegration analysis has brought time series econometrics back closer to economic theory. In the 1970s, the expansion of the Box-Jenkins analysis had generated a large gap between these two. The results mentioned in *Chapter 2* demonstrated that univariate models without (economic) theoretical underpinning led to better forecasts of the future development of economic variables. This seemed to justify the gap. These procedures did, of course, not allow for conditional forecasts, which are as important for economic policy as pure predictions. For conditional forecasts we need (empirically supported) knowledge about the basic long-run equilibrium relations. Such information can be generated much better and more precisely by using cointegration analysis rather than by employing traditional econometric methods. Thus, time series analysis and empirical investigations performed by its methods have again become much more relevant for economic policy advice than it seemed to be the case in the 1970s.

References

The idea of **cointegration** goes back to

CLIVE W.J. GRANGER, Some Properties of Time Series Data and their Use in Econometric Model Specification, *Journal of Econometrics* 16 (1981), pp. 121 – 130, as well as

CLIVE W.J. GRANGER, Developments in the Study of Co-integrated Economic Variables, *Oxford Bulletin of Economics and Statistics* 48 (1986), pp. 213 – 228.

The first basic methodological paper about cointegration was

ROBERT F. ENGLE and CLIVE W.J. GRANGER, Co-Integration and Error Correction: Representation, Estimation, and Testing, *Econometrica* 55 (1987), pp. 251 – 276.

This was one of the essential papers for which C.W.J. GRANGER received the Nobel Prize in 2003. This and the following papers,

JAMES H. STOCK, Asymptotic Properties of Least-Squares Estimators of Cointegrating Vectors, *Econometrica* 55 (1987), pp. 1035 – 1056, and

SØREN JOHANSEN, Statistical Analysis of Cointegration Vectors, *Journal of Economic Dynamics and Control* 12 (1988), pp. 231 – 254

led to the large dissemination of this approach.

An introduction to estimation and testing of cointegration in single equations is given by

UWE HASSLER, Leitfaden zum Testen und Schätzen von Kointegration in W. GAAB, U. HEILEMANN and J. WOLTERS (eds), *Arbeiten mit ökonometrischen Modellen*, Physica-Verlag, Heidelberg 2004, pp. 88 – 155.

Special textbooks covering the econometric handling of cointegrated processes are

ANINDYA BANERJEE, JUAN J. DOLADO, JOHN W. GALBRAITH and DAVID F. HENDRY, *Co-Integration, Error Correction, and the Econometric Analysis of Non-Stationary Data*, Oxford University Press, Oxford 1993; or

SØREN JOHANSEN, *Likelihood-based Inference in Cointegrated Vector Autoregressive Models*, Oxford University Press, Oxford 1995.

Based on this strongly theoretically oriented book

KATARINA JUSELIUS, *The Cointegrated VAR Model: Methodology and Applications*, Oxford University Press, Oxford 2006,

shows how to apply and interpret Vector Error Correction Models. A short review of different approaches to identify cointegrating relations and to impose restrictions on them is given by

H. PETER BOSWIJK and JURGEN A. DOORNIK, Identifying, Estimating and Testing Restricted Cointegrated Systems: An Overview, *Statistica Neerlandica* 58 (2004), pp. 440 – 465.

The problem of **spurious regressions** was first tackled in a simulation study by

CLIVE W.J. GRANGER and PAUL NEWBOLD, Spurious Regressions in Econometrics, *Journal of Econometrics* 2 (1974), pp. 111 – 120.

The corresponding **asymptotic distribution theory** is presented in

PETER C.B. PHILLIPS, Understanding Spurious Regressions in Econometrics, *Journal of Econometrics* 33 (1986), pp. 311 – 340.

Critical values of residual based tests for cointegration in single equation models are given by

ROBERT F. ENGLE and BYUNG SAM YOO, Forecasting and Testing in Cointegrated Systems, *Journal of Econometrics* 35 (1987), pp. 143 – 159;

JAMES G. MACKINNON, Critical Values for Co-Integration Tests, in: R.F. ENGLE and C.W.J: GRANGER (eds.), *Long-Run Economic Relationships*, Oxford University Press, Oxford 1991, pp. 267 – 276.

A simple correction procedure which leads to asymptotically standard normal distributed t values in static regression equations is derived by

PENTTI SAIKKONEN, Asymptotically Efficient Estimation of Cointegration Regressions, *Econometric Theory* 7 (1991), pp. 1 – 21, and

JAMES H. STOCK and MARK W. WATSON, A Simple Estimator of Cointegrating Vectors in Higher Order Integrated Systems, *Econometrica* 61 (1993), pp. 783 – 820.

Problems which might arise by neglecting the dynamic structure when using the Engle-Granger approach are shown by

ANINDYA BANERJEE, JUAN J. DOLADO, DAVID F. HENDRY and GREGOR W. SMITH, Exploring Equilibrium Relationships in Econometrics Through Static Models: Some Monte Carlo Evidence, *Oxford Bulletin of Economics and Statistics* 48 (1986), pp. 253 – 277,

Critical values for tests of cointegration in error correction models are given in

ANINDYA BANERJEE, JUAN J. DOLADO and RICARDO MESTRE, Error-Correction Mechanism Tests for Cointegration in a Single-Equation Framework, *Journal of Time Series Analysis* 19 (1998), pp. 267 – 283.

The critical values which are appropriate when the variables also include linear time trends is discussed in

UWE HASSLER, Cointegration Testing in Single Error-Correction Equations in the Presence of Linear Time Trends, *Oxford Bulletin of Economics and Statistics* 62 (2000), pp. 621 – 632.

Further test procedures for testing in single error correction equations are presented in

UWE HASSLER and JÜRGEN WOLTERS, Autoregressive Distributed Lag Models and Cointegration, *Allgemeines Statistisches Archiv* 90 (2006), pp. 59 – 74; reprinted in: O. HÜBLER and J. FROHN (eds.), *Modern Econometric Analysis*, Springer, Berlin 2006, pp. 57 – 72.

Critical values for **trace and λ_{max} tests** proposed by SØREN JOHANSEN are given by

MICHAEL OSTERWALD-LENUM, A Note with Quantiles of the Asymptotic Distribution of the Maximum Likelihood Cointegration Rank Test Statistics, *Oxford Bulletin of Economics and Statistics*, 54 (1992), pp. 461 – 472.

JAMES G. MACKINNON, ALFRED A. HAUG and LEO MICHELIS, Numerical Distribution Functions of Likelihood Ratio Tests for Cointegration, *Journal of Applied Econometrics* 14 (1999), pp. 563 – 577

present critical values which are much more accurate than those available previously and also take into account the possibility for exogenous variables in the cointegrating relation.

Tests for **hypotheses about the cointegration matrix** have been developed by

SØREN JOHANSEN and KATARINA JUSELIUS, Maximum Likelihood Estimation and Inference on Cointegration – with Applications to the Demand for Money, *Oxford Bulletin of Economics and Statistics*, 52 (1990), pp. 169 – 210.

Compared to the Johansen approach, an **alternative handling of the deterministic components** in error correction models is proposed by

HELMUT LÜTKEPOHL and PENTTI SAIKKONEN, Testing for the Cointegration Rank of a VAR Process with a Time Trend, *Journal of Econometrics* 95 (2000), pp. 177 – 198, and

PENTTI SAIKKONEN and HELMUT LÜTKEPOHL, Trend Adjustment Prior to Testing for the Cointegration Rank of a Vector Autoregressive Process, *Journal of Time Series Analysis* 21 (2000), pp. 435 – 456.

This approach can be extended to **modelling deterministic structural breaks** in the data. See for this

PENTTI SAIKKONEN and HELMUT LÜTKEPOHL, Testing for the Cointegration Rank of a VAR Process with Structural Shifts, *Journal of Business and Economic Statistics* 18 (2000), pp. 451 – 464.

Tests for cointegration in the Engle-Granger framework in the presence of structural breaks are presented in

UWE HASSLER, Dickey-Fuller Cointegration Test in the Presence of Regime Shifts at Known Time, *Allgemeines Statistisches Archiv* 86 (2002), pp. 263 – 276.

For the analysis of **structural vector error correction models** see

SØREN JOHANSEN, Cointegration in Partial Systems and the Efficiency of Single-Equation Analysis, *Journal of Econometrics* 52 (1992), pp. 389 – 402,

H. PETER BOSWIJK, Efficient Inference on Cointegration Parameters in Structural Error Correction Models, *Journal of Econometrics* 69 (1995), pp. 133 – 158, as well as

NEIL R. ERICSSON, Conditional and Structural Error Correction Models, *Journal of Econometrics* 65 (1995), pp. 159 – 171.

For the concept of **weak exogeneity** see, for example,

NEIL R. ERICSSON, Cointegration, Exogeneity, and Policy Analysis: An Overview, *Journal of Policy Modeling* 14 (1992), pp. 251 – 280, as well as

NEIL R. ERICSSON, DAVID F. HENDRY and GRAHAM E. MIZON, Exogeneity, Cointegration, and Economic Policy Analysis, *Journal of Business and Economic Statistics* 16 (1998), pp. 370 – 387.

These papers also discuss the relation between **Granger causality** and **exogeneity**. The problem of how vector error correction models with **exogenous I(1) variables** and restrictions with respect to the short-run dynamics can efficiently be estimated is discussed in

M. HASHEM PESARAN, YONGCHEOL SHIN and RICHARD J. SMITH, Structural Analysis of Vector Error Correction Models with Exogenous I(1)-Variables, *Journal of Econometrics* 97 (2000), pp. 293 – 343.

They also give the corresponding critical values of the tests for cointegration.

The problem of **Granger causality in the situation of cointegrated variables** is, for example, discussed in

CLIVE W.J. GRANGER and JIN-LUNG LIN, Causality in the Long Run, *Econometric Theory* 11 (1995), pp. 530 – 536.

Testing strategies for situations in which the question remains open whether a cointegrating relation exists or not are presented in

HIRO Y. TODA and TAKU YAMAMOTO, Statistical Inference in Vector Autoregressions with Possibly Integrated Processes, *Journal of Econometrics* 66 (1995), pp. 259 – 285, as well as in

JUAN J. DOLADO and HELMUT LÜTKEPOHL, Making Wald Tests Work for Cointegrated VAR Systems, *Econometric Reviews* 15 (1996), pp. 369 – 386.

For this, see also

HIROSHI YAMADA and HIRO Y. TODA, Inference in Possibly Integrated Vector Autoregressive Models: Some Finite Sample Evidence, *Journal of Econometrics* 86 (1998), pp. 55 – 95.

The possibilities and properties of predictions using error correction models are discussed in

PETER F. CHRISTOFFERSEN and FRANCIS X. DIEBOLD, Cointegration and Long-Horizon Forecasting, *Journal of Business and Economic Statistics* 16 (1998), pp. 450 – 458,

MICHAEL P. CLEMENTS and DAVID F. HENDRY, Forecasting with Difference-Stationary and Trend-Stationary Models, *Econometrics Journal* 4 (2001), pp. S1 – S19,

UWE HASSLER and JÜRGEN WOLTERS, Forecasting Money Market Rates in the Unified Germany, in: R. FRIEDMANN, L. KNÜPPEL and H. LÜTKEPOHL (eds.), *Econometric Studies: A Festschrift in Honour of Joachim Frohn*, Lit Verlag, Münster et al. 2001, pp. 185 – 201, as well as in

DAVID F. HENDRY and MICHAEL P. CLEMENTS, Economic Forecasting: Some Lessons from Recent Research, *Economic Modelling* 20 (2003), pp. 301 – 329.

Research on the German **money demand** is done by

JÜRGEN WOLTERS, TIMO TERÄSVIRTA and HELMUT LÜTKEPOHL, Modelling the Demand for M3 in the Unified Germany, *Review of Economics and Statistics* 80 (1998), pp. 399 – 409,

HELMUT LÜTKEPOHL, TIMO TERÄSVIRTA and JÜRGEN WOLTERS, Investigating Stability and Linearity of a German M1 Money Demand Function, *Journal of Applied Econometrics* 14 (1999), pp. 511 – 525.

HELMUT LÜTKEPOHL and JÜRGEN WOLTERS, The Transmission of German Monetary Policy in the Pre-Euro Period, *Macroeconomic Dynamics* 7 (2003), pp. 711 – 733.

The **term structure of interest rates** in the German money market is investigated by

JÜRGEN WOLTERS and UWE HASSLER, Die Zinsstruktur am deutschen Interbanken-Geldmarkt: Eine empirische Analyse für das vereinigte Deutschland, *ifo Studien* 44 (1998), pp. 141 – 160.

7 Nonstationary Panel Data

In *Chapter 4* we introduced an approach to analyse vectors of stationary time series, while *Chapter 6* was devoted to the nonstationary case. With $y_{i,t}$ we denote the ith component at time t, t = 1, ..., T. In typical time series applications the dimension of the vector is small (for instance equal to 3 in *Examples 4.4.* or *6.8*), while the time dimension is rather large (T > 100). In a panel situation the number of components or units, denoted by N, is large as well, i = 1, ..., N. There may be N price indices, N exchange rates or generally N countries or units. The unrestricted VAR(p) model from equation (4.1) allows each component to depend on its own lagged values and on the past of all other components. Hence, (4.1) includes $p \cdot N^2 + N$ parameters when modelling time series from N units, a number growing fast with the dimension N. Already with N = 10 there would be hundreds of parameters to estimate. Therefore, the VAR approach is not applicable unless the cross-sectional dimension is rather small.

Even for small N the VAR framework may not be appropriate when modelling data from N sectors or units. What we are interested in is the analysis of certain economic relationships in several units. In general, the intercepts and slope parameters may vary from one unit to the other, and in general we do not even require the vectors of explanatory regressors $x_{i,t}$ to be of equal length:

(7.1) $\quad y_{i,t} = \alpha_i + \beta_i' x_{i,t} + u_{i,t}, \quad t = 1, ..., T, \quad i = 1, ..., N,$

$\quad\quad\quad Cov[u_{i,t}, u_{j,t}] = \sigma_{ij}.$

In the most general case, the errors are allowed to be correlated at a given point in time t, $\sigma_{ij} \neq 0$, which parallels the VAR time series model.

In classical panel analysis the cross-sectional dimension N is very large (for example thousands of households or hundreds of firms) with only few so-called 'waves', i.e. with the number of time periods T being small. At the same time, regression equations like (7.1) are supposed to be independent of each other. More precisely, it is assumed that the error terms are independent across the units, which implies $Cov[u_{i,t}, u_{j,t}] = 0$ for $i \neq j$. Instead of error correlation, one typically assumes that the explanatory vectors $x_{i,t}$ are of the same length and associated with equal slope parameter vectors reflecting universal economic laws, $\beta_1 = \ldots = \beta_N = \beta$:

(7.2) $\qquad y_{i,t} = \alpha_i + \beta'x_{i,t} + u_{i,t}, \quad t = 1, \ldots, T, \quad i = 1, \ldots, N,$
$\qquad\qquad \sigma_{ij} = 0 \quad \text{for} \quad i \neq j,$

such that the equations from (7.2) are linked through this common parameter assumption.

In this chapter we deal with an intermediate case between the VAR model and the classical independent panel. It is encountered for instance when performing multi-country studies: the number of units N (for example the OECD countries) is of moderate size and often much smaller than the time dimension T. Further, the regression equations are not stochastically independent because there may be a common driving force like, for example, the business cycle shared by all countries. Moreover, with many economic or financial time series it is likely to observe nonstationarity as discussed in *Chapter 5*. Therefore, the present chapter has a special focus on unit root testing and cointegration modelling in a panel framework.

In the following, we briefly review some issues that are of general concern when analysing panel data (*Section 7.1*). In particular, we address the generalised least squares (GLS) estimation of so-called seemingly unrelated regressions (SUR) that are particularly relevant for multi-country studies. Section 7.2 is devoted to unit root testing, or more generally integration testing, with panel data. One route to panel unit root testing is the combination of individual p values. Indeed, this principle is more general and widely applicable (*Section 7.3*). In *Section 7.4* some of the cointegration techniques from *Chapter 6* are carried over to the panel framework. The chapter closes with some remarks on the virtues and limitations of nonstationary panel data tools (*Section 7.5*).

7.1 Issues with Panel Data

7.1.1 Omitted Variable Bias

Panel data (also called longitudinal data in the statistics literature) have become increasingly popular in economics. Not only that one hopes to increase efficiency of estimators and power of statistical tests when using more data; the use of panel data may also help to circumvent the problem of the omitted variable bias, which is often encountered in regressions of cross-sectional data only. Consider the case

$$y_{i,t} = \beta_0 + \beta_i' x_{i,t} + \gamma_i z_i + u_{i,t}, \quad i = 1, \ldots, N,$$

where $x_{i,t}$ is a vector of explanatory variables with marginal effects β_i. The variable z_i stands for economic or cultural attitudes that are hard to quantify (such as inflation aversion or openness with respect to new technologies, or similar). Such attitudes are country-specific and are believed to change only very slowly over decades. Consequently, such attitudes can be modelled as being constant over time: z_i does not carry a time index. Moreover, z_i is likely to be correlated with $x_{i,t}$, such that a cross-sectional OLS regression of simply $y_{i,t}$ on $x_{i,t}$, $i = 1, \ldots, N$, for a fixed time period results in an omitted variable bias. With $\alpha_i = \beta_0 + \gamma_i z_i$, however, we get (7.1), and for constant slope parameters we obtain (7.2) for each unit and each point in time, $t = 1, \ldots, T$. Although the parameter vectors and the regressors are individual-specific in (7.1), the equations may be linked through correlation of the error terms $u_{i,t}$. For that reason such systems are called seemingly unrelated regressions (SUR), as already employed in *Chapters 3 and 4*. Systems like (7.2) with constant but individual mean levels α_i are called models with *fixed effects* (FE) in the panel literature. The regression of SUR and FE will be addressed in the next subsection. In both cases the estimation of $\alpha_i = \beta_0 + \gamma_i z_i$ and β_i at N·T data points is not plagued by an omitted variable bias, and a specification and measurement of z_i is not required.

So far, we assumed an identical number of time periods for all units or individuals, which has been called a balanced panel; the case of T_i observations for unit i results in an unbalanced panel if $T_i \neq T_j$ for some $i \neq j$. We will assume balanced panels in what follows unless stated otherwise.

7.1.2 Estimation and Testing

The estimation of the FE model (7.2) is straightforward. We define temporal means

$$\bar{y}_i = \frac{1}{T}\sum_{t=1}^{T} y_{i,t} \quad \text{and} \quad \bar{x}_i = \frac{1}{T}\sum_{t=1}^{T} x_{i,t},$$

where $\bar{x}_i$ is a vector averaged for each component. They are used to compute the centred variables $\tilde{y}_{i,t} = y_{i,t} - \bar{y}_i$ and $\tilde{x}_{i,t} = x_{i,t} - \bar{x}_i$, $i = 1, \ldots, N$, and to recast (7.2) as

(7.3) $$\tilde{y}_{i,t} = \beta' \tilde{x}_{i,t} + \tilde{u}_{i,t}.$$

If the error terms are free of contemporaneous correlation, then one may simply pool all demeaned observations and run an OLS regression of (7.3) for all N·T data points. This yields an identical estimator for β as regressing (7.2) with dummy variables indicating the differing intercepts for each individual ('least squares dummy variable estimation').

Most panels of macroeconomic data exhibit cross-correlation in (7.1), i.e. $\sigma_{ij} \neq 0$. In this situation, the estimation of the SUR equations (7.1) should rely on Generalised Least Squares (GLS) in order to improve efficiency relative to OLS. This proposal dates back to ARNOLD ZELLNER (1962).

The error assumptions to perform GLS of (7.1) are the following white noise assumptions:

1. There is no serial dependence within or between the units: $E[u_{i,t} u_{j,s}] = 0$ for $t \neq s$.

2. There is homoscedasticity within the units: $E[u_{i,t}^2] = E[u_{i,s}^2] = \sigma_{ii}$ for all t and s.

3. Heteroscedasticity between the units is allowed for: $\sigma_{ii} \neq \sigma_{jj}$.

4. Contemporaneous correlation is allowed for: $E[u_{i,t} u_{j,t}] = \sigma_{ij} \neq 0$. The contemporaneous (co)variances are collected in the symmetric N×N matrix $\Sigma = (\sigma_{ij})$, i, j = 1, ..., N.

If $\beta_1, ..., \beta_N$ from (7.1) are estimated by separate OLS regressions, the information about σ_{ij} is ignored. GLS, however, employs OLS residuals from a first step in order to estimate

$$\hat{\sigma}_{ij} = \frac{1}{T}\sum_{t=1}^{T} \hat{u}_{i,t} \hat{u}_{j,t}, \quad i, j = 1, ..., N.$$

In a second step, one estimates $\hat{\Sigma}$ with $\hat{\sigma}_{ij}$ to perform the GLS estimation. To this end, $\hat{\Sigma}$ has to be inverted, which is only feasible with N smaller than T. These two steps can be iterated until the estimates converge.

GLS accounts for cross-correlation and thus uses more relevant information than OLS, which is where the improvement on OLS stems from. This GLS procedure will also be called SUR estimation in the following.

Let $\hat{\beta}_i$, i = 1, ..., N, , denote the OLS estimators obtained from separate regressions of (7.1), while $\tilde{\beta}_i$ stands for the GLS estimators. There are two important cases, where $\hat{\beta}_i = \tilde{\beta}_i$. First, if $\sigma_{ij} = 0$ for all $i \neq j$, i.e. if there is no cross-correlation, then OLS and GLS coincide notwithstanding eventual heteroscedasticity between the units. Second, if $x_{i,t} = x_{j,t}$ for all $i \neq j$ and all points in time, then again $\hat{\beta}_i = \tilde{\beta}_i$. Note that the latter case occurs when estimating an unrestricted VAR model where the same explanatory variables appear in each equation.

The null hypothesis of no contemporaneous cross-dependence is of particular interest for applied work since in this case OLS is efficient ($\hat{\beta}_i = \tilde{\beta}_i$). In terms of parameter restrictions, this hypothesis implies

$$H_0: \sigma_{ij} = 0 \quad \text{for all} \quad i \neq j, \quad i,j = 1, \ldots, N.$$

Due to the symmetry of Σ, H_0 consists of

$$(N-1) + (N-2) + \ldots + 2 + 1 = \frac{N(N-1)}{2}$$

restrictions. Under H_0 the correlation coefficients constructed from first-step OLS residuals,

$$\hat{r}_{ij} = \frac{\sum_{t=1}^{T} \hat{u}_{i,t} \hat{u}_{j,t}}{\sqrt{\sum_{t=1}^{T} \hat{u}_{i,t}^2} \sqrt{\sum_{t=1}^{T} \hat{u}_{j,t}^2}},$$

all converge to zero. Hence, TREVOR S. BREUSCH and ADRIAN R. PAGAN (1980) propose the test statistic

$$BP = T \sum_{j=2}^{N} \sum_{i=1}^{j-1} \hat{r}_{ij}^2.$$

H_0 is rejected when BP exceeds upper percentiles of a χ^2 distribution with $N(N-1)/2$ degrees of freedom.

7.1.3 Mixed Panel Evidence

We begin with an empirical example that will be leading through this chapter.

Example 7.1

In order to investigate the international interest rate linkage, following UWE HASSLER and VERENA WERKMANN (2012), we analyse 11 ten year government bond yields for Australia (AU), Canada (CA), Switzerland (CH), Germany (DE),

Denmark (DK), Japan (JP), Norway (NO), New Zealand (NZ), Sweden (SE) United Kingdom (UK) and the United States (US), or in short:

$$i \in \{AU, CA, CH, DE, DK, JP, NO, NZ, SE, UK, US\}.$$

The sample consists of monthly data from January 1990 until December 2006 (i.e. T = 204), thus not covering the period of the most recent financial and debt crises. The hypothesis of interest is whether and how strongly the US market affects the other bond yields.

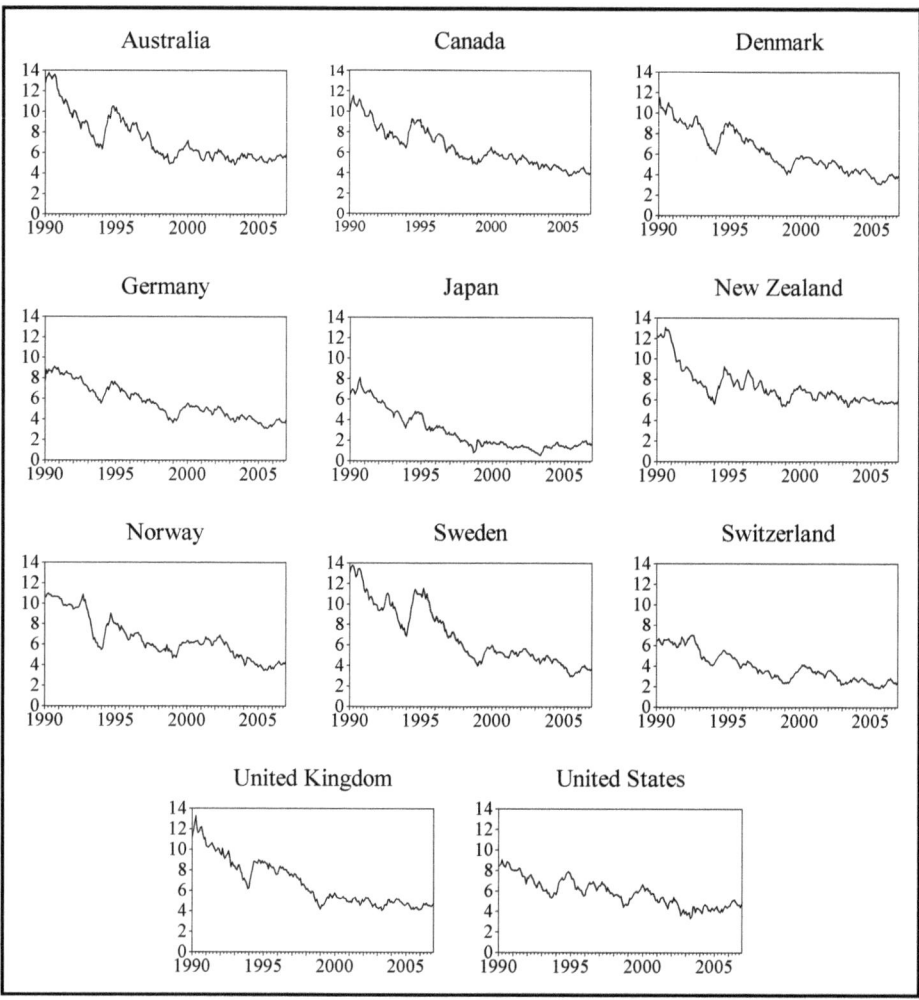

Figure 7.1: 10 year government bond yields, January 1990 – December 2006

Figure 7.1 suggests that the bond yield series $B_{i,t}$ are nonstationary. This is supported by individual ADF tests computed from regressions like (5.17'), where the lag lengths were determined according to the AIC criterion. *Table 7.1* contains the corresponding p values according to JAMES G. MACKINNON (1996). The most significant one is 0.06 for New Zealand, while the second most significant one is only 0.1503; all other values are not even significant at the 20 percent level.

Table 7.1: p values of the Augmented Dickey-Fuller Tests for 10 year government bonds

	p		p		p
Australia	0.1503	Japan	0.3254	Switzerland	0.4564
Canada	0.3168	New Zealand	0.0600	United Kingdom	0.3910
Denmark	0.3392	Norway	0.2677	United States	0.2298
Germany	0.4502	Sweden	0.2060		

The p values in *Table 7.1* vary from 0.060 over 0.1503 up to a maximum of 0.4564. This is the typical picture of mixed evidence often observed in empirical studies: Some countries are significant at the 5 percent level, some at the 10 percent level, and so on. What is the problem with summarising such mixed evidence? Why don't we simply conclude from *Table 7.1* that the unit root can be rejected at the 10 percent level for bond yields from New Zealand but not from the other countries?

Let $H_{0,i}$, i = 1, ..., N, denote N hypotheses formulated for N units. Assume for simplicity that the units are independent and that for each unit a test is performed at level α, such that the probability of a type I error individually is α. We further assume that all hypotheses are true. It then holds that the probability that $H_{0,i}$ is rejected while all other hypotheses are not is due to independence $\alpha(1-\alpha)^{N-1}$. Just as probable is under the above assumptions that $H_{0,2}$ is rejected while all other hypotheses are not. Hence, the probability that *any* of the N test statistics is significant while all the others are not becomes

$$P[\text{one false rejection}] = N \cdot \alpha (1-\alpha)^{N-1}.$$

We illustrate this point with N=11 and $\alpha = 0.1$:

$$P[\text{one false rejection}] = 1.1(0.9)^{10} = 0.384.$$

Under the hypothesis that *all* null hypotheses are true, it is coincidence that the statistic from some unit A is significant at the 10 percent level, it could just as well have been unit B, or any other unit. The probability of exactly one false rejection is 0.384, which is well above the individual nominal 10 percent level. This is the motivation to apply panel unit root tests in order to control the overall probability of a type I error when testing N hypotheses jointly.

7.2 Panel Unit Root Tests

The rate of convergence and the limiting distribution of the slope estimators of β_i from (7.1) will depend on the stochastic properties of the dependent variables $y_{i,t}$ and the regressors $x_{i,t}$, in particular on whether they are stationary or not. Therefore, panel unit root tests have been developed to establish the (non)stationarity of the data.

7.2.1 First Generation Tests

The first panel unit root tests assumed independent units, not because this assumption was believed to be met in practice, but in order to tackle the complicated distributional properties. According to JÖRG BREITUNG and M. HASHEM PESARAN (2008), these tests are said to belong to the first generation. Typically, asymptotic distributions were obtained by sequential limit theory, letting first $T \to \infty$ followed by $N \to \infty$. Such sequential limits have sometimes been interpreted as 'T should be large relative to N' for applied purposes. JOAKIM WESTERLUND and JÖRG BREITUNG (2012), however, show that such an intuition lacks theoretical grounds.

Analogously to the Augmented Dickey-Fuller regression (5.17), we consider for each of the units (i = 1, ..., N),

$$(7.4) \qquad \Delta y_{i,t} = d_{i,t} + (\rho_i - 1) y_{i,t-1} + \sum_{j=1}^{k_i} \theta_{i,j} \Delta y_{i,t-j} + u_{i,t},$$

where $d_{i,t}$ stands for a specific deterministic component.

The test by ANDREW T. LEVIN, CHIEN-FU LIN and CHIA-SHANG CHU (2002) assumes a homogeneous alternative, i.e.

$$\rho_1 = \ldots = \rho_N = \rho.$$

The null hypothesis amounts to unit roots for all individuals, while under the alternative all series are stationary:

$$H_0: \rho = 1, \quad H_1: |\rho| < 1.$$

In a first step, the Levin, Lin and Chu test individually corrects all series for deterministic terms and autocorrelation, which means that $\Delta y_{i,t}$ and $y_{i,t-1}$

are regressed separately on the deterministic components and on the lagged differences to obtain residuals. Then, one essentially runs a pooled Dickey-Fuller regression with the individually corrected series to estimate the common ρ jointly, where we omit relevant technical details here. A panel t type statistic follows a limiting normal distribution for $T \to \infty$ followed by $N \to \infty$, where normality arises from the assumption of cross-sectional independence. Note that the pooling step requires the panel to be balanced.

The test by KYUNG S. IM, M. HASHEM PESARAN and YONGCHEOL SHIN (2003) allows the ρ_i in (7.4) to be heterogeneous. The null hypothesis remains integration of all series, $\rho_1 = \ldots = \rho_N = 1$, but the alternative requires only

$$H_1: |\rho_1| < 1, \ldots, |\rho_L| < 1,$$

where $L = \ell N$, $0 < \ell < 1$; that is under the alternative, only a non-negligible fraction of individuals has to be stationary in order to reject the null hypothesis of overall nonstationarity. Further, the Im, Pesaran and Shin test does not require balanced panels since it does not pool the data but rather averages over individual test statistics. Let $t_{\rho,i}$, $i = 1, \ldots, N$, denote the t statistics from (7.4) testing for $\rho_i = 1$, and

$$\overline{t}_\rho = \frac{1}{N} \sum_{i=1}^{N} t_{\rho,i}.$$

With $T \to \infty$ the individual ADF statistics converge under $\rho_i = 1$. The mean and variance of the corresponding Dickey-Fuller distribution are known. KYUNG S. IM, M. HASHEM PESARAN and YONGCHEOL SHIN (2003) normalise the average $\overline{t}_\rho$ accordingly. Given the independence assumption, limiting normality arises upon normalisation for $N \to \infty$ due to the central limit theorem.

7.2.2 Second Generation Tests

Working paper versions of the Levin, Lin and Chu as well as Im, Pesaran and Shin tests circulated from 1993 and 1995 on, respectively. Hence, these first generation tests were subject to early critique. PAUL G. O'CONNELL (1998) provided simulation evidence that tests working under the independence assumption suffer from severe size distortions if this assumption is violated. To overcome this problem tests accounting for cross-sectional dependence have been introduced. Let Σ denote the N dimensional covariance matrix of the errors $u_t' = (u_{1,t}, ..., u_{N,t})$ from (7.4). Cross-sectional dependence is classified as weak if all eigenvalues of Σ are bounded, while strong dependence allows some of the eigenvalues of Σ to diverge with N.

Imposing again homogeneity in (7.4) i.e. $\rho_1 = ... = \rho_N = \rho$, JÖRG BREITUNG and SAMARJIT DAS (2005) discuss GLS estimation of (7.4) in the tradition of a SUR estimation discussed in the previous section. They obtain limiting normality of the t type statistic testing for $\rho = 1$ with $T \to \infty$ and $N \to \infty$, under the assumption of weak cross-sectional dependence. In practice, T has of course to be larger than N for GLS to be feasible, since GLS requires to invert the N×N dimensional matrix $\hat{\Sigma}$.

M. HASHEM PESARAN (2007) follows a different route allowing for strong cross-sectional dependence driven by a common univariate factor f_t,

$$u_{i,t} = g_i f_t + \varepsilon_{i,t}, \quad i = 1, ..., N,$$

where the so-called idiosyncratic component $\varepsilon_{i,t}$ is temporally and cross-sectionally independent. In order to account for this common factor, one simply computes cross-sectional means,

$$\bar{y}_t = \frac{1}{N} \sum_{i=1}^{N} y_{i,t},$$

and modifies the ADF regressions (7.4). The so-called cross-sectionally augmented Dickey-Fuller (CADF) regressions become

$$(7.5) \quad \Delta y_{i,t} = d_{i,t} + (\rho_i - 1)y_{i,t-1} + c_i \bar{y}_{t-1} + \sum_{j=1}^{k_i} \theta_{i,j} \Delta y_{i,t-j} + \sum_{j=0}^{k_i} \psi_{i,j} \Delta \bar{y}_{t-j} + \varepsilon_{i,t}.$$

The inclusion of $\Delta \bar{y}_{t-j}$ on the right-hand side parallels the ADF regression under structural breaks, see equation (5.23). With $t_{\rho,i}$, $i = 1, ..., N$, denoting the individual statistics testing for $\rho_i = 1$, the cumulated evidence relies on the panel average

$$C = \frac{1}{N} \sum_{i=1}^{N} t_{\rho,i}.$$

A normal approximation, however, is not valid this time. Finite sample critical values for the average C have been tabulated by M. HASHEM PESARAN (2007) for selected combinations of values of N and T. *The Tables II(b) and II(c)* in M. HASHEM PESARAN (2007) cover the cases of a constant only and a constant plus a linear time trend, respectively. The null hypothesis is rejected for too small values.

Example 7.2

From an economic point of view we expect that the interest rate differentials or spreads are stationary. For the data of *Example 7.1*, we define the interest differential or spread of country i against the U.S. dollar as:

(E7.1) $\qquad s_{i,t} = B_{i,t} - B_{US,t}, \quad i = 1, \ldots, 10.$

JEFFREY ALEXANDER FRANKEL (1992) shows, for example, that a necessary condition for the uncovered interest rate parity (UIP) to hold is that the interest rate spreads are stationary. In the presence of nonstationary bond yields this implies bivariate cointegration such that $B_i - B_{US}$ is a long-run equilibrium relation.

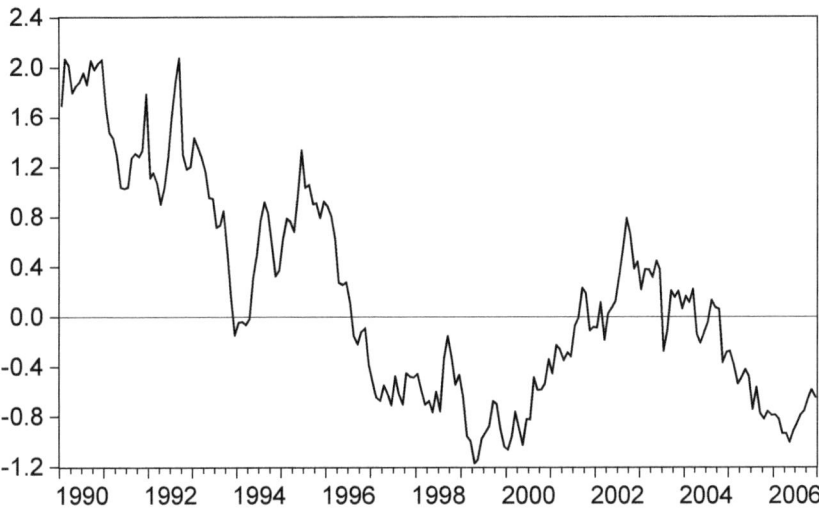

Figure 7.2: *Cross-sectional average of the U.S. spreads*

Under this cointegration assumption, there are two alternative channels for the U.S. influence through an error correction model, namely the equilibrium adjustment ($\gamma_i < 0$) and the direct short-run effect if $b_i \neq 0$ in:

(E7.2) $\Delta B_{i,t} = \alpha_i + \gamma_i s_{i,t-1} + b_i \Delta B_{US,t-1} + a_i \Delta B_{i,t-1} + u_{i,t}, \quad i = 1, \ldots, 10.$

Although the stationarity of the spreads seems to be a plausible guess, looking at *Figure 7.1*, we now formally test with the CADF test by M. HASHEM PESARAN (2007). To this end we need the cross-average $\bar{s}_t$ according to (7.5) in order to account for a common factor behind all spreads. *Figure 7.2* displays $\bar{s}_t$, which indeed seems to be characterised by a trending behaviour. In *Table 7.2* we report the lag length k_i and the t values for the coefficients in front of $y_{i,t-1}$, i.e. $t_{\rho,i}$, from (7.5) with a constant and without trend, where the number of lags was determined by the AIC of an augmented Dickey-Fuller regression without cross-sectional augmentation. The value of the panel test statistic becomes

$$C = \frac{1}{10}\sum_{i=1}^{10} t_{\rho,i} = -3.19.$$

The corresponding critical value from M. HASHEM PESARAN (2007) for N = 10 and T = 200 at the 1 percent level is -2.53. Hence, the rejection of the null hypothesis of nonstationary $s_{i,t}$ is significant at the 1 percent level.

Table 7.2: *CADF Test statistics for spreads against the U.S.*

	t_ρ	k_i		t_ρ	k_i
Australia	-2.96	0	New Zealand	-2.55	0
Canada	-3.69	0	Norway	-3.13	2
Denmark	-3.92	0	Sweden	-1.70	1
Germany	-5.05	0	Switzerland	-3.20	0
Japan	-2.80	2	United Kingdom	-2.90	2

7.2.3 The Null Hypothesis of Stationarity

In *Section 5.3.5* we introduced the so-called KPSS test for the null hypothesis of stationarity (or, more precisely, of integration of order zero). The corresponding test statistic is defined in (5.26). The panel null hypothesis reads as

$$H_0: y_{i,t} \sim I(0), \quad i = 1, \ldots, N.$$

KADDOUR HADRI (2000) proposed in a first generation framework assuming cross-sectional independence to compute the mean of individual KPSS statistics η_i:

$$\bar{\eta} = \frac{1}{N}\sum_{i=1}^{N} \eta_i.$$

With $T \to \infty$ the individual KPSS statistics converge under the null hypothesis. The corresponding mean and variance of the limiting distribution are known. Due to the independence assumption, limiting normality arises upon appropriate normalisation for $N \to \infty$.

MATEI DEMETRESCU, UWE HASSLER and ADINA I. TARCOLEA (2010) proposed a second generation version incorporating strong cross-sectional correlation into a multivariate KPSS-type statistic. In order to obtain a feasible consistent dependence estimator, they assume identical correlation between all units. KADDOUR HADRI and EIJI KUROZUMI (2012), on the contrary, allow for correlation through a common factor structure paralleling M. HASHEM PESARAN (2007).

Finally, adopting the framework of fractional integration dealt with in *Section 5.5.1*, UWE HASSLER, MATEI DEMETRESCU and ADINA I. TARCOLEA (2011) construct a panel test for arbitrary orders of integration, containing the I(0) or I(1) tests as special cases.

7.3 The Combination of Significance

So far, we have exploited the panel information by pooling the data or by combining the individual test statistics (averaging). Now, we consider a different route and combine individual p values to an overall significance level. The idea to do so can be traced back to RONALD A. FISHER (1954), and is not only applicable to panel unit root tests but more generally whenever testing a multiple null hypothesis composed of individual hypotheses for which p values are available.

Let $H_{0,i}$, $i = 1, \ldots, N$, denote again N hypotheses, and the overall null is the intersection that all hypotheses hold true:

$$H_0: \; H_{0,1} \cap H_{0,2} \cap \ldots \cap H_{0,N}.$$

Let p_i denote the p value for some statistic testing $H_{0,i}$. If the p values are independent, it then holds under H_0 that

$$F = -2 \sum_{i=1}^{N} \ln(p_i) \sim \chi^2(2N),$$

which is an exact result. Hence, RONALD A. FISHER (1954) suggested to reject H_0 for too large values of F. This approach has been proposed by GANGADHARRAO S. MADDALA and SHAOWEN WU (1999) for the purpose of unit root testing in the case of independent panels. One advantage of the p value combination is that it does not require balanced panels. For applied work, however, one has to overcome the assumption of independent units.

7.3.1 The Inverse Normal Method

With the distribution function Φ of the standard normal distribution, one defines the quantiles or so-called probits τ_i corresponding to the p values p_i:

$$\tau_i = \Phi^{-1}(p_i), \; i = 1, \ldots, N.$$

By construction the probits are standard normal; and under independence they follow a multivariate normal distribution with unit variances, such that it holds true for a linear combination with weights $\lambda_1, \ldots, \lambda_N$ where $\sum_{i=1}^{N} \lambda_i \neq 0$:

$$\sum_{i=1}^{N} \lambda_i \tau_i \sim N(0, \sum_{i=1}^{N} \lambda_i^2).$$

IN CHOI (2001) employs this property to discuss panel unit root testing under independence. JOACHIM HARTUNG (1999) assumes constant correlation between the probits $(i, j = 1, \ldots, N)$,

$$r = \text{Corr}[\tau_i, \tau_j], \quad i \neq j.$$

Under multivariate normality it then holds for finite T and N that any linear combination is again normal with variance equal to

$$\sum_{i=1}^{N} \lambda_i^2 + r \left[\left(\sum_{i=1}^{N} \lambda_i \right)^2 - \sum_{i=1}^{N} \lambda_i^2 \right].$$

Consequently, it holds for finite T and N:

$$\sum_{i=1}^{N} \lambda_i \tau_i \sim N\left(0, \sum_{i=1}^{N} \lambda_i^2 + r \left[\left(\sum_{i=1}^{N} \lambda_i \right)^2 - \sum_{i=1}^{N} \lambda_i^2 \right] \right).$$

In order to use this result for inference in practice, r has to be estimated. JOACHIM HARTUNG (1999) suggests the following rule (ensuring that the estimated correlation matrix is positive definite):

$$\hat{r}^* = \max\left(-\frac{1}{N-1}, \hat{r} \right)$$

with

$$\hat{r} = 1 - \frac{1}{N-1} \sum_{i=1}^{N} (\tau_i - \bar{\tau})^2,$$

where $\bar{\tau}$ is the mean over the probits. This estimator is consistent as $N \to \infty$. To improve the finite sample performance, JOACHIM HARTUNG (1999) introduced a tuning parameter κ, and suggested $\kappa = 0.2$ on experimental grounds. The general form of the test statistic becomes

$$\frac{\sum_{i=1}^{N}\lambda_{i}\tau_{i}}{\sqrt{\sum_{i=1}^{N}\lambda_{i}^{2}+\left[\left(\sum_{i=1}^{N}\lambda_{i}\right)^{2}-\sum_{i=1}^{N}\lambda_{i}^{2}\right]\left[\hat{r}^{*}+\kappa\sqrt{\frac{2}{N+1}}\left(1-\hat{r}^{*}\right)\right]}}.$$

In most practical applications one chooses equal weights, $\lambda_i = 1$, although with unbalanced panels the weight of a p value might be related to the length of the time dimension, for example $\lambda_i = T_i/T$. Further, the value of κ is negligible for large N (and $\hat{r}^*$ close to 1). Therefore, we work here with a simplified version of the test statistic with $\kappa = 0$ and $\lambda_i = 1$, i=1,...,N,

$$\text{Har} = \frac{\sum_{i=1}^{N}\tau_{i}}{\sqrt{N+\hat{r}^{*}\left[N^{2}-N\right]}}.$$

It is compared with quantiles from the standard normal distribution to test H_0, which is rejected for too small values. In fact, given a value of Har one may compute the p value thereof, and thus compute an overall significance from the individual p values.

MATEI DEMETRESCU, UWE HASSLER and ADINA I. TARCOLEA (2006) examine the approach by JOACHIM HARTUNG (1999) and add three aspects: First, they slightly relax the assumption of a constant r, second they provide a necessary and sufficient condition for normality to arise, and third they show experimentally that Har can be reasonably applied to ADF tests under different forms and degrees of cross-correlation even for small N.

7.3.2 Bonferroni-Type Tests

If the inverse normal method rejects the overall hypothesis H_0, the test remains silent with respect to which individual $H_{0,i}$ is to be considered as violated. To overcome this problem we order the p values in ascending order:

$$p_{(1)} \leq p_{(2)} \leq \ldots \leq p_{(N)}.$$

The Bonferroni inequality leads to a very simple test of H_0 with upper bound level α: Reject H_0 if $p_{(1)} \leq \alpha/N$. Moreover, one may not only reject the multiple null hypothesis but also detect *which* units violate it. In particular, one rejects all hypotheses $H_{0,i}$ where $p_{(i)} \leq \alpha/N$.

Since this Bonferroni test is known to be very conservative in the case of strong correlations among the p values, R. JOHN SIMES (1986) proposed the following modification: Reject H_0 at level α, if there is one sufficiently small p value with

(7.6) $$p_{(j)} \leq j\alpha/N, \quad j = 1, \ldots, N.$$

Although this proposal is clearly more powerful than the Bonferroni test, R. JOHN SIMES did not prove that it keeps the claimed level α under H_0 without independence. SANAT K. SARKAR (1998), however, establishes α as upper bound for the probability of a type I error for positively dependent multivariate distributions, for example for a multivariate normal distribution with non-negative correlations. Note that this is a finite sample result relying only on valid finite sample p values.

When rejecting H_0, JOHN SIMES suggests on heuristic grounds to consider all hypotheses $H_{0,(1)}$ till $H_{0,(j)}$ as falsified as long as

$$j = \max\{k: p_{(k)} \leq k\alpha/N\}.$$

GERHARD HOMMEL (1998), however, shows that in certain situations this may lead to an over-rejection of true null hypotheses, and he discusses an improved way to determine, which units violate $H_{0,i}$.

Example 7.3

We continue with *Examples 7.1* and *7.2*, but now wish to combine the p values of ADF regressions, like in equation (7.4), with a constant intercept. First, we investigate the 11 p values from *Table 7.1* when testing for a unit root in the bond yields. They result in Hartung's statistic

$$\text{Har} = -0.6590 \quad \text{with} \quad \hat{r}^* = 0.8189.$$

Clearly, the estimate of the correlation indicates a strong dependence over the units. Performing a one-sided test rejecting for too small values, the overall p value becomes 0.2550, which is in accordance with the nonstationarity assumption maintained in *Example 7.2*.

Next, we perform ADF tests for the spreads from (E7.1), i.e. in (7.4) we have as variable of interest $y_{i,t} = s_{i,t}$. The number of included lags according to AIC has already been reported in *Table 7.2*. The following table contains the corresponding ordered p values as well as the 10 percent bounds according to R. JOHN SIMES (1986) from equation (7.6).

From *Table 7.3* we observe that the inequality according to R. JOHN SIMES (1986) in (7.6) is never satisfied for $\alpha = 0.1$, from which we conclude that the null of I(1) spreads cannot be rejected at the 10 percent level. Unfortunately, this does not provide us with an overall significance level. To that end, we employ the Hartung test with

Table 7.3: Ordered p values for Augmented Dickey-Fuller tests for the spreads with $\alpha = 0.1$

	$p_{(j)}$	j·α/10		$p_{(j)}$	j·α/10
New Zealand	0.0298	0.01	Norway	0.1545	0.06
Switzerland	0.0406	0.02	Denmark	0.2070	0.07
Australia	0.1044	0.03	Sweden	0.2367	0.08
United Kingdom	0.1371	0.04	Japan	0.3152	0.09
Germany	0.1385	0.05	Canada	0.4717	0.10

$\text{Har} = -1.1868$ with $\hat{r}* = 0.7045$.

Here, the p value becomes 0.1176 for a one-sided test. Hence, the Hartung test, just as the Simes test, cannot establish that the U.S. spreads $s_{i,t}$ are stationary at the 10 percent level. At least, the Hartung test is significant at the 12 percent level with a p value not too distant from 10 percent. Still, a p value combination results in findings in contrast to that of the CADF test reported in *Example 7.2*. Therefore, we will come back to the issue of bivariate cointegration against the U.S. between the bond yields from *Figure 7.1* in the next section.

7.4 Panel Cointegration

If $y_{i,t}$ and the components of the vector $x_{i,t}$ are integrated (of order one), the question of cointegration naturally arises. In the case of absence of cointegration, the spurious regression problem introduced in *Example 6.1* shows up in panels, too, as has been established by HORST ENTORF (1997). In fact, due to the panel dimension, the spurious significance among independent random walks may be even increased with $N > 1$.

7.4.1 Single Equation Approaches

To justify a single equation approach, we assume that the components of the vector $x_{i,t}$ alone are not cointegrated. But we consider that the linear combination from equation (7.1) may result in stationary error terms $u_{i,t}$ that do not necessarily have to be white noise. In the case of $\beta_1 = \ldots = \beta_N = \beta$, one speaks of *homogeneous cointegration*, while $\beta_i \neq \beta_j$ for one $i \neq j$

characterises the heterogeneous case. Simple tests have been proposed to test the null hypothesis of no cointegration in all units:

$$H_0: \quad y_{i,t} - \beta_i' x_{i,t} \sim I(1), \quad i = 1, \ldots, N.$$

Since they all assume independent units, we review them only briefly here. First, CHIHWA KAO (1999) adopts the idea presented in *Section 6.2.3* and suggests panel unit root tests applied to OLS residuals from the static regressions (7.1), i = 1, ..., N. On top of cross-independence, he assumes homoskedasticity over the units and allows only to test against homogeneous cointegration. PETER PEDRONI (2004) manages to be less restrictive, while still maintaining independence. JOAKIM WESTERLUND (2007) carries the no cointegration test, based on the error correction model discussed in *Section 6.2.4*, over to nonstationary panels. In order to account for eventual cross-dependence, he switches to a computationally more involved bootstrap approach.

In the case of cointegration, PETER C.B. PHILLIPS and HYUNGSIK R. MOON (1999) establish that pooled OLS estimation $\hat{\beta}$ results in a super consistent estimation under homogeneous cointegration. The rate of convergence in case of sequential limit theory is $\sqrt{NT}$, showing that the time dimension is more informative with respect to the long-run equilibrium relation than the cross-dimension N; for N = 1 the usual time series case of *Section 6.2.3* is of course reproduced. In the case of independence, homogeneity and homoscedasticity, CHIHWA KAO and MIN-HSIEN CHIANG (2000) extend the limiting normality of the super consistent, dynamic estimator for the long-run parameters, see equation (6.7), to the panel case.

To overcome the unrealistic independence assumption when testing for no cointegration in nonstationary panels, we recommend the p value combination discussed in detail in the previous section. Since JAMES G. MACKINNON (1996) derived finite sample and asymptotic p values for residual-based cointegration tests, which are given in *Table 6.1*, it is straightforward to adopt the JOACHIM HARTUNG (1999) or R. JOHN SIMES (1986) approach to combine panel cointegration significance. Such a procedure is illustrated in the following example.

Example 7.4

Since the Hartung test did not reject nonstationarity of the bond yields at the 25 percent level (*Example 7.3*), we maintain the assumption that the bond yields from *Example 7.1* are integrated of order one. Further, in *Example 7.3* the combined significance when testing the U.S. spreads for nonstationarity was only significant at the 12 percent level, which is in contrast to the 1 percent significance found in

Example 7.2. To shed further light on this mixed evidence, we now test for cointegration of $B_{i,t}$ and $B_{US,t}$ without imposing parameter restrictions. To that end, we regress all currencies on the U.S. bond yields in levels,

(E7.3) $\qquad B_{i,t} = \alpha_i + \beta_i B_{US,t} + z_{i,t}, \quad i = 1, \ldots, 10,$

where $i = 1, \ldots, 10$ covers all the currencies except for the U.S. dollar. *Table 7.4* contains OLS estimates for β_i; note that they coincide with the SUR estimates since (E7.3) contains identical regressor observations in each equation. The estimates of β_i not only vary considerably over the currencies (from 0.96 to 1.95), they also differ substantially from 1. Next, we test the residuals $\hat{z}_{i,t}$ of the regression (E7.3) for a unit root, i.e. for no cointegration, from the regression

(E7.4) $\qquad \Delta\hat{z}_{i,t} = (\rho_i - 1)\hat{z}_{i,t-1} + \sum_{j=1}^{k_i} \theta_{i,j} \Delta\hat{z}_{i,t-j} + u_{i,t}.$

The lag length k_i was again determined with AIC. The p values testing for $\rho_i = 1$ individually from (E7.4) are given in *Table 7.4*.

Table 7.4: *Ordered p values for tests for no cointegration*

	$\hat{\beta}_j$	$p_{(j)}$		$\hat{\beta}_j$	$p_{(j)}$
Denmark	1.45	0.0748	Switzerland	0.96	0.1474
Sweden	1.95	0.0841	Australia	1.56	0.1705
United Kingdom	1.56	0.0882	Canada	1.35	0.2394
New Zealand	1.16	0.1120	Norway	1.33	0.3324
Germany	1.13	0.1200	Japan	1.26	0.4114

Although $\hat{\beta}_i$ reported in *Table 7.4* tends to be considerably larger than 1 this does not improve the cointegration evidence of $B_{i,t}$ and $B_{US,t}$: The ordered p values from *Table 7.4* tend to be even larger than those from *Table 7.3* for the spreads. Again, for all j we observe

$$p_{(j)} > \frac{j\alpha}{10}, \quad \alpha = 0.1,$$

such that the Simes procedure is not significant at $\alpha = 10$ percent. We also computed the Hartung statistic,

$$\text{Har} = -1.0800 \quad \text{with} \quad \hat{r}^* = 0.8277.$$

The panel significance, when comparing -1.08 in a one-sided test with a normal law, results in 0.1401, a p value that is larger than the one with spreads, which is

12 percent, as shown in *Example 7.3*. *Table 7.4* in view of *Table 7.3* points into the following direction: There is only weak evidence in favour of cointegration; but the evidence is stronger under the parameter restriction of the spreads.

If one has established panel cointegration at a given level, it seems advisable to perform a SUR analysis (as long as N is smaller than T) in order to estimate more precisely the adjustment parameters γ_i and the long-run parameter vectors β_i, see also equation (6.4):

$$\Delta y_{i,t} = \alpha_i + \gamma_i \left(y_{i,t-1} - \beta_i' x_{i,t-1} \right) + \sum_{j=1}^{k_i} a_{i,j} \Delta y_{i,t-j} + \sum_{j=1}^{k_i} b_{i,j}' \Delta x_{i,t-j} + u_{i,t}.$$

With $\theta_i = -\beta_i \gamma_i$ one obtains alternatively

$$(7.7) \quad \Delta y_{i,t} = \alpha_i + \gamma_i y_{i,t-1} + \theta_i' x_{i,t-1} + \sum_{j=1}^{k_i} a_{i,j} \Delta y_{i,t-j} + \sum_{j=1}^{k_i} b_{i,j}' \Delta x_{i,t-j} + u_{i,t}.$$

The components of the cointegrating vectors can be estimated individually as

$$\tilde{\beta}_{i,ec} = -\frac{\tilde{\theta}_i}{\tilde{\gamma}_i},$$

where $\tilde{\theta}_i$ and $\tilde{\gamma}_i$ are SUR estimates from (7.7). Assuming that the components of the parameter vectors vary randomly around a constant common value,

$$\beta_i = \beta + v_i, \quad v_i \sim iid(0, \sigma_v^2), \quad i = 1, \ldots, N,$$

one may consider a so-called *mean group* (MG) estimation. One simply estimates the mean of the parameters by averaging OLS or SUR estimates for each component:

$$\hat{\beta}_{MG,ec} = \frac{1}{N} \sum_{i=1}^{N} \hat{\beta}_{i,ec} \quad \text{or} \quad \tilde{\beta}_{MG,ec} = \frac{1}{N} \sum_{i=1}^{N} \tilde{\beta}_{i,ec},$$

where $\hat{\beta}_{i,ec} = -\hat{\theta}_i / \hat{\gamma}_i$ relies on the OLS estimates $\hat{\theta}_i$ and $\hat{\gamma}_i$ from (7.7). Under homogeneous cointegration, however, the MG estimators do not use the constant coefficient information. But it is straightforward to compute a SUR estimator, $\tilde{\beta}_{ec}$, or just as well an OLS estimator, $\hat{\beta}_{ec}$, imposing the homogeneous cointegration restriction by estimating

$$(7.8) \quad \Delta y_{i,t} = \alpha_i + \gamma_i \left(y_{i,t-1} - \beta' x_{i,t-1} \right) + \sum_{j=1}^{k_i} a_{i,j} \Delta y_{i,t-j} + \sum_{j=1}^{k_i} b_{i,j}' \Delta x_{i,t-j} + u_{i,t}.$$

If N is large relative to T, the SUR approach breaks down. In this case one may resort to the so-called *pooled mean group* (PMG) estimation by M. HASHEM PESARAN, YONGCHEOL SHIN and RONALD PATRICK SMITH (1999). They proposed a maximum likelihood estimation of (7.8) for large N, which comes at the price that the errors are assumed to be cross-sectionally independent. The PMG approach is a compromise between pooled estimation where all parameters are assumed to be identical over the units (except for the intercept, as in equation (7.2)) and an average of unrestricted estimation.

Example 7.5

Beyond statistical significance, good econometric practice relies on economic reasoning. Therefore, we now estimate the error correction equations (E7.2) notwithstanding ambiguous evidence with respect to the stationarity of the U.S. spreads:

$$\Delta B_{i,t} = \alpha_i + \gamma_i \, s_{i,t-1} + b_i \, \Delta B_{US,t-1} + a_i \, \Delta B_{i,t-1} + u_{i,t}.$$

Some of the OLS and SUR residuals display a mild degree of serial correlation. Since additional lags $\Delta B_{i,t-j}$, $j > 1$, turned out to be mostly insignificant, we stick to our specification with just one lag.

Table 7.5: Estimates of the error correction adjustment coefficient

	γ_i			γ_i	
	OLS	SUR		OLS	SUR
Australia	-0.045 (-2.77)	-0.067 (-5.12)	New Zealand	-0.062 (-2.75)	-0.064 (-3.63)
Canada	-0.036 (-1.56)	-0.066 (-3.68)	Norway	-0.050 (-2.97)	-0.060 (-4.64)
Denmark	-0.065 (-3.47)	-0.079 (-6.05)	Sweden	-0.032 (-3.03)	-0.043 (-4.49)
Germany	-0.063 (-2.06)	-0.083 (-5.67)	Switzerland	-0.060 (-2.03)	-0.075 (-4.27)
Japan	-0.022 (-1.13)	-0.037 (-2.54)	United Kingdom	-0.044 (-2.69)	-0.057 (-4.34)

The numbers in parentheses are the t statistics of the estimated parameters.

In *Table 7.5* we report estimates of γ_i with t statistics testing for $\gamma_i = 0$. It is interesting to compare the OLS results with those of the more efficient SUR procedure. Throughout the SUR are larger in absolute value and more significant than OLS estimates. The high significance of γ_i reduces the contribution of $\Delta B_{US,t-1}$ to the explanation of $\Delta B_{i,t}$ for all currencies. This can be seen when comparing the figures from *Table 7.6*, where SUR estimates (with t statistics) from (E7.2) are confronted with those from a – possibly misspecified – simple regression in differences:

(E7.5) $\Delta B_{i,t} = \alpha_i^* + b_i^* \Delta B_{US,t-1} + a_i^* \Delta B_{i,t-1} + u_{i,t}^*$, $i = 1, ..., 10$.

Throughout, the estimates $\tilde{b}_i^*$ from (E7.5) are larger and more significant than $\tilde{b}_i$ of (E7.2). Besides contemporaneous effects, which are not captured here, we conclude that the error correction mechanism is the dominant channel through which U.S. yields affect bonds of other currencies.

Table 7.6: Estimates of the short-run US influence

	b_i	b_i^*		b_i	b_i^*
Australia	0.224 (2.39)	0.269 (2.83)	New Zealand	0.075 (0.91)	0.114 (1.38)
Canada	0.104 (1.19)	0.151 (1.73)	Norway	0.031 (0.42)	0.092 (1.25)
Denmark	0.070 (0.98)	0.136 (1.88)	Sweden	0.010 (0.12)	0.049 (0.57)
Germany	0.127 (2.23)	0.176 (3.08)	Switzerland	0.047 (0.93)	0.100 (1.96)
Japan	0.103 (1.82)	0.123 (2.18)	United Kingdom	0.049 (0.61)	0.087 (1.07)

The numbers in parentheses are the t statistics of the estimated parameters.

Example 7.6

When estimating (7.7) without restrictions,

$$\Delta B_{i,t} = \alpha_i + \gamma_i B_{i,t-1} + \theta_i B_{US,t-1} + b_{i,1} \Delta B_{US,t-1} + a_{i,1} \Delta B_{i,t-1} + u_{i,t}.$$

we can compute the long-run parameters individually: $\tilde{\beta}_{i,ec}$. In *Table 7.7* we report those values, which can be compared with the estimates displayed in *Table 7.4*. They still vary considerably, but their mean is close to one:

$$\tilde{\beta}_{MG,ec} = \frac{1}{10}\sum_{i=1}^{10}\tilde{\beta}_{i,ec} = 1.053.$$

Next, we estimate the restricted SUR system with $\beta_1 = \ldots = \beta_N = \beta$ from (7.8), however, without imposing $\beta = 1$. The GLS estimation results in $\tilde{\beta}_{ec} = 0.919$ with a standard error of 0.103. Hence, the corresponding 95 percent confidence interval, [0.717, 1.121], covers the value 1, from which $\tilde{\beta}_{ec}$ is not significantly different at the 5 percent level. This supports the restricted error correction estimation with the U.S. spreads from *Example 7.5*. In particular, we consider the SUR adjustment parameters from *Table 7.5* as reliable.

Table 7.7: *SUR error correction estimates*

	$\tilde{\beta}_{i,ec}$		$\tilde{\beta}_{i,ec}$
Australia	0.828	New Zealand	0.588
Canada	0.957	Norway	1.312
Denmark	1.338	Sweden	1.638
Germany	1.096	Switzerland	0.938
Japan	0.590	United Kingdom	1.246

7.4.2 System Approaches

In equation (6.21) we defined the so-called trace statistic devised to test against multiple cointegration. Let r_i denote the individual cointegration ranks, which are assumed to be identical, $r_i = r < k$, $i = 1, \ldots, N$. Note that k, the number of variables in each individual system, is assumed to be constant over the sections, too. ROLF LARSSON, JOHAN LYHAGEN and MICKAEL LÖTHGREN (2001) consider trace statistics $Tr_i(r)$, $i = 1, \ldots, N$, computed individually as outlined in *Section 6.3.2*. Under the null hypothesis that the cointegration rank is r, the authors study the limit of the cross-sectional average of the trace statistics,

$$\overline{Tr}(r) = \frac{1}{N}\sum_{i=1}^{N} Tr_i(r),$$

upon appropriate normalisation. This test parallels the procedure by KYUNG S. IM, M. HASHEM PESARAN and YONGCHEOL SHIN (2003) discussed in *Section 7.2*, and unfortunately, it also does not allow for cross-correlation.

Alternatively to the maximum likelihood estimation behind $\overline{Tr}(r)$, JÖRG BREITUNG (2005) considers a two-step procedure. In a first step, all individual specific parameters are estimated. The second step assumes homogeneous cointegration across the units and estimates the long-run parameters from a pooled regression. In order to account for cross-sectional dependence, a SUR estimation could modify the first step.

A much simpler route to combine evidence from N units is of course possible with p values for the Johansen trace test available from JAMES G. MACKINNON, ALFRED A. HAUG and LEO MICHELIS (1999), in analogy to the procedures proposed by JOACHIM HARTUNG (1999) or R. JOHN SIMES (1986) described above.

So far, we did not consider the possibility of cointegration between $x_{i,t}$ and $x_{j,t}$ for $i \neq j$. Such cross-cointegration shows up naturally with certain economic models. The effect of cross-cointegration on some older panel methods has been discussed by ANINDYA BANERJEE, MASSIMILIANO MARCELLINO and CHIARA OSBAT (2004), and massive size distortions have been reported. Under cross-cointegration, the CADF test by M. HASHEM PESARAN (2007) suffers from size distortions under the null hypothesis of a unit root, too, because the included common factor in equation (7.5) will allow ρ_i to differ from one. The combination of p values, however, will be little affected by cross-cointegration under the null hypothesis, since strong dependence between the units is not ruled out. This claim is backed in particular for the Hartung test by limited simulation evidence in UWE HASSLER and ADINA I. TARCOLEA (2005). To fully account for the cross-cointegration effect, one would have to stack N vectors of k-dimensional time series. JAN J.J. GROEN and FRANK KLEIBERGEN (2003) discuss the estimation of such systems of dimension N·k, which may become intractable for practical purposes.

7.5 Concluding Remarks

In the leading example of this chapter we investigated monthly data for the period from 1990 to 2006 from 11 countries. This is a typical situation where N is small relative to T. In other multi-country studies this is not necessarily the case. The repudiated *Penn World Table* from the University of Pennsylvania, for example, provides annual data from 1950 on, but for almost 200 countries and territories such that N is about three times as large as T. We now briefly discuss, which of the above procedures remain applicable in such a large N environment.

Looking at the *Tables* in M. HASHEM PESARAN (2007), critical values for the CADF test are reported for N, T ∈ {10, 15, 20, ..., 100, 200}. This allows for large N relative to T. Similarly, the combination of p values is not restricted to the small N case. Hence, it is possible to combine significance from tests for unit roots as well as tests for the null of no cointegration from single equations for many units even if the time horizon is relatively short. Although it must be stressed that the analysis of trends is not meaningful without a sufficiently long time span.

More critical is the situation beyond (co)integration testing. When it comes to investigate cointegration in terms of error correction models, we adopted a SUR estimation procedure. It not only accounts for correlation between the errors, but also allows to easily incorporate, for example, long-run parameter restrictions. This approach, however, breaks down when N is not much smaller than T. For large values of N one may resort to pooled mean group estimation or the VAR approach by MICHAEL BINDER, CHENG HSIAO, and M. HASHEM PESARAN (2005) for short panels. Further, if we wish to model cross-cointegration, a full panel approach may be inaccessible. A VAR approach with N very small will be preferable, see for instance JÜRGEN WOLTERS (2002) or RALF BRÜGGEMANN and HELMUT LÜTKEPOHL (2005) for N = 2 (U.S. and Europe) when reporting evidence with respect to the uncovered interest rate parity hypothesis and the expectations hypothesis of the term structure.

References

The estimation of **SUR equations** accounting for cross-correlation has been proposed by

ARNOLD ZELLNER, An Efficient Method of Estimating Seemingly Unrelated Regression Equations and Tests of Aggregation Bias, *Journal of the American Statistical Association* 57 (1962), pp. 500 – 509.

The test for the null hypothesis of no cross-sectional error correlation is by

TREVOR S. BREUSCH and ADRIAN R. PAGAN, The Lagrange Multiplier Test and Its Applications to Model Specification in Econometrics, *Review of Economic Studies* 47 (1980), pp. 239 – 253.

The first two **panel unit root tests** that became popular are by

ANDREW T. LEVIN, CHIEN-FU LIN and CHIA-SHANG J. CHU, Unit Root Tests in Panel Data: Asymptotic and Finite-Sample Properties, *Journal of Econometrics* 108 (2002), pp. 1 – 24,

KYUNG S. IM, M. HASHEM PESARAN and YONGCHEOL SHIN, Testing for Unit Roots in Heterogeneous Panels, *Journal of Econometrics* 115 (2003), pp. 53 – 74.

For application and critique see also

PAUL G. J. O'CONNELL, The Overvaluation of Purchasing Power Parity, *Journal of International Economics* 44 (1998), pp. 1 – 19,

JOAKIM WESTERLUND and JÖRG BREITUNG, Lessons from a Decade of IPS and LLC, *Econometric Reviews* 31 (2012), forthcoming.

Tests **allowing for cross-correlation** have been introduced among others by

JÖRG BREITUNG and SAMARJIT DAS, Panel Unit Root Tests under Cross Sectional Dependence, *Statistica Neerlandica* 59 (2005), pp. 414 – 433,

M. HASHEM PESARAN, A Simple Panel Unit Root Test in the Presence of Cross Section Dependence, *Journal of Applied Econometrics* 22 (2007), pp. 265 – 312,

UWE HASSLER, MATEI DEMETRESCU and ADINA I. TARCOLEA, Asymptotic Normal Test for Integration in Panels with Cross-Dependent Units, *Advances in Statistical Analysis* 95 (2011), pp. 187 – 204.

Tests for the **null hypothesis of stationarity** can be found in

KADDOUR HADRI, Testing for Stationarity in Heterogeneous Panel Data, *Econometrics Journal* 3 (2000), pp. 148 – 161,

KADDOUR HADRI and EIJI KUROZUMI, A Simple Panel Stationarity Test in the Presence of Serial Correlation and a Common Factor, *Economics Letters* 115 (2012), pp. 31 – 34,

MATEI DEMETRESCU, UWE HASSLER and ADINA I. TARCOLEA, Testing for Stationarity in Large Panels with Cross-Dependence, and US Evidence on Unit Labor Cost, *Journal of Applied Statistics* 37 (2010), pp. 1381 – 1397.

The idea to **combine p values** goes back to

RONALD A. FISHER, *Statistical Methods for Research Workers*, Oliver & Bond, Berlin, 12th edition (1954),

and has first been applied to independent panels by

GANGADHARRAO S. MADDALA and SHAOWEN WU, A Comparative Study of Unit Root Tests with Panel Data and A New Simple Test, *Oxford Bulletin of Economics and Statistics* 61 (1999), pp. 631 – 652, and

IN CHOI, Unit Root Tests for Panel Data, *Journal of International Money and Finance* 20 (2001), pp. 249 – 272.

The case of constant correlation between the units has been discussed in

JOACHIM HARTUNG, A Note on Combining Dependent Tests of Significance, *Biometrical Journal* 41 (1999), pp. 849 – 855, and

MATEI DEMETRESCU, UWE HASSLER and ADINA I. TARCOLEA, Combining Significance of Correlated Statistics with Application to Panel Data, *Oxford Bulletin of Economics and Statistics* 68 (2006), pp. 647 – 663.

An EViews programme to execute the Hartung test is available from the homepage of UWE HASSLER.

Alternative ways of p value combination under dependence are from

R. JOHN SIMES, An Improved Bonferroni Procedure for Multiple Tests of Significance, *Biometrika* 73 (1986), pp. 751 – 754,

GERHARD HOMMEL, A Stagewise Rejective Multiple Test Procedure based on a Modified Bonferroni Test, *Biometrika* 75 (1988), pp. 383 – 386, and

SANAT K. SARKAR, Probability Inequalities for Ordered MTP_2 Random Variables: A Proof of the Simes Conjecture, *The Annals of Statistics* 26 (1998), pp. 494 – 504.

For a comparison of the different procedures including experimental evidence see

CHRISTOPH HANCK, An Intersection Test for Panel Unit Roots, *Econometric Reviews* 31 (2012), forthcoming.

A related but different approach relying on the false discovery rate has been advocated by

HYUNGSIK R. MOON and BENOIT PERRON, Beyond Panel Unit Roots Tests: Using Multiple Testing to Determine the Nonstationarity Properties of Individual Series in a Panel, *Journal of Econometrics* (2012), forthcoming.

For unit root and cointegration tests one may take p values from

JAMES G. MACKINNON, Numerical Distribution Functions for Unit Root and Cointegration Tests, *Journal of Applied Econometrics* 11 (1996), pp. 601 – 618,

JAMES G. MACKINNON, ALFRED A. HAUG and LEO MICHELIS, Numerical Distribution Functions of Likelihood Ratio Tests for Cointegration, *Journal of Applied Econometrics* 14 (1999), pp. 563 – 577.

Spurious regressions and **tests for no cointegration** have been discussed in

HORST ENTORF, Random walks with Drifts: Nonsense Regressions and Spurious Fixed-effect Estimation, *Journal of Econometrics* 80 (1997), pp. 287 – 296,

CHIHWA KAO, Spurious Regression and Residual-based Tests for Cointegration in Panel Data, *Journal of Econometrics* 90 (1999), pp. 1 – 44,

PETER PEDRONI, Panel Cointegration: Asymptotic and Finite Sample Properties of Pooled Time Series Tests with Application to the PPP Hypothesis, *Econometric Theory* 20 (2004), pp. 597 – 625,

JOAKIM WESTERLUND, Testing for Error Correction in Panel Data, *Oxford Bulletin of Economics and Statistics* 69 (2007), pp. 709 – 748.

For an **analysis of cointegrated panels** see

PETER C. B. PHILLIPS and HYUNGSIK R. MOON, Linear Regression Limit Theory for Nonstationary Panel Data, *Econometrica* 67 (1999), pp. 1057 – 1111,

CHIHWA KAO and MIN-HSIEN CHIANG, On the Estimation and Inference of a Cointegrated Regression in Panel Data, in: B.H. BALTAGI (ed.), *Advances in Econometrics*, Elsevier Science, Oxford, vol. 15 (2000), pp. 179 – 222,

ROLF LARSSON, JOHAN LYHAGEN and MICKAEL LÖTHGREN, Likelihood-Based Cointegration Tests in Heterogeneous Panels, *The Econometrics Journal* 4 (2001), pp. 109 – 142,

JAN J. J. GROEN and FRANK KLEIBERGEN, Likelihood-Based Cointegration Analysis in Panels of Vector Error-Correction Models, *Journal of Business & Economic Statistics* 21 (2003), pp. 295 – 318,

MICHAEL BINDER, CHENG HSIAO and M. HASHEM PESARAN, Estimation and Inference in Short Panel Vector Autoregressions with Unit Roots and Cointegration, Econometric Theory 21 (2005), pp. 795 – 8 37,

ANINDYA BANERJEE, MASSIMILIANO MARCELLINO and CHIARA OSBAT, Some Cautions on the Use of Panel Methods for Integrated Series of Macroeconomic Data, *Econometrics Journal* 7 (2004), pp. 322 – 340,

JÖRG BREITUNG, A Parametric Approach to the Estimation of Cointegration Vectors in Panel Data, *Econometric Reviews* 24 (2005), pp. 151 – 173.

The **(pooled) mean group estimation** has been discussed in

M. HASHEM PESARAN and RONALD PATRICK SMITH, Estimation of Long-run Relationships from Dynamic Heterogeneous Panels, *Journal of Econometrics* 68 (1995), pp. 79 – 113,

M. HASHEM PESARAN, YONGCHEOL SHIN and RONALD PATRICK SMITH, Pooled Mean Group Estimation of Dynamic Heterogeneous Panels, *Journal of the American Statistical Association* 94 (1999), pp. 621 – 634.

For a **general review** on nonstationary panels we recommend

JÖRG BREITUNG and M. HASHEM PESARAN, Unit Roots and Cointegration in Panels, in: L. MÁTYÁS and P. SEVESTRE (eds.), *The Econometrics of Panel Data:*

Fundamentals and Recent Developments in Theory and Practice, Kluwer Academic Publishers, Dordrecht 3rd edition (2008), pp. 279 – 322.

Finally, for a theoretical exposition on the uncovered interest rate parity hypothesis, and for more recent empirical evidence on the **international interest rate linkage** and the expectations hypothesis of the term structure, see

JEFFREY ALEXANDER FRANKEL, Measuring International Capital Mobility: A Review, *American Economic Review* 82 (1992), pp. 197 – 202,

UWE HASSLER and ADINA I. TARCOLEA, Combining Multi-Country Evidence on Unit Roots: The Case of Long-Term Interest Rates, *Applied Economics Quarterly* 51 (2005), pp. 181 – 189,

UWE HASSLER and VERENA WERKMANN, New Panel Evidence on International Interest Rate Linkage, *mimeo*, 2012,

RALF BRÜGGEMANN and HELMUT LÜTKEPOHL, Uncovered Interest Rate Parity and the Expectations Hypothesis of the Term Structure: Empirical Results for the U.S. and Europe, *Applied Economics Quarterly* 51 (2005), pp. 143 – 154,

JÜRGEN WOLTERS, Uncovered Interest Rate Parity and the Expectations Hypothesis of the Term Structure: Empirical Results for the U.S. and Europe, in: I. KLEIN and S. MITTNIK (eds.), *Contributions to Modern Econometrics: From Data Analysis* to *Economic Policy. In Honour of Gerd Hansen*, Kluwer Academic Publishers (2002), pp. 271 – 282.

8 Autoregressive Conditional Heteroscedasticity

All models discussed so far use the conditional expectation to describe the mean development of one or more time series. The optimal forecast, in the sense that the variance of the forecast errors will be minimised, is given by the conditional mean of the underlying model. Here, it is assumed that the residuals are not only uncorrelated but also homoscedastic, i.e. that the unexplained fluctuations have no dependencies in the second moments. However, BENOIT MANDELBROT (1963) already showed that financial market data have more outliers than would be compatible with the (usually assumed) normal distribution and that there are 'volatility clusters': small (large) shocks are again followed by small (large) shocks. This may lead to 'leptokurtic distributions', which – as compared to a normal distribution – exhibit more mass at the centre and at the tails of the distribution. This results in 'excess kurtosis', i.e. the values of the kurtosis are above three.

Example 8.1

As an example, we take the German Stock Market Index (DAX). We use daily observations from 2 January 1996 to 19 May 1999, i.e. we have 842 observations. *Figure 8.1a* shows the time series, *Figure 8.1b* the continuous returns, i.e. the first differences of the logarithms of this series. 'Clusters' for the returns appear. While the development of this series is relatively quiet at the beginning, i.e. the amplitude is small; more pronounced fluctuations can be observed in the second half of the observation period. This leads to the excess kurtosis which can be seen in *Figure 8.1c*: The kurtosis of the returns is 6.344, i.e. far above the value of 3.0, which would be expected if the variable were normally distributed. Thus, we get a value of 456.051 (p = 0.000) for the Jarque-Bera statistic. The null hypothesis of normal distribution has to be rejected at any conventional significance level.

The correlogram of the returns indicates second order autocorrelation. If we estimate an AR(2) model (with the modulus of t values in parentheses) for this series we get:

$$\Delta\ln(DAX_t) = \underset{(2.07)}{0.001} - \underset{(2.62)}{0.090}\,\Delta\ln(DAX_{t-2}) + \hat{\varepsilon}_t,$$

$$\bar{R}^2 = 0.007,\ SE = 0.015,\ Q(9) = 5.947\ (p = 0.745).$$

8 Autoregressive Conditional Heteroscedasticity

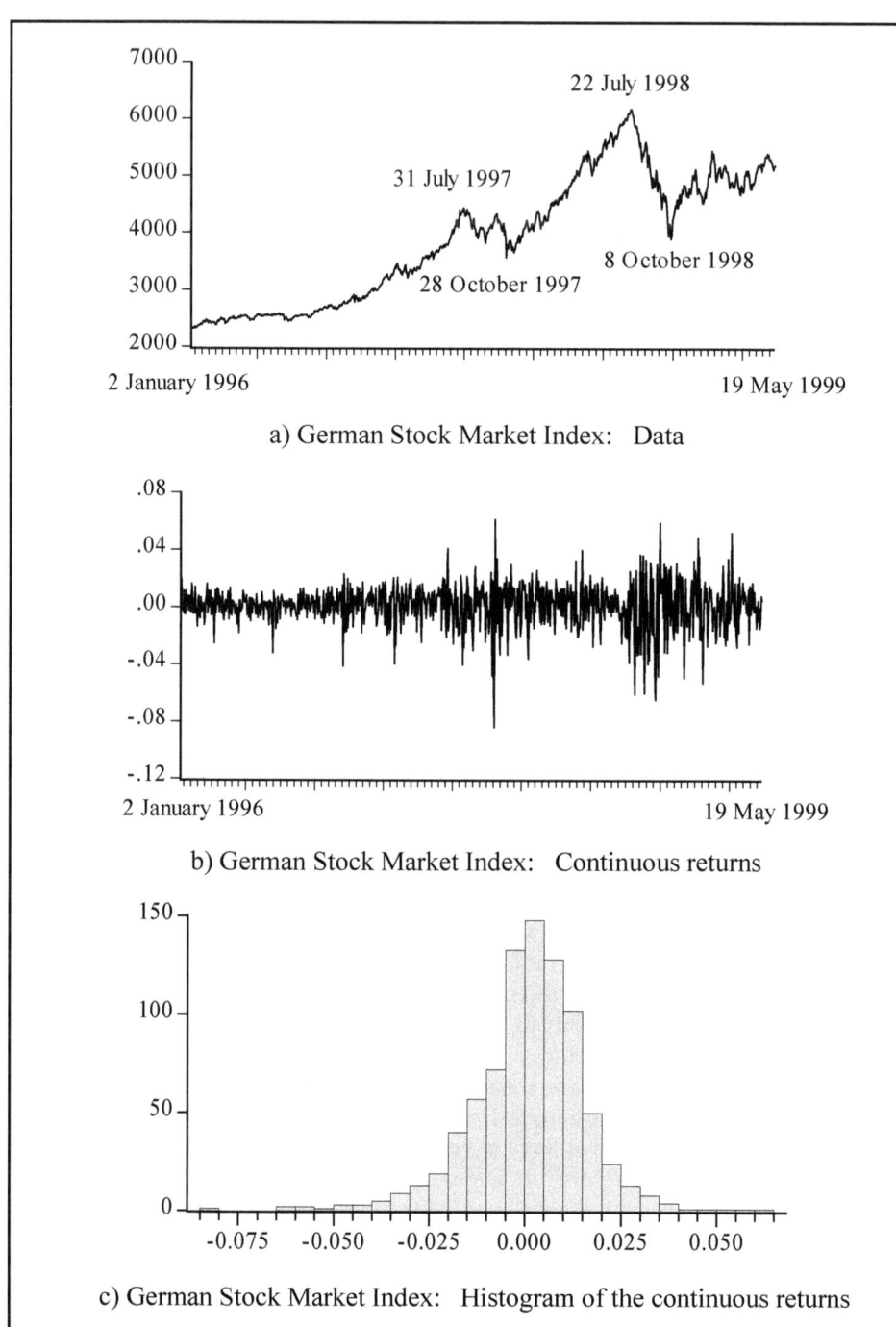

Figure 8.1: German Stock Market Index, 2 January 1996 until 19 May 1999, 842 observations

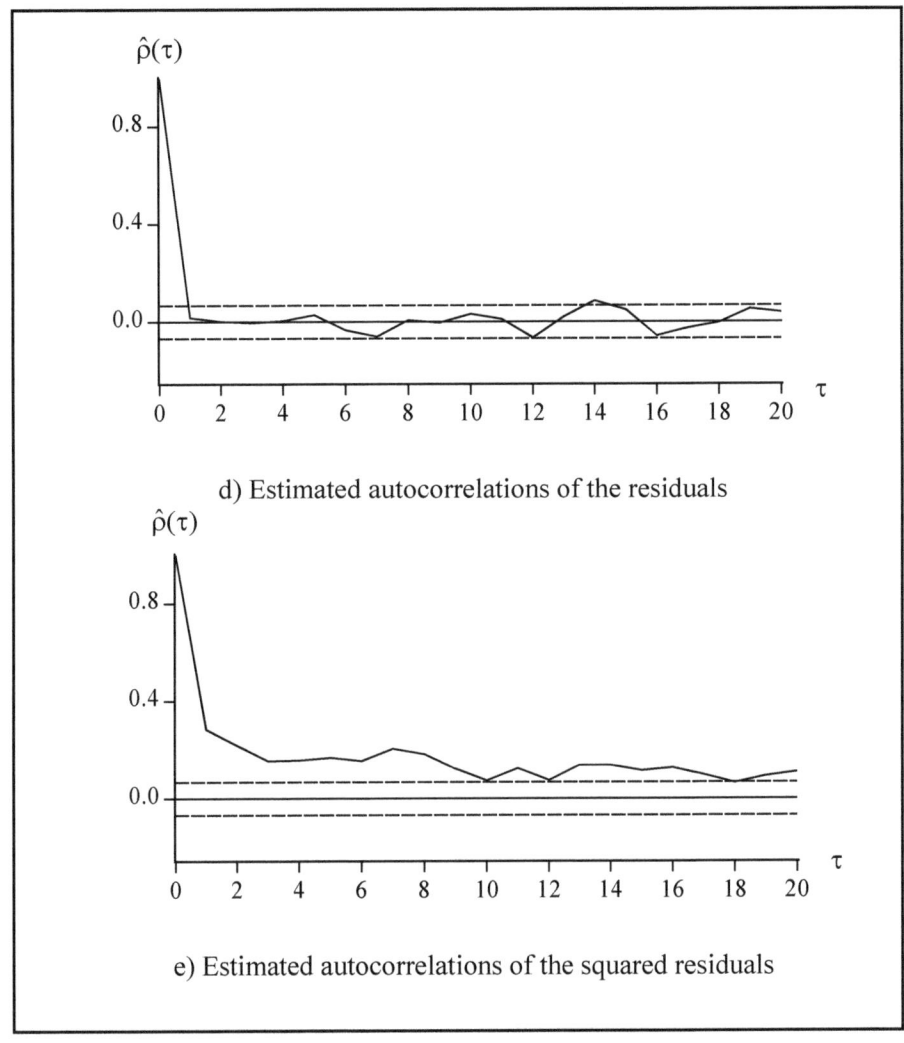

Figure 8.1: German Stock Market Index, 2 January 1996 until 19 May 1999, 842 observations (continued)

Figure 8.1d indicates that the residuals of this model no longer exhibit any significant autocorrelation. On the other hand, *Figure 8.1e* shows highly significant autocorrelation between the squares of these residuals. This indicates dependency in the second moments of the residuals, which contradicts the assumption of a constant, time-invariant variance. Thus, ε is not pure white noise.

In order to capture such problems by extending the models, we first present the conditional and unconditional means and variances of an AR(1) process. As shown in *Section 2.1.1*, for the process (2.1)

holds
$$x_t = \delta + \alpha x_{t-1} + u_t, \text{ with } |\alpha| < 1,$$

$$E[x_t] = \frac{\delta}{1-\alpha} \text{ and } V[x_t] = \frac{\sigma^2}{1-\alpha^2}.$$

Contrary to this, the conditional mean
$$E[x_t | x_{t-1}, \ldots] = E_{t-1}[x_t] = \delta + \alpha x_{t-1}$$
is not constant but depends on the observation of the previous period. However, for the conditional variance it holds that
$$V[x_t | x_{t-1}, \ldots] = E[(x_t - E_{t-1}[x_t])^2 | x_{t-1}, \ldots]$$
$$= E[u_t^2 | x_{t-1}, \ldots] = \sigma^2.$$

It is constant, just like the unconditional variance. Thus, phenomena like volatility clusters cannot be described by this model. We need different distributional assumptions to allow for 'fat tails', i.e. for values of the kurtosis above three.

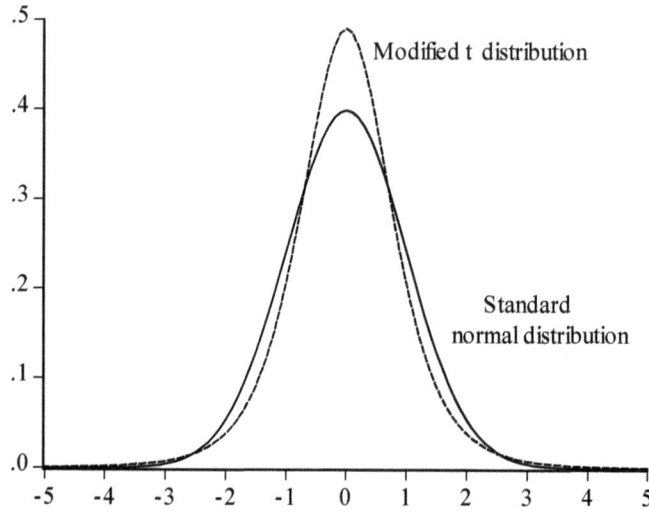

Figure 8.2: Density functions of a normalised t distribution with 5 degrees of freedom, variance one and a standard normal distribution

One possibility is to leave the normal distribution and to use, for example, a t distribution. *Figure 8.2* shows a t distribution with five degrees of freedom which is transformed so that it has a variance of one, i.e. the same variance as the standard normal distribution. Its kurtosis is nine. It can clearly be seen that the sides are steeper compared to the normal distribution also presented in *Figure 8.2*. (In 'stable distributions', the density functions are shaped similarly to the t distribution.)

On the other hand, in his paper on inflation in Great Britain, ROBERT F. ENGLE (1982) retained the normal distribution assumption but allowed the conditional variance of the residuals to vary linearly with the lagged squared residuals. This leads to models with autoregressive conditional heteroscedastic residuals, the ARCH models. The residuals of these models are also leptokurtic. The idea behind this approach is that the same models which are used to represent the conditional mean of a variable, i.e. AR, MA or ARMA models, can be applied to the squared residuals of equations. *Section 8.1* will present these ARCH models. Generalisations will be discussed in *Section 8.2*, and problems of estimation and testing in *Section 8.3*. Multivariate models are introduced in *Section 8.4*. We will conclude this chapter with examples of the application of ARCH/GARCH models in financial market analysis (*Section 8.5*).

8.1 ARCH Models

In the following, we will first discuss dependencies of the squared residuals by using autoregressive models. The main properties of such models will be presented. In addition, we will show that it largely depends on the frequency of data collection whether autoregressive conditional heteroscedasticity occurs.

8.1.1 Definition and Representation

Let us assume that the variable y can be explained in a linear model with the predetermined variables X and the parameter vector β,

(8.1) $$y_t = X_t' \beta + \varepsilon_t.$$

Along with truly exogenous deterministic and stochastic variables, the vector X might also contain lagged endogenous variables. The error term ε has zero mean, $E[\varepsilon_t] = 0$, and a constant unconditional variance, $E[\varepsilon_t^2] = \sigma^2$. It also holds that ε is not autocorrelated whereas ε^2 is allowed to be autocorrelated. It is assumed that this autocorrelation can be captured by an AR(q) process,

(8.2) $$\varepsilon_t^2 = \alpha_0 + \alpha_1 \varepsilon_{t-1}^2 + \alpha_2 \varepsilon_{t-2}^2 + \ldots + \alpha_q \varepsilon_{t-q}^2 + v_t,$$

were v_t is white noise. The information set I_t contains all information which is available at time t (as in *Chapter 3*), thus $I_{t-1} = \{y_{t-1}, y_{t-2}, \ldots, X_{t-1}, X_{t-2}, \ldots\}$. If the parameter vector β is known, this information set also contains all residuals up to time $t - 1$ because of $\varepsilon_{t-i} = y_{t-i} - X'_{t-i}\beta$, $i = 1, 2, \ldots$.

The conditional variance of ε_t, h_t^2, can be written as

(8.3) $$h_t^2 := V[\varepsilon_t | I_{t-1}] = E[\varepsilon_t^2 | I_{t-1}].$$

Because of (8.2) we get the ARCH(q) model

(8.4) $$h_t^2 = \alpha_0 + \sum_{i=1}^{q} \alpha_i \varepsilon_{t-i}^2$$

with $\alpha_0 > 0$ and $\alpha_i \geq 0$ for $i = 1, \ldots, q - 1$, as well as $\alpha_q > 0$. These conditions ensure that the conditional variance is always positive.

If a large shock occurs in equation (8.1), i.e. if there is a large positive or negative value of ε, this leads, according to relation (8.4), to a series of large values for the conditional variance, as the latter is a monotonically increasing function of lagged squared realised values of ε. If the occurring shock is only small, further small shocks are assumed to occur in the near future. The higher the value of q, the more extended are the volatility clusters.

ARCH effects can, for example, result from random coefficients, as shown by ANIL K. BERA and MATTHEW L. HIGGINS (1993). Let ε be a time dependent autoregressive process of order q (in contrast to the assumption above),

$$\varepsilon_t = \sum_{i=1}^{q} \phi_{it} \varepsilon_{t-i} + u_t,$$

with

$u_t \sim (0, \alpha_0)$, $\phi_{it} = \phi_i + \eta_{it}$, $\eta_{it} \sim (0, \alpha_i)$, $i = 1, 2, \ldots, q$,

$Cov[\eta_{it}, \eta_{jt}] = 0$ for $i \neq j$, $Cov[\eta_{it}, u_{t+j}] = 0$ for all i and j.

Then the conditional variance of the residuals leads to

$$V[\varepsilon_t | I_{t-1}] = \alpha_0 + \alpha_1 \varepsilon_{t-1}^2 + \alpha_2 \varepsilon_{t-2}^2 + \ldots + \alpha_q \varepsilon_{t-q}^2,$$

i.e. the residuals do not follow an AR(q) but an ARCH(q) process.

Example 8.2

Assume that the residuals follow the AR(1) process with random coefficient

$$\varepsilon_t = \phi_t \varepsilon_{t-1} + u_t,$$

with

$$\phi_t \sim (\phi, \alpha_1) \quad \text{and} \quad u_t \sim (0, \alpha_0),$$

where ϕ_t and u_t are independently generated. Then the conditional mean of the residuals results in

$$E[\varepsilon_t \mid I_{t-1}] = \phi \varepsilon_{t-1},$$

and their conditional variance in

$$V[\varepsilon_t \mid I_{t-1}] = \alpha_0 + \alpha_1 \varepsilon_{t-1}^2,$$

i.e. the residuals do not only follow an AR(1) but also an ARCH(1) process. This allows, for example, to model time dependent risk premia.

Large values of q demand models with many parameters, which contradicts the parsimony principle of univariate time series analysis. Therefore, ROBERT F. ENGLE (1982) proposed the following model with distributed lags where only two parameters have to be estimated:

(8.5) $$h_t^2 = \alpha_0 + \alpha_1 \sum_{i=1}^{q} w_i \varepsilon_{t-i}^2$$

with

$$w_i = \frac{2(q+1-i)}{q(q+1)}, \quad i = 1, 2, ..., q.$$

These weights decrease linearly and sum up to one.

For estimating and testing, assumptions on the conditional distribution of ε have to be made. Following ROBERT F. ENGLE (1982), it is often assumed that the residuals follow a conditional normal distribution,

(8.6) $$\varepsilon_t \mid I_{t-1} \sim N(0, h_t^2).$$

The assumption of a conditional univariate normal distribution implies that neither the joint nor the marginal distributions are normal. It is, however, possible to approximate leptokurtic distributions.

8.1.2 Unconditional Moments

In the following, we use a special version of the *law of iterated expectations*

(8.7) $$E[Z] = E[E[Z \mid I]],$$

where Z is a random variable and I the relevant information as a set of conditioning random variables.

Due to (8.6), it holds that $E[\varepsilon_t \mid I_{t-1}] = 0$. Thus, because of (8.7) $E[\varepsilon_t] = 0$ also holds. Due to (8.7) and (8.3), we get

$$\sigma^2 = E[\varepsilon_t^2] = E[E[\varepsilon_t^2 \mid I_{t-1}]] = E[h_t^2]$$

for the unconditional variance of the residuals.

Because of (8.4) we get

$$\sigma^2 = E[h_t^2] = \alpha_0 + \sum_{i=1}^{q} \alpha_i E[\varepsilon_{t-i}^2] = \alpha_0 + \sigma^2 \sum_{i=1}^{q} \alpha_i.$$

This leads to

(8.8) $$\sigma^2 = \frac{\alpha_0}{1 - \sum_{i=1}^{q} \alpha_i}, \quad \text{if } \sum_{i=1}^{q} \alpha_i < 1.$$

If this condition is violated, this process does not possess a finite variance.

For the kurtosis of an ARCH(1) process, ROBERT F. ENGLE (1982) derived the following expression assuming normally distributed ε's:

(8.9) $$K[\varepsilon_t] = \frac{E[\varepsilon_t^4]}{(E[\varepsilon_t^2])^2} = 3 \frac{1 - \alpha_1^2}{1 - 3\alpha_1^2}.$$

Thus, the kurtosis only exists if $3\alpha_1^2 < 1$. It is larger than three, i.e. than its value in case of a normal distribution. We get this value if α_1 tends towards zero. Compared to a normal distribution with the same variance, the ARCH(1) process has more mass in the centre of the distribution and fatter tails. As shown above, these are the properties often exhibited by financial market data if they are measured in short time distances.

For the autocovariances, we get

$$E[\varepsilon_t \varepsilon_{t-\tau}] = E[E[\varepsilon_t \varepsilon_{t-\tau} \mid I_{t-1}]]$$
$$= E[\varepsilon_{t-\tau} E[\varepsilon_t \mid I_{t-1}]] = 0$$

for $\tau \geq 1$.

As the ARCH(q) process has zero mean and is not autocorrelated, it is weakly stationary if its variance is finite, i.e. if the above shown condition that the sum of the α_i, $i = 1, \ldots, q$, is smaller than one is fulfilled.

The fact that ε is not autocorrelated does, of course, not imply that it is distributed independently. After all, the autocorrelation of ε^2 is modelled in relation (8.2). This prevents higher moments from disappearing.

Example 8.3

For the time series of the German Stock Market Index used in *Example 8.1*, we can estimate the following model:

$$\Delta\ln(DAX_t) = \underset{(3.41)}{0.0012} - \underset{(-1.97)}{0.072} \Delta\ln(DAX_{t-2}) + \hat{\varepsilon}_t,$$

$$\hat{h}_t^2 = \underset{(3.96)}{2.52 \cdot 10^{-5}} + \underset{(3.79)}{0.163 \hat{\varepsilon}_{t-1}^2} + \underset{(3.29)}{0.150 \hat{\varepsilon}_{t-2}^2} + \underset{(2.11)}{0.107 \hat{\varepsilon}_{t-3}^2} + \underset{(1.72)}{0.063 \hat{\varepsilon}_{t-4}^2}$$

$$+ \underset{(2.54)}{0.120 \hat{\varepsilon}_{t-5}^2} + \underset{(2.85)}{0.139 \hat{\varepsilon}_{t-6}^2} + \underset{(2.62)}{0.139 \hat{\varepsilon}_{t-7}^2} + \underset{(2.20)}{0.085 \hat{\varepsilon}_{t-8}^2},$$

SE = 0.015, Q(9) = 5.794 (p = 0.760), $Q^2(9)$ = 2.838 (p = 0.970)
JB = 65.652.

Looking at the t values given in parentheses, we can conclude that, with one-sided tests, all estimated parameters prove to be positive significant at least at the 5 percent level. Thus, they satisfy the conditions for a non-negative variance. The sum of the ARCH coefficients is 0.964 (< 1). Therefore, the unconditional variance exists and has a value of $7.2 \cdot 10^{-4}$. The value of the Jarque-Bera statistic indicates that the null hypothesis of a normal distribution can still be rejected at any conventional significance level, but now it is much smaller than before. The reason for this is that the kurtosis is now only 3.806 compared with the kurtosis 6.344 of the data themselves. Thus, the kurtosis of the estimated residuals, standardised with $\hat{h}_t$, comes quite close to the one of a normal distribution. In addition, as the Q^2 statistic shows, the squared standardised residuals do no longer exhibit significant autocorrelation. (The estimation and testing of such models is discussed in *Section 8.3*.)

8.1.3 Temporal Aggregation

In the following, we will derive the behaviour of the conditional variance of an ARCH(q) process if the series can only be observed over time intervals that are larger than the frequency of the data generating process. For example, only monthly, quarterly or annual data might be available instead of daily observations. (See for this also *Section 2.2.2.*) We consider the case of temporal aggregation where only every m-th observation is taken into account. This is, for example, the case if, instead of (available) daily data, only end-of-month or end-of-quarter data are used for interest rate or exchange rate data.

We consider an ARCH(1) process with $\alpha_1 = \alpha$. By repeated substitution with q = 1 in relation (8.2), we get:

$$\begin{aligned}
\varepsilon_t^2 &= \alpha_0 + \alpha\, \varepsilon_{t-1}^2 + v_t, \\
&= \alpha_0 + \alpha\, (\alpha_0 + \alpha\, \varepsilon_{t-2}^2 + v_{t-1}) + v_t, \\
&= \alpha_0 (1 + \alpha) + \alpha^2\, \varepsilon_{t-2}^2 + v_t + \alpha\, v_{t-1}, \\
&= \alpha_0 (1 + \alpha) + \alpha^2 (\alpha_0 + \alpha\, \varepsilon_{t-3}^2 + v_{t-2}) + v_t + \alpha\, v_{t-1}, \\
&= \alpha_0 (1 + \alpha + \alpha^2) + \alpha^3\, \varepsilon_{t-3}^2 + v_t + \alpha\, v_{t-1} + \alpha^2\, v_{t-2}, \\
&= \ldots,
\end{aligned}$$

and, finally, for arbitrary m,

(8.10) $$\varepsilon_t^2 = \alpha_0 \sum_{j=0}^{m-1} \alpha^j + \alpha^m\, \varepsilon_{t-m}^2 + \sum_{j=0}^{m-1} \alpha^j\, v_{t-j}.$$

The conditional variance in the original relation leads to

$$h_t^2 = E[\varepsilon_t^2 | I_{t-1}] = \alpha_0 + \alpha\, \varepsilon_{t-1}^2$$

for t = 1, 2, ..., T.

If only every second value is observed, i.e. the information set changes to $I_{t-2} = \{ y_{t-2}, y_{t-4}, \ldots, X_{t-2}, X_{t-4}, \ldots \}$ for t = 2, 4, ..., T, and due to (8.10) we get

$$h_{t,2}^2 = E[\varepsilon_t^2 | I_{t-2}] = \alpha_0 (1 + \alpha) + \alpha^2\, \varepsilon_{t-2}^2$$

for the conditional variance and m = 2.

In the general situation when only every m-th value is observed, we get, according to relation (8.10):

(8.11) $$h_{t,m}^2 = E[\varepsilon_t^2 | I_{t-m}] = \alpha_0 \frac{1 - \alpha^m}{1 - \alpha} + \alpha^m\, \varepsilon_{t-m}^2$$

for t = m, 2m, 3m, ..., T.

The conditional variance of the temporally aggregated data again follows an ARCH(1) process. Due to $0 < \alpha < 1$, however, the ARCH effect becomes the weaker the longer the observational intervals are. If m increases above all limits we get

$$\lim_{m \to \infty} h^2_{t,m} = \frac{\alpha_0}{1-\alpha}.$$

Here, the temporally aggregated process has a constant conditional variance. Because of (8.8) it coincides with the unconditional variance of the ARCH(1) process. This effect was detected by FRANCIS X. DIEBOLD (1988, pp. 12ff.) when modelling temporally aggregated exchange rates. If, in addition, the distributional assumption (8.6) holds, not only the conditional distribution is normal but also the unconditional one, i.e. the fat tails disappear.

Example 8.4

Let the following ARCH(1) model be given:

$$h^2_t = 0.1 + 0.5\,\varepsilon^2_{t-1}, \quad t = 1, 2, \ldots, T.$$

This process has the unconditional variance of

$$\sigma^2 = \frac{0.1}{1 - 0.5} = 0.2$$

and due to (8.9) the kurtosis of

$$K = 3\,\frac{1 - 0.25}{1 - 0.75} = 9.$$

If we observe only every second value, i.e. for $t = 2, 4, \ldots$, the conditional variance changes to

$$h^2_{t,2} = 0.15 + 0.25\,\varepsilon^2_{t-2}$$

because of (8.11).

The unconditional variance of the temporally aggregated process is still 0.2, while the kurtosis is reduced to 3.4615. Thus, ARCH effects can hardly be noticed. If we aggregate once again and consider only every fourth observation, i.e. if $t = 4, 8, \ldots$, we get the following process:

$$h^2_{t,4} = 0.1875 + 0.0625\,\varepsilon^2_{t-4}.$$

The variance is still 0.2, but the kurtosis has become 3.0237. Thus, the ARCH effect has disappeared almost completely.

Example 8.5

We consider the first differences of the logarithm of the exchange rate between the Swiss Franc and the U.S. Dollar, as used in *Example 1.3* of *Chapter 1*. For the period from January 1980 to December 2011, with 384 observations, we get a kurtosis of 3.451 for the end-of-month data shown in *Figure 1.8*. The value of the Jarque-Bera statistic is 3.857 (p = 0.145). Thus, the null hypothesis of a normal distribution cannot be rejected at any conventional significance level. If we use daily data for the same period, we have 7900 observations and the value of the Jarque-Bera statistic is 4772.721. This extremely high value is exclusively determined by a kurtosis of 6.805, as the value of the skewness of -0.074 is – in absolute terms – even slightly smaller than the value of -0.097 which is based on monthly data.

8.2 Generalised ARCH Models

Modelling the dependencies between the squared residuals by ARMA models, we get parsimonious parameterisations. These approaches can be extended to represent asymmetric effects, i.e. to allow for different impacts of positive and negative shocks.

8.2.1 GARCH Models

If the maximum lag in ARCH(q) models becomes too large, problems with the non-negativity constraints might occur if the estimates are not restricted appropriately. To get more parsimoniously parameterised models in which such problems occur less frequently but which are nevertheless capable of dealing with long-lasting volatility clusters, the approach of relation (8.5) was applied. Its disadvantage is, however, that possible dynamics of ARCH processes are captured only restrictively, i.e. with given, linearly declining weights.

Independently of each other, TIM BOLLERSLEV (1986) and STEPHEN J. TAYLOR (1986) developed a generalisation of the ARCH approach, the *Generalised Autoregressive Conditional Heteroscedasticity* (GARCH) model which is more flexible than the approach (8.5). They additionally included p lagged values of the conditional variance into relation (8.4). This leads to a GARCH(p,q) process:

(8.12) $\quad h_t^2 = \alpha_0 + \alpha_1 \varepsilon_{t-1}^2 + \ldots + \alpha_q \varepsilon_{t-q}^2 + \beta_1 h_{t-1}^2 + \ldots + \beta_p h_{t-p}^2.$

Sufficient conditions for the non-negativity of the conditional variance of this process are $\alpha_0 > 0$, $\alpha_i \geq 0$, $i = 1, \ldots, q - 1$, $\alpha_q > 0$, $\beta_i \geq 0$, $i = 1, \ldots, p - 1$, $\beta_p > 0$.

Using the lag polynomials

$$\alpha(L) := \alpha_1 L + \ldots + \alpha_q L^q, \quad \beta(L) := \beta_1 L + \ldots + \beta_p L^p,$$

(8.12) can be written as

(8.13) $$h_t^2 = \alpha_0 + \alpha(L)\varepsilon_t^2 + \beta(L) h_t^2,$$

or, if all roots of $1 - \beta(L) = 0$ are outside the unit circle, as

(8.13') $$h_t^2 = \frac{\alpha_0}{1-\beta(1)} + \frac{\alpha(L)}{1-\beta(L)} \varepsilon_t^2.$$

If the rational function of the lag operator is expanded into a series as, for example, in *Section 2.1.2*, we get the ARCH(∞) process

(8.14) $$h_t^2 = \alpha_0^* + \sum_{i=1}^{\infty} \delta_i \varepsilon_{t-i}^2,$$

with $\alpha_0^* > 0$ and $\delta_i \geq 0$, $i = 1, 2, \ldots,$. Thus, GARCH(p,q) models allow the parsimonious parameterisation for conditional variances in the same way as ARMA(p,q) models for conditional means.

The non-negativity conditions of the δ_i are sufficient for the conditional variances to be strictly positive. Thus, they are less restrictive than the conditions placed on α_i and β_i for equation (8.12).

In the following way we can show that ε_t^2 really follows an ARMA process: Due to (8.2) and (8.3), $v_t = \varepsilon_t^2 - h_t^2$ and

$$E[v_t \mid I_{t-1}] = E[\varepsilon_t^2 - h_t^2 \mid I_{t-1}] = 0.$$

Thus, v has zero mean and is uncorrelated. It satisfies the conditions of white noise. If we insert $h_t^2 = \varepsilon_t^2 - v_t$ into (8.12) we get

$$\varepsilon_t^2 = \alpha_0 + \alpha_1 \varepsilon_{t-1}^2 + \ldots + \alpha_q \varepsilon_{t-q}^2$$
$$+ \beta_1 (\varepsilon_{t-1}^2 - v_{t-1}) + \ldots + \beta_p (\varepsilon_{t-p}^2 - v_{t-p}) + v_t.$$

It follows that

(8.15) $$\varepsilon_t^2 = \alpha_0 + \sum_{i=1}^n (\alpha_i + \beta_i) \varepsilon_{t-i}^2 + v_t - \sum_{i=1}^p \beta_i v_{t-i},$$

with n = max(p, q). Relation (8.15) shows that the structure of dependence of the squared residuals ε^2 of a GARCH(p,q) process is given by an ARMA(n,p) process.

The considerations to calculate the unconditional variance and the autocorrelation function of ε for a GARCH process are the same as for the ARCH process in *Section 8.1.2*. Thus, the residuals are uncorrelated. According to (8.13), we get

$$(8.16) \qquad V[\varepsilon_t] = E[\varepsilon_t^2] = \frac{\alpha_0}{1 - \alpha(1) - \beta(1)}$$

for the variance.

Thus, it is necessary for the existence of the variance of a GARCH(p,q) process that

$$\alpha(1) + \beta(1) = \sum_{i=1}^{q} \alpha_i + \sum_{i=1}^{p} \beta_i < 1.$$

Together with the non-negativity constraints given above this condition is also sufficient. If the above condition holds, the GARCH(p,q) process is weakly stationary.

8.2.2 The GARCH(1,1) Process

For the empirical modelling of financial market data, a GARCH(1,1) model is often sufficient. It is given by

$$(8.17) \qquad h_t^2 = \alpha_0 + \alpha\, \varepsilon_{t-1}^2 + \beta\, h_{t-1}^2 ,$$

with $\alpha_0 > 0$, $\alpha > 0$ and $\beta > 0$. Due to (8.15), the squared residuals follow the ARMA(1,1) process

$$(8.18) \qquad \varepsilon_t^2 = \alpha_0 + (\alpha + \beta)\varepsilon_{t-1}^2 + v_t - \beta v_{t-1} ,$$

which is stable for $0 < \alpha + \beta < 1$. Then, the unconditional variance also exists:

$$(8.19) \qquad V[\varepsilon_t] = \frac{\alpha_0}{1 - \alpha - \beta}.$$

According to JÜRGEN FRANKE, WOLFGANG HÄRDLE and CHRISTIAN HAFNER (2004, p. 221), the kurtosis also exists if $3\alpha^2 + 2\alpha\beta + \beta^2 < 1$:

$$\text{(8.20)} \qquad K[\varepsilon_t] \;=\; 3 + \frac{6\alpha^2}{1-\beta^2-2\alpha\beta-3\alpha^2}.$$

It is always above three, the value of the normal distribution, since $\alpha > 0$ holds. Thus, the GARCH(1,1) process can be used to model distributions with fat tails. If α tends towards zero, the heteroscedasticity disappears and the value of the kurtosis tends towards three. It depends more strongly on α than on β. Correspondingly, in order to reach high values of the kurtosis, high values of α are always more effective than high values of β.

By transforming (8.17), we can show that the GARCH(1,1) model is really able to represent long-lasting effects:

$$(1-\beta L)\, h_t^2 \;=\; \alpha_0 + \alpha\, \varepsilon_{t-1}^2,$$

$$h_t^2 \;=\; \frac{\alpha_0}{1-\beta} + \frac{\alpha}{1-\beta L}\varepsilon_{t-1}^2,$$

$$\text{(8.21)} \qquad h_t^2 \;=\; \frac{\alpha_0}{1-\beta} + \alpha \sum_{j=1}^{\infty} \beta^{j-1}\varepsilon_{t-j}^2.$$

Due to $\alpha > 0$, $\beta > 0$ and $\alpha + \beta < 1$, the GARCH(1,1) process is transformed into an ARCH(∞) process with geometrically declining weights. The larger β, the longer is the effect of the shocks. Even if $\alpha + \beta = 1$, i.e. if we have an *Integrated GARCH process* (IGARCH), representation (8.21) is still valid for the conditional variance whereas the unconditional variance does not exist in this case.

To forecast the conditional variances of a GARCH(1,1) process, we use the ARMA(1,1) representation in (8.18). Following the considerations in Section 2.4.1, we get the optimal forecasts for the period $t+\tau$ with $\tau > 0$ as

$$h_{t+\tau|t}^2 \;=\; E[\varepsilon_{t+\tau}^2 \mid I_t].$$

(8.18) results in

$$\varepsilon_{t+\tau}^2 \;=\; \alpha_0 + (\alpha+\beta)\varepsilon_{t+\tau-1}^2 + v_{t+\tau} - \beta v_{t+\tau-1}.$$

Thus, for the one step ahead forecast we get

$$h_{t+1|t}^2 \;=\; E[\varepsilon_{t+1}^2 \mid I_t] \;=\; \alpha_0 + (\alpha+\beta)\varepsilon_t^2 - \beta v_t$$

$$\;=\; \alpha_0 + \alpha \varepsilon_t^2 + \beta h_t^2.$$

For $\tau = 2$ we get

$$h^2_{t+2|t} = E[\varepsilon^2_{t+2} | I_t] = \alpha_0 + (\alpha + \beta) E[\varepsilon^2_{t+1} | I_t]$$

and, therefore,

$$h^2_{t+2|t} = \alpha_0 + (\alpha + \beta) h^2_{t+1|t}.$$

Iteration leads to

$$h^2_{t+\tau|t} = \alpha_0 \frac{1-(\alpha+\beta)^{\tau-1}}{1-\alpha-\beta} + (\alpha+\beta)^{\tau-1} h^2_{t+1|t}.$$

If the forecast horizon grows above all limits, if $\alpha + \beta < 1$ and when taking (8.19) into account, we have

$$\lim_{\tau \to \infty} h^2_{t+\tau|t} = \frac{\alpha_0}{1-\alpha-\beta} = V[\varepsilon_t].$$

Thus, the conditional variance of ε converges towards its unconditional variance. This is no longer true for an IGARCH process. In this case we have $\alpha + \beta = 1$, implying that the conditional variance grows linearly with the forecast horizon. The conditional variance for period t, which defines the information set for the forecasts, has a permanent influence.

Example 8.6

If we apply an AR(2) process for the mean and a GARCH(1,1) process for the conditional variance of the DAX returns used in *Examples 8.1* and *8.3*, the AR(2) parameter is no longer significantly different from zero even at the 10 percent significance level. Thus, the correspondingly reduced model is

$$\Delta \ln(DAX_t) = \underset{(3.27)}{0.0012} + \hat{\varepsilon}_t,$$

$$\hat{h}^2_t = \underset{(3.22)}{3.69 \cdot 10^{-6}} + \underset{(6.23)}{0.164 \hat{\varepsilon}^2_{t-1}} + \underset{(33.35)}{0.829 \hat{h}^2_{t-1}},$$

SE = 0.015, Q(10) = 5.686 (p = 0.841), Q^2(10) = 3.183 (p = 0.977)
JB = 75.310,

with t values given in parentheses.

The simple as well as the partial autocorrelations of the squared residuals are no longer significantly different from zero.

Because of $\alpha + \beta = 0.993$ the unconditional variance is 0.00053. The high persistence that was already apparent in *Example 8.3*, where a pure ARCH process was applied, becomes obvious again if the estimated GARCH(1,1) model is, according to (8.21), transformed into an ARCH representation:

$$\hat{h}_t^2 = 0.0000215 + 0.164\,\hat{\varepsilon}_{t-1}^2 + 0.136\,\hat{\varepsilon}_{t-2}^2 + 0.113\,\hat{\varepsilon}_{t-3}^2$$
$$+ 0.093\,\hat{\varepsilon}_{t-4}^2 + 0.077\,\hat{\varepsilon}_{t-5}^2 + 0.064\,\hat{\varepsilon}_{t-6}^2 + \ldots\,.$$

The significant value of the Jarque-Bera statistic is caused by the still existing excess kurtosis. Although the kurtosis has been reduced drastically, it is still 3.953.

8.2.3 Nonlinear Extensions

A problem arises especially when estimating higher order ARCH models without restrictions: the estimated coefficients violate the non-negativity constraints. To avoid this problem, JOHN GEWEKE (1986) suggested to use a multiplicative approach for the conditional variance:

$$h_t^2 = e^{\alpha_0} \cdot \varepsilon_{t-1}^{2\alpha_1} \cdot \varepsilon_{t-2}^{2\alpha_2} \cdot \ldots \cdot \varepsilon_{t-q}^{2\alpha_q}.$$

This expression is always positive, regardless of whether the parameters are positive or negative. By taking logarithms, we get the estimating equation

(8.22) $$\ln(h_t^2) = \alpha_0 + \alpha_1 \ln(\varepsilon_{t-1}^2) + \ldots + \alpha_q \ln(\varepsilon_{t-q}^2).$$

All models discussed so far have the disadvantage that positive and negative shocks exert the same impact on the conditional variance as the signs disappear due to squaring. On the other hand, it is well known that the reaction of volatility of share prices is different if the shocks are negative, i.e. if they result from bad news, than if they are positive, i.e. if they result from good news. This *leverage effect* leads to higher volatility as a result of negative shocks as compared to positive ones. In the following, two extensions of the symmetric GARCH(1,1) model are presented which are capable to treat such asymmetric effects.

The *Threshold ARCH model* (TARCH), developed by LAWRENCE R. GLOSTEN, RAVI JAGANNATHAN and DAVID E. RUNKLE (1993) assumes different GARCH models for positive and negative shocks. Thus, the TARCH(1,1) model can be written as

(8.23) $$h_t^2 = \alpha_0 + \alpha\,\varepsilon_{t-1}^2 + \gamma\,\varepsilon_{t-1}^2\,d_{t-1} + \beta\,h_{t-1}^2,$$

with

$$d_t = \begin{cases} 1 & \text{if } \varepsilon_t < 0 \\ 0 & \text{otherwise} \end{cases}.$$

If $\gamma > 0$, a leverage effect is observed as the impulse $\alpha + \gamma$ of negative shocks is larger than the impulse α of positive shocks.

By presenting an *Exponential GARCH model* (EGARCH), DANIEL B. NELSON (1991) not only captures asymmetries but also ensures that the conditional variance is always positive. The EGARCH(1,1) model can be written as

$$(8.24) \qquad \ln(h_t^2) = \alpha_0 + \alpha \left| \frac{\varepsilon_{t-1}}{h_{t-1}} \right| + \gamma \frac{\varepsilon_{t-1}}{h_{t-1}} + \beta \ln(h_{t-1}^2).$$

Here, the standardised residuals ε/h are used. The ARCH effect is produced by the absolute value of the standardised residuals and not by their squares. The asymmetry is also captured by the standardised residuals. For $\gamma \neq 0$ we find an ARCH effect of $\alpha + \gamma$ for positive residuals and one of $\alpha - \gamma$ for negative residuals. If a leverage effect exists, we expect γ to be negative.

Example 8.7

To investigate whether the leverage effect plays a role for the DAX returns, the data of *Example 8.1* are taken to estimate a TARCH(1,1) as well as an EGARCH(1,1) model. The results of the TARCH model are:

$$\Delta\ln(DAX_t) = \underset{(2.89)}{0.0011} + \hat{\varepsilon}_t,$$

$$\hat{h}_t^2 = \underset{(3.20)}{3.75 \cdot 10^{-6}} + \underset{(4.34)}{0.146\hat{\varepsilon}_{t-1}^2} + \underset{(0.85)}{0.032\hat{\varepsilon}_{t-1}^2 d_{t-1}} + \underset{(33.30)}{0.830\hat{h}_{t-1}^2},$$

SE = 0.015, Q(10) = 5.911 (p = 0.823), Q²(10) = 3.173 (p = 0.977), JB = 74.492,

where t values are given in parentheses. For the EGARCH model we get:

$$\Delta\ln(DAX_t) = \underset{(2.46)}{0.0009} + \hat{\varepsilon}_t,$$

$$\ln(\hat{h}_t^2) = \underset{(-5.78)}{-0.501} + \underset{(7.00)}{0.281} \left| \frac{\hat{\varepsilon}_{t-1}}{\hat{h}_{t-1}} \right| - \underset{(-2.99)}{0.059} \frac{\hat{\varepsilon}_{t-1}}{\hat{h}_{t-1}} + \underset{(120.55)}{0.968 \ln(\hat{h}_{t-1}^2)},$$

SE = 0.015, Q(10) = 5.147 (p = 0.881), Q²(10) = 3.639 (p = 0.962), JB = 75.000,

with t values given in parentheses.

The main difference between these two approaches is that the leverage effect is significant in the EGARCH but not in the TARCH model. In the former, the short-run reaction to positive shocks is 0.222 and 0.340 on negative shocks. This difference is highly significant. In both models, the remaining deviation from a normal distribution of the residuals is again due to the existing excess kurtosis: The estimated kurtosis is 3.953 in the TARCH and 3.931 in the EGARCH model. In both approaches the squared standardised residuals are no longer autocorrelated.

Usually, it is assumed that higher returns of a financial asset imply a higher risk. Therefore, mean and variance tend to go into the same direction. If we assume the risk premium to be time-dependent, this can be represented by applying the ARCH-in-mean (ARCH-M) approach developed by ROBERT F. ENGLE, DAVID M. LILIEN and RASSEL P. ROBINS (1987). Relation (8.1) is extended to

(8.25) $$y_t = X_t' \beta + \delta h_t^2 + \varepsilon_t,$$

with

$$\varepsilon_t | I_{t-1} \sim N(0, h_t^2),$$

where the variance h_t^2 might be generated by an ARCH or GARCH process. As this variance is part of model (8.25), the residuals of the original model (8.1), ζ,

$$\zeta_t = y_t - X_t' \beta = \delta h_t^2 + \varepsilon_t,$$

are now autocorrelated.

8.3 Estimation and Testing

We consider model (8.1)

$$y_t = X_t' \beta + \varepsilon_t,$$

and allow for a time-dependent conditional variance of ε_t, i.e. we assume

(8.26) $$\varepsilon_t | I_{t-1} \sim f(0, h_t^2),$$

where f is a distribution function and the conditional variance h_t^2 possibly follows a (G)ARCH process.

If the residuals in (8.1) are independent, as is assumed in the classical model, autocorrelation appears neither in the estimated residuals nor in their squares.

Usually, a model for the mean is regarded as appropriate if the estimated residuals do not exhibit significant autocorrelation and if the null hypothesis of normally distributed residuals cannot be rejected. If the Jarque-Bera test (described in *Section 1.3*) indicates that the normality assumption has to be rejected because the value of the kurtosis is larger than three, this can be seen as evidence for the existence of (G)ARCH effects. If such effects exist, the simple as well as the partial autocorrelation functions of the squared residuals should have values significantly different from zero. This can be checked by applying the Q and Q* statistics described in *Section 1.4* on the squared residuals, denoted as Q^2 and Q^2*. Under the null hypothesis of no autocorrelation these statistics are asymptotically χ^2 distributed, and the number of degrees of freedom is (as in the linear case) equal to the considered number of autocorrelation coefficients (of the squared residuals) minus the number of estimated parameters in the equation for the mean.

It can also be checked by using Lagrange Multiplier tests whether autoregressive conditional heteroscedasticity exists. The squared residuals are in an auxiliary regression regressed on a constant and their own lagged values up to order q,

$$\hat{\varepsilon}_t^2 = \alpha_0 + \alpha_1 \hat{\varepsilon}_{t-1}^2 + \ldots + \alpha_q \hat{\varepsilon}_{t-q}^2 + v_t .$$

The test statistic is $T \cdot R^2$, i.e. the product of the number of observations, T, and the multiple correlation coefficient of the auxiliary regression, R^2. Under the null hypothesis of homoskedasticity this statistic is χ^2 distributed with q degrees of freedom. Alternatively, an F statistic can be performed for the combined null hypothesis $H_0: \alpha_1 = \alpha_2 = \ldots = \alpha_q = 0$.

In these tests, it is possible to employ the OLS residuals of equation (8.1), as they are consistently estimated despite the existence of (G)ARCH effects. These estimates are, however, not efficient. If such effects exist, relations (8.1) and (8.2) (or other (G)ARCH specifications) are therefore usually estimated simultaneously using maximum likelihood methods. For the conditional distribution in (8.26) a normal distribution is mostly supposed, i.e. it is assumed that the standardised residuals ε/h follow a standard normal distribution. This does, of course, not imply that the unconditional distribution is normal, too, because h^2 is also a random variable under this assumption. The above ARCH(1) and GARCH(1,1) models exemplified that the tails of the unconditional distribution are typically fatter than those of the normal distribution.

Normally, when estimating such processes, the stationarity conditions are not imposed as this would be numerically too complex. To avoid the risk of these conditions being violated, one should choose rather small values of p and q. The standard programme systems employ two procedures with respect to the non-negativity constraints. The first one is to use no restrictions at all. If negative values of α_i or β_i are estimated, it has to be checked whether all composite parameters δ_i in (8.14) are positive. The alternative is to impose the sufficient conditions directly on the α_i and β_i. This often leads to corner solutions which do not necessarily represent the maximum of the likelihood function.

Even if the assumption of the normal distribution of standardised residuals does not hold, the maximum likelihood estimator is still providing consistent results despite the misspecification of the likelihood function, if at least the first two moments are specified correctly. However, these *quasi maximum likelihood estimates* demand corrections for the consistent estimation of the standard errors. Such a procedure is to be found, for example, in JAMES D. HAMILTON (1994, p. 663).

For (8.26), TIM BOLLERSLEV (1987) assumes a conditional t distribution with a small number of degrees of freedom. As shown above, for a finite number of degrees of freedom the t distribution has fatter tails than the normal distribution. With an increasing number of degrees of freedom, however, it converges to the latter. (From 100 degrees of freedom on, there is practically no longer any difference from the normal distribution.) This provides the possibility to check whether a conditional normal distribution is appropriate.

8.4 Multivariate Models

Since many volatilities move together over time across assets and markets, multivariate models should be valuable tools for decision makers working, for example, on portfolio selection, risk management or option pricing. In the case of a volatility spillover, one would like to know, which market is leading the volatility of the other markets; the transmission channel could be direct from one conditional variance to the others, or indirect through the conditional covariances. We now assume that ε_t is a k-dimensional vector with zero mean and free of serial correlation. Let I_{t-1} denote again the past information, then the conditional (co)variance matrix is defined as

(8.27) $\qquad H_t = E(\varepsilon_t \varepsilon_t' | I_{t-1}) = E_{t-1}(\varepsilon_t \varepsilon_t')$.

Several proposals have been made to model H_t in terms of past ε_{t-j}. For such a proposal to be sensible it must be guaranteed that H_t is positive definite in the whole sample space. Often, the square matrices H_t and $\varepsilon_t \varepsilon_t'$ are vectorised for modelling purposes. Let vech denote the so-called vector-half operator, stacking the lower triangular portion of a (symmetric) square matrix into a vector of length $k(k+1)/2$. For $k = 2$ we have

$$\text{vech} \begin{pmatrix} h_{11} & h_{12} \\ h_{12} & h_{22} \end{pmatrix} = \begin{pmatrix} h_{11} \\ h_{12} \\ h_{22} \end{pmatrix}.$$

With this notation we define

$$h_t = \text{vech}(H_t), \quad \eta_t = \text{vech}(\varepsilon_t \varepsilon_t').$$

8.4.1 VAR-Type Models

At first glance, the most natural way to extend the univariate GARCH(p, q) model from equation (8.12) to the multivariate case is similar to the VAR system from equation (4.1). It is often called the VEC model (since it is vectorised with the vech operator):

(8.28) $h_t = c + A_1 \eta_{t-1} + \ldots + A_q \eta_{t-q} + B_1 h_{t-1} + \ldots + B_p h_{t-p},$

where the matrices A_i, $i = 1, \ldots, q$, and B_j, $j = 1, \ldots, p$, are all square of dimension $k(k+1)/2$, and c is a vector of length $k(k+1)/2$. Without further restrictions, equation (8.28) contains a huge number of parameters, the number growing with k^4. Even with $q = p = 1$ and only $k = 3$ series, there are 78 parameters. In order to obtain a more parsimonious model, TIM BOLLERSLEV, ROBERT F. ENGLE, and JEFFREY M. WOOLDRIDGE (1988) advocated the so-called diagonal VEC model where A_i and B_j are diagonal matrices, such that variances depend only on own past squared realizations, and covariances depend only on own past cross-products. With the diagonality restriction the number of parameters in (8.28) is reduced to the order k^2; for $q = p = 1$ and $k = 3$ we have 18 parameters left. For illustrative purposes we spell out an example of the diagonal VEC model for $k = 2$ series with $q = p = 1$:

$$h_{11,t} = c_1 + a_{11} \varepsilon_{1,t-1}^2 + b_{11} h_{11,t-1},$$

$$h_{12,t} = c_2 + a_{22} \varepsilon_{1,t-1} \varepsilon_{2,t-1} + b_{22} h_{12,t-1},$$

$$h_{22,t} = c_3 + a_{33} \varepsilon_{2,t-1}^2 + b_{33} h_{22,t-1}.$$

The estimation of a diagonal VEC model is computationally less demanding than the general case since each of the $k(k+1)/2$ equations can be treated separately. But the simple diagonal VEC model has two serious drawbacks in applied work. First, positive definiteness of H_t is not guaranteed, and second, the exclusion of interaction between different conditional (co)variances is not reasonable for practical purposes.

A different restriction of the general VEC model is the so-called BEKK model (named after YOSHI BABA, ROBERT F. ENGLE, DENNIS KRAFT and KENNETH F. KRONER (1990)) that has been popularized in the paper by ROBERT F. ENGLE and KENNETH F. KRONER (1995). To ensure positive definiteness the BEKK model assumes a quadratic form for the parameter matrices:

$$(8.29) \qquad H_t = C^{*'}C^* + \sum_{j=1}^{q} A_j^{*'} \varepsilon_{t-j} \varepsilon_{t-j}' A_j^* + \sum_{j=1}^{p} B_j^{*'} H_{t-j} B_j^*,$$

where C^* is an upper triangular matrix of dimension k, and A_j^* and B_j^* are k×k matrices without restrictions. Equation (8.29) produces positive definite matrices H_t for all possible ε_t as long as C^* or B_j^* are of full rank. The link between the BEKK model and the VEC representations is not at all obvious but spelled out in the paper by ROBERT F. ENGLE and KENNETH F. KRONER (1995).

For k = 2, q = 1 and p = 0 the BEKK model reads as

$$h_{11,t} = c_1 + a_{11}^{*2} \varepsilon_{1,t-1}^2 + 2 a_{11}^* a_{21}^* \varepsilon_{1,t-1} \varepsilon_{2,t-1} + a_{21}^{*2} \varepsilon_{2,t-1}^2,$$

$$h_{12,t} = c_2 + a_{11}^* a_{12}^* \varepsilon_{1,t-1}^2 + (a_{21}^* a_{12}^* + a_{11}^* a_{22}^*) \varepsilon_{1,t-1} \varepsilon_{2,t-1} + a_{21}^* a_{22}^* \varepsilon_{2,t-1}^2,$$

$$h_{22,t} = c_3 + a_{12}^{*2} \varepsilon_{1,t-1}^2 + 2 a_{12}^* a_{22}^* \varepsilon_{1,t-1} \varepsilon_{2,t-1} + a_{22}^{*2} \varepsilon_{2,t-1}^2,$$

with

$$c_1 = c_{11}^{*2}, \quad c_2 = c_{12}^* c_{11}^*, \quad c_3 = c_{12}^{*2} + c_{22}^{*2}.$$

The number of parameters in equation (8.29) is as in the diagonal VEC of order k^2, for example equal to 24 for k = 3 with p = q = 1. Hence, the estimation of BEKK models is still numerically a bit demanding, and algorithms may not converge since (8.29) is nonlinear in the parameters. A straightforward simplification is to assume diagonal matrices A_j^* and B_j^* in (8.29), but then again the interaction between different conditional (co)variances is ruled out by assumption.

8.4.2 Correlation Models

Conditional correlation models rely on a two-step approach. First, one employs univariate GARCH-type models for each conditional variance. Based on this, in the second step the conditional correlation matrix is modelled. Such a procedure is less greedy in parameters than a VEC or BEKK model and hence less troublesome when it comes to estimation.

Define the diagonal matrix D_t as in (4.6),

$$(8.30) \qquad D_t = \text{diag}(\sqrt{h_{11,t}}, ..., \sqrt{h_{kk,t}})$$

such that $e_t = D_t^{-1}\varepsilon_t$ is a standardised vector. Then the conditional correlation matrix of ε_t becomes

$$R_t = E(\varepsilon_t \varepsilon_t' \mid I_{t-1}) = D_t^{-1} H_t D_t^{-1},$$

with typical element $\rho_{ij,t}$ and $\rho_{ii,t} = 1$, or

$$(8.31) \qquad H_t = D_t R_t D_t.$$

The model of constant conditional correlation (CCC) has been proposed by TIM BOLLERSLEV (1990). While $R_t = R$ is constant with a typical element ρ_{ij}, the conditional variances and covariances are time-dependent:

$$h_{ij,t} = \rho_{ij} \sqrt{h_{ii,t}} \sqrt{h_{jj,t}}.$$

As long as D_t contains positive entries, the positive definiteness of H_t follows from that of R. In practice, ρ_{ij} can be estimated simply by the usual sample correlations. Such a model is very easy to estimate and very parsimonious. Assume k = 3 series with 3 correlation coefficients; if each $h_{ij,t}$ is estimated as univariate GARCH (1, 1), then this amounts to only 12 parameters. At the same time, the assumption of constant correlation may be too restrictive for some empirical concerns.

ROBERT F. ENGLE (2002) introduced the dynamic conditional correlation (DCC) model. Again, D_t from (8.30) can be estimated from k univariate GARCH-type models. H_t is computed from (8.31) with time-varying R_t:

$$R_t = (Q_t^*)^{-1} Q_t (Q_t^*)^{-1}$$

with the diagonal matrix $Q_t^* = \text{diag}(\sqrt{q_{11,t}}, ..., \sqrt{q_{kk,t}})$ containing the square root main diagonal entries of Q_t. In order to guarantee positive definiteness it is assumed that all conditional correlations obey the same dynamic structure governed by two scalar parameters a and b only:

$$(8.32) \qquad Q_t = (1-a-b)\bar{Q} + ae_{t-1}e'_{t-1} + bQ_{t-1},$$

with $e_t = D_t^{-1}\varepsilon_t$ and $\bar{Q} = E(e_t e'_t)$ which is the unconditional covariance matrix of the standardised residuals and hence straightforward to estimate

$$\hat{\bar{Q}} = T^{-1}\sum_{t=1}^{T} e_t e'_t.$$

As long as a, b > 0 with a + b < 1, the DCC process is stationary producing positive definite conditional covariance matrices H_t.

Note that the DCC model is remarkably parsimonious with only two parameters accounting for conditional correlation. Hence, the estimation is not complicated with growing k. At the same time, the assumption of common dynamics in (8.32) becomes debatable for a large number of series. For more flexible time-varying correlation approaches we refer to the literature.

8.5 ARCH/GARCH Models as Instruments of Financial Market Analysis

To evaluate the risk of different portfolio strategies is one of the basic tasks of financial market analysis. As mentioned in the introduction of this chapter, when modelling asset returns, it has long been known that the residuals of the estimated models are not homoscedastic but that their variances partly show strong variations over time. A possibility to reflect this in the models is provided by the ARCH and GARCH approaches.

The estimated conditional standard deviations of the residuals can, for example, be used to construct more precise intervals for the forecasts of asset returns. Point forecasts of returns modelled according to equation (8.1) are the same, regardless of whether the residuals follow a (G)ARCH process or not. In both cases, the conditional expectation given all information up to period t is an optimal forecast (compare *Section 2.4*).

If the residuals are homoscedastic, the forecast error variance only depends on the length of the forecast horizon but not on the elements of the information set I_t. In case of heteroscedastic residuals, we use, according to (2.57), the information set dependent conditional variances for the construction of forecast error variances. These conditional variances can be derived from the ARMA representation (8.15) of the squared residuals which are assumed to follow a GARCH process.

Moreover, estimates of conditional variances to capture volatilities are, for example, necessary for the following approaches:

- The approach of FISCHER BLACK and MYRON S. SCHOLES (1973) is often employed to evaluate options. Besides the basic price, the expiry date, the share price and the riskless interest rate, an estimate of the volatility is necessary. All of these quantities can usually be observed directly except for the last one.

- The *Value at Risk* (VaR) has recently been applied to capture market risks. It is defined as the maximum loss to be expected over a fixed time horizon (holding period) with a specified confidence level. Typically, a normal distribution is assumed to calculate a VaR for holding periods of one day or ten days and confidence levels of 95 or 99 percent. This implies that the probability that losses are larger than calculated by the VaR is five or one percent.

  Statistically, the VaR is an α-quantile of the left edge of a distribution for the **change of the value** of a portfolio. To calculate this quantile, besides other quantities, the conditional standard deviation of the portfolio returns, which cannot be observed directly, is necessary.

A variety of models exists for estimation VaR (see in particular PHILLIPPE JORION (2001)). Here, we will focus on approaches which estimate volatilities by time series methods.

Traditionally, 'historical volatilities', i.e. the standard deviations of the last n price changes, are used to estimate this conditional heteroscedasticity. If Δx denotes the change of the logarithm of an asset price, it holds that

$$(8.33) \quad \hat{\sigma}_t = \sqrt{\frac{1}{n}\sum_{i=1}^{n}\left(\Delta x_{t-i} - \overline{\Delta x_{(t)}}\right)^2} \quad \text{with} \quad \overline{\Delta x_{(t)}} = \frac{1}{n}\sum_{i=1}^{n}\Delta x_{t-i}.$$

To give current observations a higher weight, exponentially weighted moving averages are used:

$$(8.34) \quad \hat{\sigma}_t = \sqrt{\frac{1-\lambda}{1-\lambda^n}\sum_{i=1}^{n}\lambda^{i-1}\left(\Delta x_{t-i} - \overline{\Delta x_{(t)}}\right)^2}, \quad 0 < \lambda < 1.$$

We get (8.33) as the limit of (8.34) for $\lambda = 1$ since

$$\lim_{\lambda \to 1} \frac{1-\lambda}{1-\lambda^n} = \frac{1}{n}.$$

If Δx is a conditionally normal distributed zero mean return, $\Delta x_t = r_t \sim N(0, \sigma_t^2)$, we get from (8.34) for $n \to \infty$

$$\sigma_t^2 = (1-\lambda)\sum_{i=1}^{\infty}\lambda^{i-1} r_{t-i}^2 = \frac{1-\lambda}{1-\lambda L} r_{t-1}^2, \text{ or}$$

(8.35) $$\sigma_t^2 = (1-\lambda) r_{t-1}^2 + \lambda \sigma_{t-1}^2.$$

This approach is recommended by RISK METRICS GROUP (1996). For daily data $\lambda = 0.94$ and for monthly data $\lambda = 0.97$ is suggested. In (8.35), the coefficients $(1 - \lambda)$ and λ add up to one. Thus, we get an IGARCH(1,1) with the constant term restricted to zero. ROBERT F. ENGLE (2001) shows for the example of a GARCH(1,1) model how such models can be used to calculate the VaR.

Two different other applications have already been mentioned. Firstly, the ARCH approach can be used to model time-dependent risk premia. Secondly, the ARCH-M model allows to represent the possibility that assets with higher expected returns imply higher risk. At least risk neutral and risk averse investors will only buy assets with higher risk if they can expect a higher return.

References

The first to mention that changes of speculative markets are **not normally distributed** was

BENOIT MANDELBROT, The Variation of Certain Speculative Prices, *Journal of Business* 36 (1963), pp. 394 – 419.

In this context, he discussed Pareto distributions. The **ARCH model** was developed by

ROBERT F. ENGLE, Autoregressive Conditional Heteroscedasticity with Estimates of the Variance of U.K. Inflation, *Econometrica* 50 (1982), pp. 987 – 1008.

In 2003, ROBERT F. ENGLE received the Nobel prize for this important paper. The **GARCH model** was introduced by

TIM BOLLERSLEV, Generalized Autoregressive Conditional Heteroskedasticity, *Journal of Econometrics* 31 (1986), pp. 307 – 327, and

STEPHEN J. TAYLOR, *Modelling Financial Time Series*, John Wiley, Chichester (U.K.) 1986,

independently of each other. The **IGARCH** approach was discussed by

ROBERT F. ENGLE and TIM BOLLERSLEV, Modelling the persistence of Conditional Variances, *Econometric Reviews* 5 (1986), pp. 1 – 87.

The **TARCH model** was developed by

LAWRENCE R.GLOSTEN, RAVI JAGANNATHAN and DAVID E. RUNKLE, On the Relation between the Expected Value and the Volatility of the Nominal Excess Return of Stocks, *Journal of Finance* 48 (1993), pp. 1779 – 1801,

while the **EGARCH model** goes back to

DANIEL B. NELSON, Conditional Heteroscedasticity in Asset Returns: A New Approach, *Econometrica* 59 (1991), pp. 347 – 370,

and the **ARCH-M model** to

ROBERT F. ENGLE, DAVID M. LILIEN and RASSEL P. ROBINS, Estimating Time Varying Risk Premia in the Term Structure: The ARCH-M Model, *Econometrica* 55 (1987), pp. 391 – 407.

The **multiplicative model** which guarantees the non-negativity of the estimated conditional variances was proposed by

JOHN GEWEKE, Modelling the Persistence of Conditional Variances: Comment, *Econometric Reviews* 5 (1986), pp. 57 – 61.

Surveys are, for example, given by

ANIL K. BERA and MATTHEW L. HIGGINS, ARCH Models: Properties, Estimation and Testing, *Journal of Economic Surveys* 7 (1993), pp. 305 – 366, and

TIM BOLLERSLEV, ROBERT F. ENGLE and DANIEL B. NELSON, ARCH Models, in: R.F. ENGLE and D.L. MCFADDEN (eds.), *Handbook of Econometrics*, volume IV, Elsevier, Amsterdam et al. 1994, pp. 2959 – 3038,

as well as, for example, in the textbook of

JAMES D. HAMILTON, *Time Series Analysis*, Princeton University Press, Princeton 1994.

The **t distribution** to model leptokurtic behaviour was proposed by

TIM BOLLERSLEV, A Conditionally Heteroscedastic Time Series Model for Speculative Prices and Rates of Return, *Review of Economics and Statistics* 69 (1987), pp. 542 – 547.

The effects of **temporal aggregation** are discussed in

FRANCIS X. DIEBOLD, *Empirical Modeling of Exchange Rate Dynamics*, Springer, New York et al. 1988, as well as in

FEIKE C. DROST and THEO E. NIJMAN, Temporal Aggregation of GARCH Processes, *Econometrica* 61 (1993), pp. 909 – 927.

A **first discussion of multivariate GARCH (VEC) models** can be found in

TIM BOLLERSLEV, ROBERT F. ENGLE and JEFFREY M. WOOLDRIDGE, A Capital Asset Pricing Model with Time-varying Covariances, *Journal of Political Economy* 96 (1988), pp. 116 – 131.

The **BEKK model** was introduced by

ROBERT F. ENGLE and KENNETH F. KRONER, Multivariate Simultaneous Generalized GARCH, *Econometric Theory* 11 (1995), pp. 122 – 150,

with reference to an earlier, unpublished paper by

YOSHI BABA, ROBERT F. ENGLE, DENNIS Kraft and KENNETH F. KRONER, Multivariate Simultaneous Generalized ARCH, mimeo, University of California San Diego, 1990.

Some recent **survey papers** are

CHRIS BROOKS, SIMON P. BURKE and GITA PERSAND, Multivariate GARCH Models: Software Choice and Estimation Issues, *Journal of Applied Econometrics* 18 (2003), pp. 725 – 734,

LUC BAUWENS, SÉBASTIEN LAURENT and JEROEN V.K. ROMBOUTS, Multivariate GARCH Models: A Survey, *Journal of Applied Econometrics* 21 (2006), pp. 79 – 109, or

ANNASTIINA SILVENNOINEN and TIMO TERÄSVIRTA, Multivariate Garch Models, in: T.G. ANDERSEN, R.A. DAVIS, J.-P. KREISS, and TH. MIKOSCH (eds.), *Handbook of Financial Time Series*, Springer, Heidelberg et al. 2009, pp. 201 – 229.

Models on **conditional correlation** were introduced by

TIM BOLLERSLEV, Modeling the Coherence in Short-run Nominal Exchange Rates: A Multivariate Generalized ARCH Model, *Review of Economics and Statistics* 72 (1990), pp. 498 – 505, and

ROBERT F. ENGLE, Dynamic Conditional Correlation: A Simple Class of Multivariate Generalized Autoregressive Conditional Heteroscedasticity Models, *Journal of Business and Economics Statistics* 20 (2002), pp. 339 – 350.

The modern **analysis of option prices** was founded by

FISCHER BLACK and MYRON S. SCHOLES, The Pricing of Options and Corporate Liabilities, *Journal of Political Economy* 81 (1973), pp. 637 – 659.

The **Value at Risk** was discussed extensively by

PHILIPPE JORION, Value at Risk: *The New Benchmark for Managing Financial Risk*, McGraw Hill Trade, 2nd edition 2001.

How the value at risk of an asset can be calculated using a GARCH(1,1) model is described in

ROBERT F. ENGLE, GARCH 101: The Use of ARCH/GARCH Models in Applied Econometrics, *Journal of Economic Perspectives* 15 (2001), pp. 157 – 168.

See for this also

JAMES CHONG, Value at Risk from Econometric Models and Implied from Currency Options, *Journal of Forecasting* 23 (2004), pp. 603 – 620.

Generally, for the **econometric analysis of financial market data** see

ADRIAN PAGAN, The Econometrics of Financial Markets, *Journal of Empirical Finance* 3 (1996), pp. 15 – 102,

TIM BOLLERSLEV, RAY Y. CHOU and KENNETH F. KRONER, ARCH Modelling in Finance: A Review of the Theory and Empirical Evidence, *Journal of Econometrics* 52 (1992), pp. 5 – 59,

TERENCE C. MILLS, *The Econometric Modelling of Financial Time Series*, Cambridge University Press, Cambridge (U.K.), 2nd edition 1999, or

JÜRGEN FRANKE, WOLFGANG HÄRDLE and CHRISTIAN HAFNER, *Einführung in die Statistik der Finanzmärkte*, Springer, Berlin et al., 2nd edition 2004.

The approaches of the RISK METRICS GROUP are described in

J.P. MORGAN/REUTERS, *RiskMetrics: Technical Document*, Morgan Trust, New York, 4th edition 1996.

Index of Names and Authors

A

Akaike, Hirotugu 56, 91
Amisano, Gianni 154
Andrews, Donald W. 175, 200

B

Baba, Yoshi 303, 309
Babbage, Charles 3
Baillie, Richard T. 202
Banerjee, Anindya 217, 222-23, 245-46, 274, 278
Bartlett, Maurice Stevenson 16, 24, 174, 199
Bauwens, Luc 309
Bera, Anil K. 19, 25, 286, 308
Beveridge, Stephen 183-88, 201
Binder, Michael 275, 278
Black, Fischer 306, 309
Blanchard, Oliver 152, 154
Bollerslev, Tim 292, 301-02, 304, 307-10
Boswijk, H. Peter 238, 245, 247
Box, George E.P. 1, 4, 17, 23-24, 27, 90, 207
Breitung, Jörg 154, 192, 202, 258, 260, 274, 276, 278
Breusch, Trevor S. 17, 25, 255, 275
Brooks, Chris 309
Brown, Bryan W. 87, 91
Brockwell, Peter J. 90
Brüggemann, Ralf 154, 275, 279
Buiter, Willem H. 124
Burke, Simon P. 309

C

Carvalho, José L. 22
Chan, K. Hung 201
Chiang, Min-Hsien 268, 278
Choi, In 264, 277
Chong, James 310
Chou, Ray Y. 310
Chu, Chia-Shang J. 258, 276
Christoffersen, Peter F. 242, 248
Clements, Michael P. 242, 248
Cochrane, Donald 1, 23, 27, 91
Cochrane, John H. 185-86, 202

D

Das, Samarijt 260, 276
Davidson, James E.H. 153
Davis, Richard A. 90
Demetrescu, Matei 192, 202, 262-63, 265, 276-77
Dickey, David A. 167-68, 172, 175, 178, 199-200
Diebold, Francis X. 92, 242, 248, 291, 308
Dolado, Juan J. 217, 222-23, 241, 245-46, 248
Doornik, Jürgen A. 245
Drost, Feike C. 308
Durbin, James 1, 23

E

Enders, Walter 24, 92, 153
Engle, Robert F. 195, 199, 201-02, 207-10, 216, 242, 244-45, 285, 287-88, 299, 302-04, 307-10

Entorf, Horst 267, 277
Ericsson, Neil R. 238, 247

F

Feige, Edgar L. 105, 108, 111, 123
Fisher, Roland A. 263, 276
Franke, Jürgen 294, 310
Frankel, Jeffrey Alexander 261, 279
Friedman, Milton 39, 93
Friedmann, Ralph 248
Fuller, Wayne A. 167-68, 172, 175, 199

G

Galbraith, John W. 245
Galilei, Galileo 95
Geweke, John 297, 308
Giannini, Carlo 154
Glosten, Lawrence R. 297, 308
Godfrey, Leslie G. 17, 25
Gomez, Victor 23
Granger, Clive W.J. 1, 24, 92, 95-97, 104, 120, 122, 124, 128, 154, 195, 199, 202, 205, 208-10, 216, 240, 244-45, 248
Groen, Jan J. 274, 278
Grether, David M. 22
Griffiths, William E. 153

H

Hadri, Kaddour 262-63, 276
Hafner, Christian 294, 310
Hagen, Hanns Martin 88, 92
Hamilton, James D. 24, 209, 301, 308
Hanck, Christoph 277
Hannan, Edward J. 56, 91
Hansen, Bruce E. 200
Härdle, Wolfgang 294, 310
Hartley, Peter R. 203
Hartung, Joachim 264-65, 268, 274, 277
Harvey, Andrew C. 202

Hassler, Uwe 179, 181-82, 192, 201-02, 217, 222-23, 242, 244, 246-49, 255, 262-63, 265, 274, 276-77, 279
Hatanaka, Michio 24
Haug, Alfred 199, 246, 274, 277
Haugh, Larry D. 99, 104, 108, 122-23
Hayya, Jack C. 201
Hendry, David F. 128, 152-53, 217, 242, 245-48
Higgins, Matthew L. 286, 308
Hill, R. Carter 153
Hodrick, Robert J. 186, 188, 202
Hommel, Gerhard 266, 277
Hsiao, Cheng 104, 112, 123, 275, 278
Hume, David 95-96, 120
Hylleberg, Svend 195-96, 202

I

Im, Kyung S. 259, 273, 276

J

Jagannathan, Ravi 297, 308
Jarque, Carlos M. 19, 25
Jenkins, Gwilym M. 1, 4, 23, 27, 90, 207
Jeong, Jinook 87, 91
Jevons, William Stanley 3
Johansen, Søren 208, 212, 225, 228-29, 231-32, 238, 244-47
Jorion, Philippe 306, 309
Joyeux, Roselyne 202
Judge, George G. 153
Juselius, Katarina 152, 243, 245-46

K

Kang, Heejoon 201
Kao, Chihwa 268, 278
Kepler, Johannes 2
Kirchgässner, Gebhard 37, 88, 92-93, 96, 119, 123-25, 145, 201
Kleibergen, Frank 274, 278

Kraft, Dennis 303, 309
Krämer, Jörg W. 124
Krätzig, Markus 24, 154
Kroner, Kenneth F. 303, 309-10
Kurozumi, Eiji 263, 276
Kuzin, Vladimir 192, 202
Kwiatkowski, Denis 180-81, 200
Kydland, Finn E. 203

L

Langfeld, Enno 124
Larsson, Rolf 273, 278
Laurent, Sébastien 309
Lee, Tsoung-Chao 153
Levin, Andrew T. 258, 276
Lilien, David M. 299, 308
Lin, Chien-Fu 258, 276
Lin, Jin-Lung 240, 248
Ljung, Greta M. 17, 24
Löthgren, Mickael 273, 278
Lütkepohl, Helmut 24, 124, 135, 153-54, 241, 247-49, 275, 279
Lyhagen, Johan 273, 278

M

MacKinnon, James G. 168, 170, 172, 175, 178, 199, 216-17, 245-46, 257, 268, 274, 277
Maddala, Gangadharrao S. 87, 91, 263, 276
Maital, Shlomo 87, 91
Mandelbrot, Benoit 281, 307
Maravall, Augustin 23
Marcellino, Massimiliano 274, 278
Mariano, Roberto S. 92
Mestre, Ricardo 222-23, 246
Michelis, Leo 199, 246, 274, 277
Mills, Terence C. 90, 310
Mincer, Jacob 86, 91
Mishkin, Frederic S. 124
Mizon, Graham E. 247
Moon, Hyungsik Roger 268, 277-78
Müller, Ulrich K. 200

N

Nelson, Charles R. 183, 186, 188, 201
Nelson, Daniel B. 298, 308
Nerlove, Marc 22
Newbold, Paul 1, 24, 96, 124, 205, 208, 245
Nijman, Theo E. 308

O

O'Connell, Paul G. 259, 276
Orcutt, Guy H. 1, 23, 27, 91
Ord, J.-Keith 201
Osbat, Chiara 274, 278
Osterwald-Lenum, Michael 232, 246

P

Pagan, Adrian 153, 255, 275, 310
Palm, Franz C. 89, 92
Parzen, Emanuel 24
Pearce, Douglas K. 105, 108, 111, 123
Pedroni, Peter 268, 278
Perron, Benoit 277
Perron, Pierre 170-71, 173-74, 179-80, 199-200, 203
Persand, Gita 309
Persons, Warren M. 3, 23
Pesaran, M. Hashem 247, 258-63, 271, 273-76, 278
Phillips, Peter C.B. 173, 180-81, 199-201, 203, 245, 268, 278
Pierce, David A. 17, 24, 104, 108, 111, 123-24
Pindyck, Robert S. 90
Plosser, Charles I. 201
Prescott, Edward C. 186, 188, 202
Price, J. Michael 123

Q

Quah, Danny 152, 154
Quinn, Barry G. 56, 91

R

Richard, Jean-François 152
Robertson, Donald 153
Robins, Rassel P. 299, 308
Rombouts, Jeroen V.K. 309
Rubinfeld, Daniel 90
Rudebusch, Glenn D. 203
Runkle, David E. 297, 308

S

Saïd, Saïd E. 172, 178, 200
Saikkonen, Pentti 218, 245, 247
Sargan, J. Dennis 128, 153
Sargent, Thomas J. 104, 122, 124
Sarkar, Sanat K. 266, 277
Savioz, Marcel R. 125, 145
Schmidt, Peter 180-81, 200
Scholes, Myron S. 306, 309
Schumpeter, Joseph A. 198, 203
Schwarz, Gideon 56, 91
Schwert, G. William 97, 108, 123, 172, 178, 200
Shin, Yongcheol 180-81, 200, 247, 259, 271, 273, 276, 278
Silvennoinen, Annastiina 309
Simes, R. John 266, 268, 274, 277
Sims, Christopher A. 95, 104-05, 119, 122-23, 127, 151-52
Slutzky, Evgenij Evgenievich 4
Smith, Gregor W. 217, 246
Smith, Richard J. 247
Smith, Ronald Patrick 271, 278
Spanos, Aris 24
Srba, Frank 153
Stadler, George W. 203
Stock, James H. 153, 198, 201-03, 208, 218, 223, 244-45

T

Tarcolea, Adina I. 262, 263, 265, 274, 276-77, 279
Taylor, Stephen J. 292, 307
Temple, Jonathan 203
Teräsvirta, Timo 248, 309

Theil, Henry 88, 92
Tiao, George C. 64-65, 91, 166, 199
Tinbergen, Jan 1, 22, 90, 92
Toda, Hiro Y. 241, 248
Tsay, Ruey S. 166, 199

U

Uhlig, Harald 152, 154

W

Watson, Geoffrey S. 1, 23
Watson, Mark W. 153, 198, 202-03, 218, 245
Werkmann, Verena 255, 279
Westerlund, Joakim 258, 268, 276, 278
Whitt, Joseph A. 203
Wickens, Michael 153
Wold, Herman 4, 21, 23, 90
Wolters, Jürgen 124, 179, 181-82, 192, 201-02, 223, 242, 244, 246, 248-49, 275, 279
Wooldridge, Jeffrey M. 302, 309
Working, Holbrook 63, 91
Wu, Shaowen 263, 276

X

Xiao, Zhijie 201

Y

Yamada, Hiroshi 248
Yamamoto, Taku 241, 248
Yeo, Stephen 153
Yoo, Byung Sam 195, 202, 207, 242, 245
Yule, George Udny 4

Z

Zarnowitz, Victor 86, 91
Zellner, Arnold 89, 92, 112, 123, 254, 275

Subject Index

A

ARCH models 281-310
  ARCH-M 299, 307-08
  EGARCH 298-99, 308
  GARCH 285, 292-310
  IGARCH 295-96, 307
  TARCH 297-99, 308
ARFIMA process 191-92
ARIMA process 23, 160, 166, 171, 183-84, 186-87, 191, 199-200, 213
Autocorrelogram 16, 33, 61, 65-66
Autocorrelation 1-2, 16-18, 22-25, 33, 39-40, 43, 50-51, 59, 63-67, 69, 72, 74, 76, 130, 158, 161, 165-66, 173, 175-78, 180-81, 191, 200, 219, 222, 258, 285, 289, 294, 299-300
  estimation 16
  partial autocorrelation 52-54, 61-62, 69, 72, 74, 76, 300
Autocovariance 13, 15-16, 22, 31-32, 39, 43, 50, 59, 66, 71-72, 76, 129-30, 174, 288
  autocovariance matrix 129-30

B

Bartlett window 174, 199
Beveridge-Nelson decomposition 183, 186, 201
Bonferroni type test 265-66, 277
Box-Jenkins approach 1, 27, 78, 89-90, 165, 243
Breusch-Godfrey test 17, 25

C

Choleski decomposition 138, 140
Cochrane measure of persistence 185-86, 202
Cointegration 199-200, 202, 205-49, 252, 267-75, 277-78
  cointegration rank 202, 209-10, 215, 225-26, 229-30, 232, 237, 246-47, 273
  cointegration test 199, 200, 209, 215, 221-23, 246-47, 268, 277-78
  cointegration vector 209, 211-12, 214-15, 225-26, 228, 232, 238-39, 244, 278
  definition 209-10
  Engle-Granger test 216-17, 244-45, 247
  Johansen approach 208, 225, 228-232, 238, 244-47
  λ_{max} test 230, 246
  trace test 230, 246, 274
Combination of significance 263-67
  inverse normal method 263-65
Conditional correlation 304-05, 309
  constant conditional correlation 304
  dynamic conditional correlation 304, 309
Consistency, consistent estimation 13, 16, 54, 56-57, 59, 76, 86-87, 91, 135, 139, 166, 174, 199-200, 209, 216-18, 223, 232, 239, 243, 262, 264, 268, 300-01

315

super consistency 217-18, 221, 223, 239, 243, 268
Cross-correlation 103-04, 108, 116, 139-41, 207, 254, 265, 274-76
Breusch-Pagan test 255, 275

D

Deterministic 3-4, 21, 28, 52, 99, 120, 155, 157, 160-61, 165, 170-71, 174-75, 179-81, 195-97, 202, 217, 222, 225, 228, 230-31, 237, 242, 246-47, 258-59, 285
   deterministic trend 155, 157, 160-61, 170-71, 174-75, 179-81, 196, 202, 222, 231
Deterministic rule 120-21
Durbin-Watson statistic 1, 18, 23, 161, 165, 205

E

Equation, system of 50, 53, 66-67, 89, 128, 135, 139
   reduced form 96, 135
   structural form 139
Equilibrium error 211-12, 216
Ergodicity 12-13, 16, 24
Error correction 128, 132, 153, 210, 212, 221-23, 225, 230, 232, 237-48, 268, 275, 277-78
   conditional error correction 223, 238-39
   structural error correction 239, 247
   vector error correction 225-26, 230, 232, 237, 240, 245, 247, 278
Exogeneity, weak 212, 221, 223, 232, 237, 239, 247
Expectation 2-3, 13, 31-32, 40, 342, 50, 58, 66, 71, 79-81, 91, 97, 127, 131, 146, 155, 160, 211, 214, 218, 275, 279, 281, 288, 305
   conditional expectation 80, 131, 146, 281, 305

rational expectations 81, 97, 121, 127

F

Fixed effects 253, 277
Forecast (see also prediction) 78-89
   forecast function (*see* prediction function)
   forecast error (*see* also prediction error) 79-82, 84, 86-87, 91, 97-99, 128, 132, 146-47, 157, 281, 305
   forecast error variance 81, 97-98, 146, 305
   mean absolute forecast error 87
   mean squared forecast error 78-81
   root mean squared forecast error 87

G

Granger causality 95-125, 128, 137-40, 151, 207, 210, 213, 237, 240-41, 247-48
   and policy rules 120-22
   and rational expectations 121
   definition 97-98
   direct Granger procedure 104-08, 112, 116-17
   feedback 98, 104-05, 108, 117, 119-20, 123
   Haugh-Pierce test 108, 111, 116, 119, 122-24
   Hsiao procedure 104, 112, 118, 123
   instantaneous causality 97-100, 103-05, 108, 112, 117, 119-23, 129, 138-40, 151, 223, 239
   in VAR models 138-40
   in vector error correction models 240-41
   with cointegrated variables 210, 213
Granger representation theorem 210, 212, 221

H

Hodrick-Prescott filter 186, 202

I

Identification 91, 95-96, 139, 152, 154, 214, 228
Impulse response 128, 137, 140-46, 151, 153-54, 237, 241
Information criteria 56-57, 76, 91, 105, 112, 135-36, 218
   Akaike criterion 56, 91, 135-36
   final prediction error 56, 112, 135
   Hannan-Quinn criterion 56-57, 91, 135
   Schwarz criterion 56, 91, 135
Innovations 97, 123, 139-41, 146-47, 151-52, 179, 185
Integration 159-60, 182, 191-96, 202, 241, 259, 262-63, 275-76
   definition 159-60
   fractional integration 191-193, 202, 263
   seasonal integration 193-96, 202
Invertible, Invertibility 60, 62, 64, 67, 71-72, 75-76, 78, 83-85, 99, 102, 156, 159, 161, 171, 191, 215, 231

J

Jarque-Bera test 19, 25, 300

K

Kurtosis 18-19, 281, 284-85, 288, 294-95, 300

L

Lag operator 2, 10-12, 29, 40, 42, 49, 59, 65, 67, 69, 75, 99-100, 102, 237, 293
Lagrange multiplier test 17, 25, 275, 300

Least squares estimation 23, 56, 76, 87, 91, 135, 166, 168, 174, 193, 199, 205, 209, 216, 222-23, 228, 244, 252, 254
   generalised least sqares (GLS) 87, 135, 252, 254, 260
Leptokurtosis 281, 285, 287, 308
Leverage effect 297-98
LSE approach 128, 153

M

Maximum likelihood estimation 56, 76, 228, 246, 271, 274, 300-01
   quasi maximum likelihood estimation 301
Mean group estimation 270-71, 275, 278
Method of undetermined coefficients 41, 50, 67
Multivariate ARCH/GARCH 301-05, 308-09
   VEC representation 302-03, 308-09
   BEKK representation 303, 309

N

Nonparametric approach 175, 177-78
Normal distribution 13, 16, 18-19, 25, 54, 56, 152, 169, 176-77, 180, 193, 198, 217-19, 225, 245, 259, 263-66, 281, 284-85, 287-89, 291, 295, 300-01, 306-07
   conditional normal distribution 287, 291, 301, 307
   multivariate normal distribution 13, 152, 264, 266

P

Panel cointegration 268-73, 278
   homogeneous cointegration 267-68, 270, 274

heterogeneous cointegration 268, 276, 278
Panel unit root test 252, 258-264, 268, 275-77
   Levin-Lin-Chu test 258-59, 276
   Im-Pesaran-Shin test 259, 273, 276
   CADF test 260, 274-75
Popularity function 93
Popularity series 37-38
Prediction (*see* also forecast) 23, 56, 78-84, 86-87, 89, 91, 97, 104, 112, 128, 135, 244, 248
   prediction error 56, 80, 82-83, 112, 135
   prediction function 78-80, 82, 97, 104
Predictability 86, 96, 120, 122

Q

Q statistic 17-18, 108

R

Random walk 14-15, 63, 122, 158-165, 167-70, 175, 180, 183, 194-95, 199, 201-03, 205, 211, 267, 277
Real business cycle 197, 203
Residuals, standardised 298, 300-01, 305

S

Seemingly unrelated regressions (SUR) 112, 123, 135, 252-54, 260, 270-71, 274-75
Skewness 18-19
Stability condition 39, 41-43, 49-50, 52, 56, 70, 75, 157, 166, 248
Stable distribution 285
Stationarity
   covariance stationarity 14, 21-22, 122, 156
   definition 13-14

difference stationarity 159, 161, 167, 248
mean stationarity 14-15, 156, 158
stationarity test (see also unit root test) 180-82, 200, 276
trend stationarity 156, 160-61, 167-68, 170-71, 178, 180, 248
variance stationarity 14
weak stationarity 13-16, 28-29, 31, 90, 97, 99, 112, 128, 159-60, 191, 207, 289, 294
Structural break 6, 9, 178-80, 200-01, 247, 260

T

t distribution 19, 152, 168-69, 205, 284-85, 301, 308
Temporal aggregation 62-65, 91, 120, 123-24, 253, 289-92, 308
Term structure of interest rates 106, 124, 232, 236, 249, 275, 279, 308
Theil's U 88, 92
Trend elimination 161-65, 201

U

Unit root test (*see* also stationarity test as well as panel unit root test) 165-82, 192, 196, 199, 201-02, 215-17, 222, 252, 258-64, 268, 275-79
   Augmented Dickey-Fuller (ADF) test 170-73, 175, 178-79, 192, 217, 258-60, 265
   Dickey-Fuller test 167-71, 175, 192, 199-200, 202, 216-17,
   HEGY test 195-96
   Phillips-Perron test 173-78, 181, 199-200

V

Value at risk 306, 309-10
Variance decomposition 81, 128, 137, 146-51, 241

Vector Autoregression (VAR) 95, 118, 127-54, 225-242, 245, 246-48, 251-52, 255, 275, 278, 302-03
   structural VAR 139, 152, 154, 247
Volatility 281, 284, 286, 292, 297, 301, 306, 308

W

White noise 14, 16, 56-57, 63-64, 76, 87, 99, 102-03, 108, 120-21, 147, 161, 181, 194-95, 210-11, 225, 254, 267, 286, 293
Wold decomposition (representation) 2, 21-22, 27-29, 41-42, 49, 66, 58, 68-70, 75-76, 78, 81-82, 99, 101, 140-41, 146, 183

Y

Yule-Walker equations 50-51, 53, 56